Riding the Airwaves with Alpha & Zulu

Preparation for the Novice
and
No-Code Technician
Amateur Radio License Examinations

by
John Abbott, K6YB

Graphics & additional material

by
Bill Smith, N6MQS
John Mitchell

artsci

Artsci, Inc.
P.O. Box 1428
Burbank, CA 91507 U.S.A.
(818) 843-4080
(818) 846-2298 FAX

Riding the Airwaves with Alpha & Zulu

Copyright © 1993 artsci, inc.
Copyright © 1992, 1993 John Abbott

All rights reserved. No part of this book may be reproduced or utilized in any form or by any means, electronic or mechanical, including photocopying, recording or by any information storage and retrieval system, without permission in writing from the Publisher, artsci inc., P.O. Box 1428, Burbank, CA 91507.

93 94 95 96 97 98 99   First Edition   10 9 8 7 6 5 4 3 2 1

ISBN 0-917963-14-8 $14.95

Notice of Liability

The information in this book is distributed on an "AS IS" basis, without warranty. Neither the author nor artsci inc shall have any liability to the customer or any other person or entity with respect to any liability, loss, or damage caused or alleged to be caused directly or indirectly by the modifications contained herein. This includes, but is not limited to, interruption of service, loss of use, loss of business or anticipatory profits, or consequential damages from the information in this book.

artsci inc, books are available for bulk sales at quantity discounts. For information, please contact Marketing Manager, artsci inc, P.O. Box 1428, Burbank, CA 91507 FAX: (818) 846-2298

Printed in the United States of America

Acknowledgements

Special Thanks to the friends and associates who helped in the developement and proof reading of this book

Doug Wynn, KB6YZD
Steve Blumenfeld, WA6PQD
Helen Bratton
Marc Cohen
Kris Smith

"To the young in age and the young at heart" - Bill Smith

"To better communication for the people of the world" - John Mitchell

"To My XYL, Teri" - John Abbott

Riding the Airwaves with Alpha & Zulu

artsci inc

Riding the Airwaves with Alpha & Zulu

Contents

Preface.. i
What is Ham Radio?... iii
Information for the Teacher... iv
Which license is for you ?... v
To Morse or not to Morse.. vi
Where can you take the test ?.. vii

Novice License Section

Learning the Phonetic Aphabet.. 1
Alpha Wants to Talk to Someone.. 2
Word Search Game - Phonetico Finder... 4
The "RST" Signal Report... 5
Identification.. 7
Purpose and Rules.. 9
Definitions.. 11
Station / Operator License... 13
License Possession and Classes... 15
Eligibility, Exam Elements, and U.S. Call Signs.. 17
Mailing Addresses, Lost and Renewed Licenses... 19
Emergency Operation.. 21
Operating a Station.. 23
Word Search Game - Antenna Terms... 26
Operating Another Station.. 27
U.S. Operation, Non-Amateur Contacts, & Space Stations....................... 29
Unauthorized Persons.. 31
Business Use of Amateur Radio.. 33
Third Party & Foreign Contacts.. 35
Word Search Game - Equipment & Controls... 38
Broadcasting Music, Codes, & Harmful Interference................................. 39
Unidentified & False Signals.. 41
Respect For Others.. 43
FM Repeaters... 45
Morse Code Emissions & Connections... 49
Morse Code Operation.. 51
Voice Emissions & Connections.. 53
RTTY Emissions & Connections... 55
Packet Radio & Connections... 57
RTTY & Packet Operation.. 59
Direct Current, Alternating Current, and Frequency.................................. 61
Audio and Radio Frequencies.. 63
Hertz, Kilohertz, and Megahertz.. 65
Frequency and Wavelength.. 67
CW Only Frequencies.. 69
CW Only Frequency Emissions... 71
CW, Voice, and Data Frequencies... 73

CW, Voice, and Data Frequency Emissions	75
Word Search Game - Suppliers	78
Energy, Power, and Watts	79
High Frequency Power	81
Very and Ultra High Frequency Power	83
Ground Wave, Sky Wave, and Sun Spots	85
Phonetico Coloring Page	88
Line of Sight Communications	89
Voltage	91
Insulators, Conductors, Current, and Resistance	93
Ohm's Law, Open & Short Circuits	96
Measures	99
Morse Code Practice Toon #1	102
Resistor, Fuse, and Battery Symbols	103
Switch Symbols	105
Transistor Symbols	107
Tube Symbols	109
Antenna and Ground Symbols	111
Power Supplies	113
Receiver Overload	115
Harmonics	117
Harmonic Calculations	119
Spurious Emissions	121
Splatter	123
Term & Description Matching Game	126
Standing Wave Ratio	127
Unusual SWR Readings, Power Meters	129
Station Diagrams	131
Antenna Feed Lines, Tuners, and SWR Meters	133
Half Wavelength Antenna Length	135
Quarter Wavelength Vertical Antennas	137
Changing Antenna Length	139
The "Yagi" Beam Antenna	141
Vertical and Dipole Antennas	143
Coaxial Cable	145
Coaxial Cable Connections	147
Parallel Conductor Feed Lines	149
Health Risk and Antenna Location	151
Safety	153
Grounding	155
Lightning Protection	157
Morse Code	159
How to build a Morse Code Practice Oscillator	160
Morse Code Practice Toon #2	161
Call Sign map with State game	162
Connect the Dots	163

Technician - NO-CODE Section

The "RST" Signal Report	165
Identification	167
Emergency Operation	169
Radio Amateur Civil Emergency Service	173
Control Point and License Renewal	175
Broadcasting Third Party & Non-Amateur Contacts	177
Respect for Others	179
Dummy Antennas	181
FM Repeater Operation	183
FM Repeater Coordination, Open & Closed Repeaters	187
FM Repeater Frequencies & Simplex Operation	189
CW & Data Modulation & Emissions	191
Voice Modulation & Emissions	193
Digital Symbol Rate	195
RTTY and Data Frequency Shift & Bandwidth	197
High and Very High Frequencies	199
Frequency Use and Emissions	201
Beacon Stations & Model Craft	203
Transmitter Power	205
RF Filters & Signal Bandwidth	207
Term & Description Matching Game	210
Detectors, VFO's & FM Circuits	211
Signal & Marker Generators, Crystal Calibrations & WWV	213
The Ionosphere	215
Ionospheric Absorption	217
Ionospheric Changes	219
Troposphere VHF Communications	221
Scatter & VHF Skip	223
Ohm's Law	225
Resistors	229
Inductors	233
Inductor Cores & Symbols	235
Capacitors	237
Voltmeters & Ammeters	241
Wattmeters	243
Standing Wave Ratio & Reflectometers	245
Beam Antennas	247
Non-Directional Antennas & Polarization	249
Feed Line Losses	251
Baluns, Connectors & Coax	253
Health Risk & Antenna Location	255
Crossword Puzzle - Radio Crosstalk	258
Electric Wiring	259
Safety	261
U.S. Amateur Bands	264
X-RAY'S Common Radio Equations	265
Games Solution Page	266

RIDING THE AIRWAVES WITH ALPHA & ZULU

Study for the Novice and No-Code Technician license tests with the newest comic book characters the Phoneticos.
112 comic strips review all the questions and answers.
288 pages, 8 1/2 X 11" format

U.S. REPEATER MAPBOOK #2

A repeater guide that shows where in each state principal open amateur repeaters are located. The Maps also show the important highways in each state. Tables showing the popular repeater in the states major cities are also presented. 2 meter, 200, 440 MHz and 1.2 GHz repeaters are shown. 144 pages, 6 x 9" format

FEDERAL ASSIGNMENTS Vol 3

The Frequency assignment master file.
The complete listing of all U.S. government used frequencies listed by agency and in frequency order. Frequencies for Departments of: Agriculture, Air Force, Army, Commerce, Defence, Energy, Health and Human Services, Housing and Urban Developement, Interior, Justice, Labor, Navy, State, Treasury, Transportation and 29 Independent agencies & Commissions.
Over 350 pages, 8 1/2 X 11" format

AMATEUR HAMBOOK

Equipment & Log Sheets, Charts, Tables showing: worldwide callsigns, world times, shortwave listening frequencies, coax losses, CTCSS details, conversions, construction plans, emergency information, etc.
This book contains all the useful information a amateur radio operator needs to reference.
133 pages, 6 X 9" format.

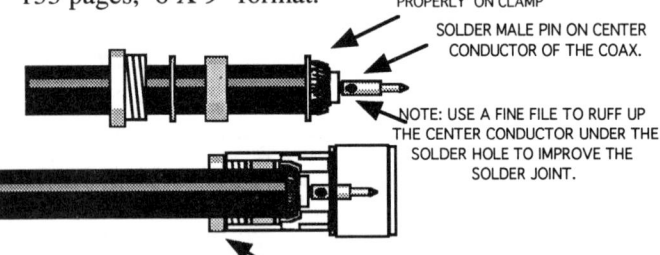

QUICK-N-EASY SHORTWAVE LISTENING

What kind of radio to Buy? Whats a good antenna, What is there to listen to? This book contains pictures of radios, antenna construction and frequency lists. 6 X 9" format

P.O. Box 1428
Burbank, CA 91507
(818) 843-4080
FAX: (818) 846-2298

RADIO/TECH MODIFICATIONS # 5A

Modifications and alignment controls for ICOM & KENWOOD amateur radios, and UNIDEN, RADIO SHACK & REGENCY Scanners. Over 200 pages, 8 1/2" X 11"

RADIO/TECH MODIFICATIONS # 5B

Modifications and alignment controls for ALINCO, YAESU, STANDARD, AZDEN radios and 10 meter & CB radios.

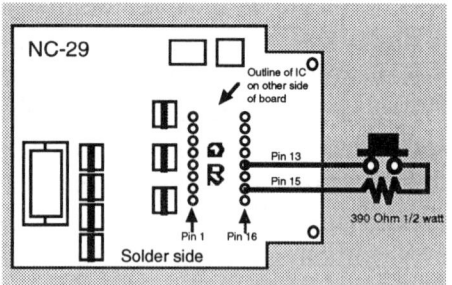

HAM RADIO RESOURCE GUIDE

For Southern California only. A booklet of all the information an amateur radio operator needs. Listings of clubs, testing centers, sario stores and surplus dealers. Maps of repeaters, store and swap meet locations. Listing of Packet repeaters, phone BBS and node lists. 64 pages 8 1/2" X 5 1/2"

NORTH AMERICAN SHORTWAVE FREQUENCY GUIDE

Accurate and complete listing of all English and Spanish broadcasts on the 0 - 30 MHz shortwave bands. Listing are presented in frequency order.
Over 200 pages, 8 1/2" X 11"

LOST USER MANUALS

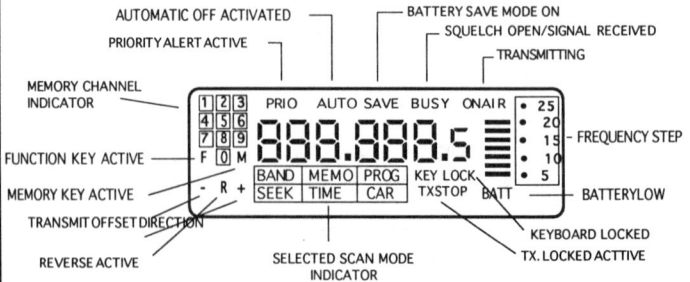

Lost the manual for your HT or Mobile rig? Did you purchase a used radio and it did not come with a manual? Do you have the manual but still can work the radio quickly?
"LOST USERS MANUALS" contains operating instructions for all the popular amateur radios and scanners. ICOM, Yaesu, Kenwood, Alinco, Standard, Uniden and other manufactors radios. Each radio is given 2 to 5 pages of drawing, charts and programming instructions. Over 140 Pages, 8 1/2 X 11" format.

Preface

This book prepares you to take two different amateur radio license exams. Both of these examinations will test your knowledge of radio theory and its proper use.

Currently over 500,000 Americans have their amateur radio license. There are over 1,500,000 radio amateurs worldwide who communicate with each other using Voice, Morse code, RTTY, Packet and Video. Friends in Japan, Africa, Europe and Australia are waiting for you to "get your ticket" and get on the air.

There are probably amateur radio clubs in your town who will be more than glad to help you study for the test. They may even have a testing session every month. Men, women, boys and girls are all welcome. It does not matter who you are, what you look like, or where you are from. Ham radio is a Hobby, a fun hobby for everyone.

You would be surprised to hear the names of some amateur radio operators. Astronaut Owen Garriott, W5LFL, who was the first to operate from space in 1983, Senator Barry Goldwater, K7UGA, actor Gary Shandling and Priscilla Presley are but a few of the people you might meet on the air.

This book uses a family of cartoon characters, called Phoneticos, to review all the questions and answers in the Examination pool. Alpha and Zulu are the stars of the book and with them you will be introduced to all the aspects of amateur radio. If you take a close look at the "Phonetico" characters, you may notice that their bodies are made up of Morse code "Dits" and "Dahs". Each Phonetico's body is the proper Morse code symbol for the characters letter.

Following each cartoon page is a testing page. If you read the cartoon, you should be able to answer each question correctly. On each question page there is an answer area for you to fill in. When you have completed the test page, you can look ahead to the answer block contained on the bottom, left side of the next page to check your answers. If you missed any, go back read the cartoon once more, and try again.

Probably the best method of studying for the exam is with a friend. Have a friendly competition to see who can complete each cartoon page and answer the questions perfectly. This method works great for Morse code also.

Don't try to read the book in one sitting. Take some time and think about what you have read. You may even want to try the test page first and see how you do and then jump back and read the cartoon. Ham radio is fun, don't make the learning a chore.

The cartoon artists have included some interesting sub-plots into the cartoon strips,. Don't be in such a hurry that you miss some of these puns and story lines. Scattered in the book you will also find some game pages. The word search games and crossword puzzles help to exercise your mind while you are studying for the test.

For those who are reading this, I am sure you always read the instructions before you start a game, puzzle or project. One of the games we have included is a Connect the Dots page. If you read the instructions first, you will have no problems; if not, then have the eraser handy. There is a valuable lesson to be learned, and we hope this page will help you learn it.

In the back of this book you will find a number of reference pages that can be used over and over again. Your amateur radio library begins here and will expand as you select which area of amateur radio interests you.

If you have half the fun reading this book as the artists had creating the characters and the cartoon strips, then go out and pass the test and send us a QSL card or letter and let us know which character you like best, what cartoon was your favorite and any suggestions you may have.

Good luck and 73

WHAT IS HAM RADIO?

Ham radio is a fun hobby! It allows people all over the world to communicate continuously with each other, and is available to anyone, anywhere, who has obtained an amateur radio license. You will always find a friend to talk to .

Ham Radio becomes a part of you. Whether you are 8 or 80, it doesn't matter who you are when you are on the air! Those who obtain their licenses will have this experience for the rest of their lives, and may be able to provide communications to help people during emergencies. Ham Radio can bring you many advantages, and lead to a career in science or communications! Even if you do not obtain a license, a Ham Radio involvement will give you:

1) <u>The ability to talk over a microphone to anyone</u>. Once this ability is acquired, public speaking becomes a natural. Never again will you suffer "stage fright" or feel ill at ease when speaking. You will be able to "think as you talk"!

2) <u>A knowledge of the Morse code</u>. Morse code was the first method of radio communication, and was the beginning of today's digital communications. It uses the least amount of radio spectrum, can "get through" when voice cannot, and permits contact between people with different languages. Contact with deep space satellites is maintained using very slow speed digital signals, very similar to Morse code. If we contact another civilization in space, it will be by use of slow speed digital communications!

3) <u>A knowledge of geography.</u> Ham Radio crosses all borders and time zones. You will talk to people worldwide and become aware of "Universal Time", measured in Greenwich, England. When a station is contacted, it can be located on a map of the U.S.A. or the world, and you and the ham at the other end can share interesting details about your location.

4) <u>A beginning technical knowledge of radio, electronics, and an application of mathematics</u>. Ham Radio requires a knowledge of basic mathematics. You will see how electrical currents flow in a circuit, begin to understand how a radio wave travels around the earth, learn that all phenomenon in the universe are related to some form of electromagnetic wave, such as the light and heat from the stars, planets, and the sun. You will experience the sun's effect on radio waves, caused by eruptions on the sun called "sun spots". *MOST OF ALL YOU WILL HAVE FUN!* 73 (Best Regards) K6YB

INFORMATION FOR THE TEACHER

"Riding the Air Waves with ALPHA and ZULU" was written to simplify the experience of studying for the Novice and No-Code Technician amateur radio licenses! The No-Code Technician examination substitutes 25 technical questions for the 5 word-per-minute code test of the 30 question Novice examination, for a total of 55 questions. Of course you can take the code test and the No-Code technician examination, for even greater operating privileges as a Technician Plus.

The first five Novice topics can be taught as a general introduction to amateur radio, and permit the students to get on the air in a "professional" manner. Students are urged to use phonetics for their names, and call "CQ" on the air, using the format in the first Novice topic. An amateur radio station should be installed at the school, containing the equipment referred to in the examination questions. A licensed amateur must of course be the control operator!

Those students who do well on the first five Novice topics, and enjoy Morse code, should be given an accelerated class to cover the rest of the topics. Those who have difficulty with the code should be encouraged to study for the more advanced No-Code Technician license. Everyone should be encouraged to study at their own pace, and be given the opportunity to get on the air frequently. Often, a slow student will suddenly get the amateur radio "bug"!

A typical class of one hour may be organized in the following manner: The first quarter hour, code practice; The second quarter hour, an explanation and discussion of one theory topic from this book; The last half hour, on the air time, practically demonstrating the theory and code! Frequent code and theory examinations should be given, and the students separated into groups according to their progress.

Students should be encouraged to study at home. If a computer is available to the instructor, a print out of an actual examination, using available software, should be given as a take home test. The students can refer to this book for the answers to reinforce the classroom sessions, and to provide material for additional home study.

The primary object of this book is to provide the student with the minimum amount of technical information required to pass the Novice and No-Code Technician examinations. A detailed technical explanation of many of the subjects tends to confuse and discourage many potential young amateurs. The information is available elsewhere, should the student desire to go further.

Artsci Inc. P.O.Box 1428, Burbank, CA 91507, the National Amateur Radio Association (NARA), P.O. Box 201407, Arlington, TX 76006, and the American Radio Relay League (ARRL), 225 Main St., Newington, CT 06111, publish many excellent technical books, have computer examination software available, and provide free material for beginners. Write to ARTSCI, NARA and ARRL for further information.

Which license is for you ?

The first license available to you will be the "Novice" license. To receive a Novice class license, you must pass two tests; a 30 question multiple choice test on the theory of radio and its proper operation and a 10 question test proving your ability to copy a 5 word per minute Morse conversation.

The second license available is the new "No-Code" Technician license. This license has no Morse code requirement. However, it does have a 55 question multiple choice test which includes all the information used in the Novice license exam along with 25 additional technical questions.

The "No-Code" License is the highest class license available to you without a Morse code requirement. If you elect to study Morse Code (It's more fun than you think), you can continue up the licensing ladder. The licenses available from the FCC are: Novice, Technician, Technician Plus, General, Advanced and Extra. Each license has specific testing and most have Morse code knowledge as a requirement. The table below shows the requirement for each license.

License Class	Test Element	Type of Examination
Novice	Element 2 Element 1A	30-Question Written Exam 5 - Word per minute Code test
Technician	Element 2 Element 3A	55-Question Written Exam (in 2 parts, 30-question element 2 & 25 question element 3A)
Technician Plus	Element 2 Element 3A Element 1A	30-Question Written Exam 25-Question Written Exam 5 - Word per minute Code test
General	Element 3B Element 1B	25-Question Written Exam 13 - Word per minute Code test A Technician License
Advanced	Element 4A	50-Question Written Exam A General Class License
Extra	Element 4B Element 1C	40-Question Written Exam 20 - Word per minute Code test An Advanced Class License

To Morse or not to Morse

You don't need to answer this question right away. You could make the decision after you have finished the first section of this book. That will be the time to decide which license you will work towards. In fact you can decide to learn Morse code at any time after your have your Technician license.

If you are studying with a friend or in a class room, learning Morse Code will help to break up the study routine. It is a great deal easier to learn Morse code with another person. You can even practice Morse code over the telephone at night!!

Give Morse a try before you decide. It may turn out to be more fun and easier that you think.

If you do decide to study Morse Code, below is a suggested sequence:

```
Lesson #1    T H O S E
Lesson #2    G R I P W
Lesson #3    M A N L F
Lesson #4    D U C K Q
Lesson #5    X Z B J X V
Lesson #6    1 2 3 4 5
Lesson #7    6 7 8 9 0
Lesson #8    (.) (,) (?) (/) (AR) (SK) (BT)
```

There are many cassette tapes available that will help you learn the Morse code. You may be able to borrow one from someone in an amateur radio club.

To learn Morse code in this book, we make use of words that suggest the sound of the Morse code symbols. We use the words Dit and Dah to show the Morse code components of each letter. We do this in place of the symbol "•" and "-".

The letter "A" in Morse code is Di Dah. It sounds just like "Di Dah" over the radio. It does not sound like "• -", you do not transmit "• -", Don't learn "• -". Learn the letter as it sounds.

Most people have made the mistake of learning all the letters in a series of "•" and "-". You can learn the code at 5 words per minute this way, but it takes too much time and brain power to hear "Di Dah" and turn the Di into a "•" and the Dah into a "-" and then reassemble the "•" and "-" into "• -" and then remember "• -" is an "A".

Learn the sound of the Morse code letter, and pretend Morse code is another language. If you do learn this way, you can make the jump from 5 words per minute to 13 and 20 words per minute without much effort.

Where can you take the test ?

Once you have studied for your test, there are testing groups in every state who will be happy to let you take the test. A minor fee may be charged by the testing group. The amount is limited by the FCC and is currently below $6.00.

To find a local testing group in your area, contact one of the Volunteer-Examiner Coordinators near you to get a list of the testing sites.

U.S. Volunteer-Examiner Coordinators in Amateur Service

Anchorage Amateur Radio Club
2628 Turnagain Parkway
Anchorage, AK 99503
(907) 243-2221, 344-5401

ARRL/VEC
225 Main Street
Newington, CT 06111
(203) 666-1541

Central Alabama VEC, Inc.
606 Tremont Street
Selma, AL 36701
H: (205) 872-1166, 872-5450
O: (205) 874-1688

Charlotte VEC
227 Bennett Lane
Charlotte, NC 28213
(704) 596-2168

Golden Empire Amateur Radio Society
P.O. Box 508
Chico, CA 95927

Greater Los Angeles Amateur Radio Group
9737 Noble Avenue
Sepulveda, CA 91343
(818) 892-2068

Jefferson Amateur Radio Club
P.O. Box 73665
Metairie, LA 70033

Koolau Amateur Radio Club
45-529 Nakuluai Street
Kaneohe, HI 96744
(808) 235-4132

Laurel Amateur Radio Club
P.O. Box 3039
Laurel, MD 20709-0039
(301) 434-6087, 572-5124

Mountain Amateur Radio Club
P.O. Box 234
Cumberland, MD 21502
(304) 289-3576

PHD Amateur Radio Association, Inc.
P.O. Box 11
Liberty, MO 64068
(816) 781-7313

Triad Emergency Amateur Radio Club
3504 Stonehurst Place
High Point, NC 27260
(919) 841-7576

Sandarc-VEC
P.O. Box 2456
La Mesa, CA 92044
(619) 465-3926

Sunnyvale VEC Amateur Radio Club
P.O. Box 60307
Sunnyvale, CA 94088-0307
(408) 255-9000

The Milwaukee Radio Amateurs Club, Inc.
1737 N. 116th St.
Wauwatosa, WI 53226
(414) 774-6999

Western Carolina Amateur Radio Society
5833 Clinton Hwy, Suite 203
Knoxville, TN 37912-2545
(615) 688-7771

W5YI-VEC
P.O. Box 565101
Dallas, TX 75356-5101
(817) 461-6443

CALL FOR OTHER LOCATIONS
(800) 669- W5YI

Notes

Riding the Airwaves with Alpha & Zulu

Riding the Airwaves with Alpha & Zulu

SEE IF YOU CAN ANSWER THESE QUESTIONS WITHOUT LOOKING BACK TO THE LAST 'TOON'. WRITE YOUR ANSWERS IN THE ANSWER BOX BELOW. GOOD LUCK.

PRACTICE TEST PAGE N01

N2A20
To make your call sign better understood when using voice transmissions, what should you do?
- A. Use Standard International Phonetics for each letter of your call
- B. Use any words which start with the same letters as your call sign for each letter of your call
- C. Talk louder
- D. Turn up your microphone gain

N2A19
How should you answer a voice CQ call?
- A. Say the other station's call sign at least ten times, followed by "this is," then your call sign at least twice
- B. Say the other station's call sign at least five times phonetically, followed by "this is," then your call sign at least once
- C. Say the other station's call sign at least three times, followed by "this is," then your call sign at least five times phonetically
- D. Say the other station's call sign once, followed by "this is," then your call sign given phonetically

N2A11
What is meant by the term "DX"?
- A. Best regards
- B. Distant station
- C. Calling any station
- D. Go ahead

N2A12
What is the meaning of the term "73"?
- A. Long distance
- B. Best regards
- C. Love and kisses
- D. Go ahead

N2A08
What is the meaning of the procedural signal "CQ"?
- A. "Call on the quarter hour"
- B. "New antenna is being tested" (no station should answer)
- C. "Only the called station should transmit"
- D. "Calling any station"

N2A18
What is the correct way to call CQ when using voice?
- A. Say "CQ" once, followed by "this is," followed by your call sign spoken three times
- B. Say "CQ" at least five times, followed by "this is," followed by your call sign spoken once
- C. Say "CQ" three times, followed by "this is," followed by your call sign spoken three times
- D. Say "CQ" at least ten times, followed by "this is," followed by your call sign spoken once

YOUR ANSWERS TO THIS TEST

N2A20	_____
N2A11	_____
N2A19	_____
N2A12	_____
N2A08	_____
N2A18	_____

artsci inc

PHONETICO FINDER

THE PUZZLE BELOW HAS THE NAMES OF ALL THE PHONETICOS YOU HAVE MET, PLUS OTHERS THAT ARE STILL TO COME. THE FULL LIST OF NAMES IS AT THE BOTTOM OF THE PAGE. CAN YOU FIND THEM ALL?

	1	2	3	4	5	6	7	8	9	10	11	12	13	14	15	16
A	X	R	A	Y	S	F	O	X	T	R	O	T	T	E	M	R
B	T	A	T	A	N	G	O	B	J	U	L	I	E	T	T	A
C	A	C	S	N	S	B	U	W	I	K	I	L	O	I	Z	B
D	C	Y	H	K	A	G	O	L	F	L	U	Y	S	N	U	O
E	L	Q	U	E	B	E	C	O	I	S	L	L	C	T	L	U
F	R	O	M	E	O	U	H	S	I	E	R	R	A	P	U	T
G	K	H	A	L	P	H	A	H	L	T	A	B	R	A	V	O
H	R	V	I	C	T	O	R	O	O	L	M	U	S	P	L	C
I	I	S	I	R	W	R	L	T	U	I	E	N	T	A	S	A
J	S	D	N	A	O	E	I	E	D	M	M	I	K	E	M	R
K	S	F	D	N	W	D	E	L	T	A	E	F	M	I	A	R
L	W	H	I	S	K	E	Y	L	A	T	J	O	H	N	R	O
M	L	I	A	A	A	C	L	U	C	K	Y	R	O	S	T	T
N	A	H	L	S	B	H	P	N	O	V	E	M	B	E	R	J
O	M	V	T	F	O	O	R	U	T	A	C	T	I	N	G	A
P	R	B	A	P	R	I	L	S	H	O	W	E	R	S	R	N

MY HAT'S OFF TO YOU IF YOU CAN FIND US ALL.

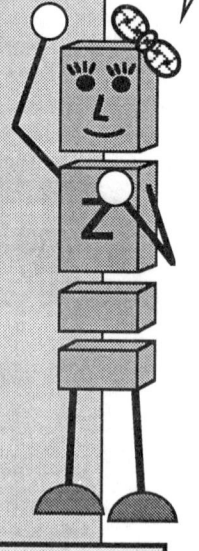

ALPHA MAY BE CORNY, BUT HE'S A LOT OF FUN TO BE WITH.

ALPHA	JULIETT	SIERRA
BRAVO	KILO	TANGO
CHARLIE	LIMA	UNIFORM
DELTA	MIKE	VICTOR
ECHO	NOVEMBER	WHISKEY
FOXTROT	OSCAR	XRAY
GOLF	PAPA	YANKEE
HOTEL	QUEBEC	ZULU
INDIA	ROMEO	

Riding the Airwaves with Alpha & Zulu

Riding the Airwaves with Alpha & Zulu

SEE IF YOU CAN ANSWER THESE QUESTIONS WITHOUT LOOKING BACK TO THE LAST 'TOON'. WRITE YOUR ANSWERS IN THE ANSWER BOX BELOW. GOOD LUCK.

PRACTICE TEST PAGE N02

N2A13
What are RST signal reports?
- A. A short way to describe ionospheric conditions
- B. A short way to describe transmitter power
- C. A short way to describe signal reception
- D. A short way to describe sunspot activity

N2A14
What does RST mean in a signal report?
- A. Recovery, signal strength, tempo
- B. Recovery, signal speed, tone
- C. Readability, signal speed, tempo
- D. Readability, signal strength, tone

N2A16
What is one meaning of the Q signal "QTH"?
- A. Time here is
- B. My name is
- C. Stop sending
- D. My location is

N2A17
What is a QSL card?
- A. A letter or postcard from an amateur pen pal
- B. A Notice of Violation from the FCC
- C. A written proof of communication between two amateurs
- D. A postcard reminding you when your license will expire

N2A15
What is one meaning of the Q signal "QRS"?
- A. Interference from static
- B. Send more slowly
- C. Send RST report
- D. Radio station location is

ANSWERS TO PREVIOUS TEST

N2A20	A
N2A11	B
N2A19	D
N2A12	B
N2A08	D
N2A18	C

YOUR ANSWERS TO THIS TEST

N2A13 _____

N2A14 _____

N2A16 _____

N2A17 _____

N2A15 _____

artsci inc

Riding the Airwaves with Alpha & Zulu

SEE IF YOU CAN ANSWER THESE QUESTIONS WITHOUT LOOKING BACK TO THE LAST 'TOON'. WRITE YOUR ANSWERS IN THE ANSWER BOX BELOW. GOOD LUCK.

PRACTICE TEST PAGE N03

N1H06 [97.119a]
How often must an amateur station be identified?
- A. At the beginning of a contact and at least every ten minutes after that
- B. At least once during each transmission
- C. At least every ten minutes during and at the nd of a contact
- D. At the beginning and end of each transmission

N1H11 [97.119a]
What is the longest period of time an amateur station can operate without transmitting its call sign?
- A. 5 minutes
- B. 10 minutes
- C. 15 minutes
- D. 20 minutes

N1H07 [97.119a]
What do you transmit to identify your amateur station?
- A. Your "handle"
- B. Your call sign
- C. Your first name and your location
- D. Your full name

N1H08 [97.119a]
What identification, if any, is required when two amateur stations begin communications?
- A. No identification is required
- B. One of the stations must give both stations' call signs
- C. Each station must transmit its own call sign
- D. Both stations must transmit both call signs

N1H09 [97.119a]
What identification, if any, is required when two amateur stations end communications?
- A. No identification is required
- B. One of the stations must transmit both stations' call signs
- C. Each station must transmit its own call sign
- D. Both stations must transmit both call signs

ANSWERS TO PREVIOUS TEST	
N2A13	C
N2A14	D
N2A16	D
N2A17	C
N2A15	B

YOUR ANSWERS TO THIS TEST	
N1H06	
N1H11	
N1H07	
N1H08	
N1H09	

Riding the Airwaves with Alpha & Zulu

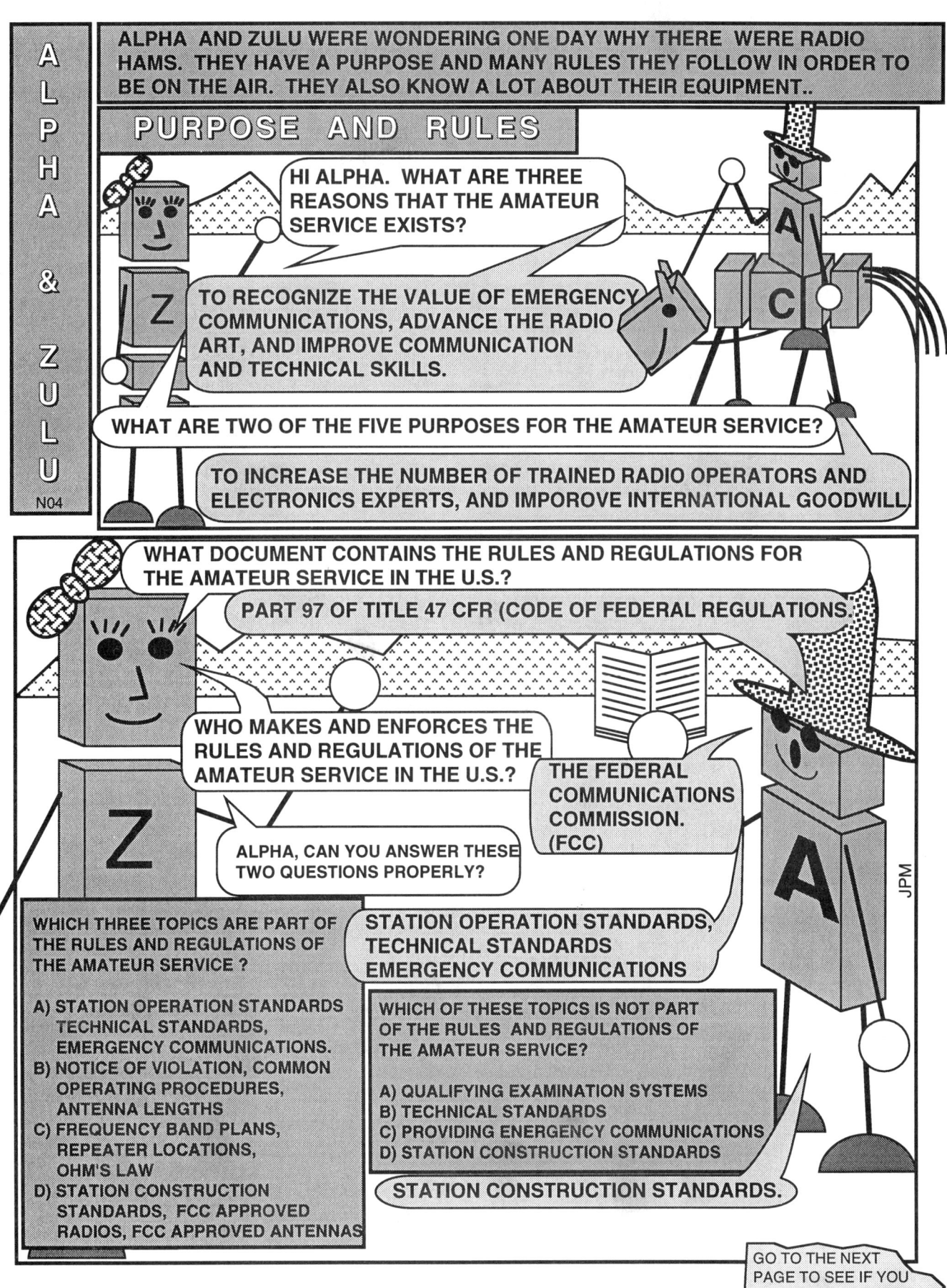

Riding the Airwaves with Alpha & Zulu

SEE IF YOU CAN ANSWER THESE QUESTIONS WITHOUT LOOKING BACK TO THE LAST 'TOON'. WRITE YOUR ANSWERS IN THE ANSWER BOX BELOW. GOOD LUCK.

PRACTICE TEST PAGE N04

N1A05 [97.1]
What are three reasons that the amateur service exists?
A. To recognize the value of emergency communications, advance the radio art, and improve communication and technical skills
B. To learn about business communications, increase testing by trained technicians, and improve amateur communications
C. To preserve old radio techniques, maintain a pool of people familiar with early tube-type equipment, and improve tube radios
D. To improve patriotism, preserve nationalism, and promote world peace

N1A06 [97.1]
What are two of the five purposes for the amateur service?
A. To protect historical radio data, and help the public understand radio history
B. To help foreign countries improve communication and technical skills, and encourage visits from foreign hams
C. To modernize radio schematic drawings, and increase the pool of electrical drafting people
D. To increase the number of trained radio operators and electronics experts, and improve international goodwill

N1A01 [97]
What document contains the rules and regulations for the amateur service in the US?
A. Part 97 of Title 47 CFR (Code of Federal Regulations)
B. The Communications Act of 1934 (as amended)
C. The Radio Amateur's Handbook
D. The minutes of the International Telecommunication Union meetings

N1A02 [97]
Who makes and enforces the rules and regulations of the amateur service in the US?
A. The Congress of the United States
B. The Federal Communications Commission (FCC)
C. The Volunteer Examiner Coordinators (VECs)
D. The Federal Bureau of Investigation (FBI)

N1A04 [97]
Which of these topics is NOT part of the rules and regulations of the amateur service?
A. Qualifying examination systems
B. Technical standards
C. Providing emergency communications
D. Station construction standards

N1A03 [97]
Which three topics are part of the rules and regulations of the amateur service?
A. Station operation standards, technical standards, emergency communications
B. Notice of Violation, common operating procedures, antenna lengths
C. Frequency band plans, repeater locations, Ohm's law
D. Station construction standards, FCC approved radios, FCC approved antennas

ANSWERS TO PREVIOUS TEST
- N1H06 C
- N1H11 B
- N1H07 B
- N1H08 A
- N1H09 C

YOUR ANSWERS TO THIS TEST
- N1A05 ____
- N1A06 ____
- N1A01 ____
- N1A02 ____
- N1A04 ____
- N1A03 ____

Riding the Airwaves with Alpha & Zulu

SEE IF YOU CAN ANSWER THESE QUESTIONS WITHOUT LOOKING BACK TO THE LAST 'TOON'. WRITE YOUR ANSWERS IN THE ANSWER BOX BELOW. GOOD LUCK.

PRACTICE TEST PAGE N05

N1A07 [97.3a1]
What is the definition of an amateur operator?
- A. A person who has not received any training in radio operations
- B. A person who has a written authorization to be the control operator of an amateur station
- C. A person who has very little practice operating a radio station
- D. A person who is in training to become the control operator of a radio station

N1A09 [97.3a5]
What is the definition of an amateur station?
- A. A station in a public radio service used for radio communications
- B. A station using radiocommunications for a commercial purpose
- C. A station using equipment for training new radio communications operators
- D. A station in an Amateur Radio service used for radio communications

N1A08 [97.3a4]
What is the definition of the amateur service?
- A. A private radio service used for profit and public benefit
- B. A public radio service for US citizens which requires no exam
- C. A personal radio service used for self-training, communication, and technical studies
- D. A private radio service used for self-training of radio announcers and technicians

N1A10 [97.3a11]
What is the definition of a control operator of an amateur station?
- A. Anyone who operates the controls of the station
- B. Anyone who is responsible for the station's equipment
- C. Any licensed amateur operator who is responsible for the station's transmissions
- D. The amateur operator with the highest class of license who is near the controls of the station

N1A11 [97.513a]
What is a Volunteer Examiner (VE)?
- A. An amateur who volunteers to check amateur teaching manuals
- B. An amateur who volunteers to teach amateur classes
- C. An amateur who volunteers to test others for amateur licenses
- D. An amateur who volunteers to examine amateur station equipment

ANSWERS TO PREVIOUS TEST
N1A05
A
N1A06
D
N1A01
A
N1A02
B
N1A03
A
N1A04
D

YOUR ANSWERS TO THIS TEST
N1A07
N1A08
N1A09
N1A10
N1A11

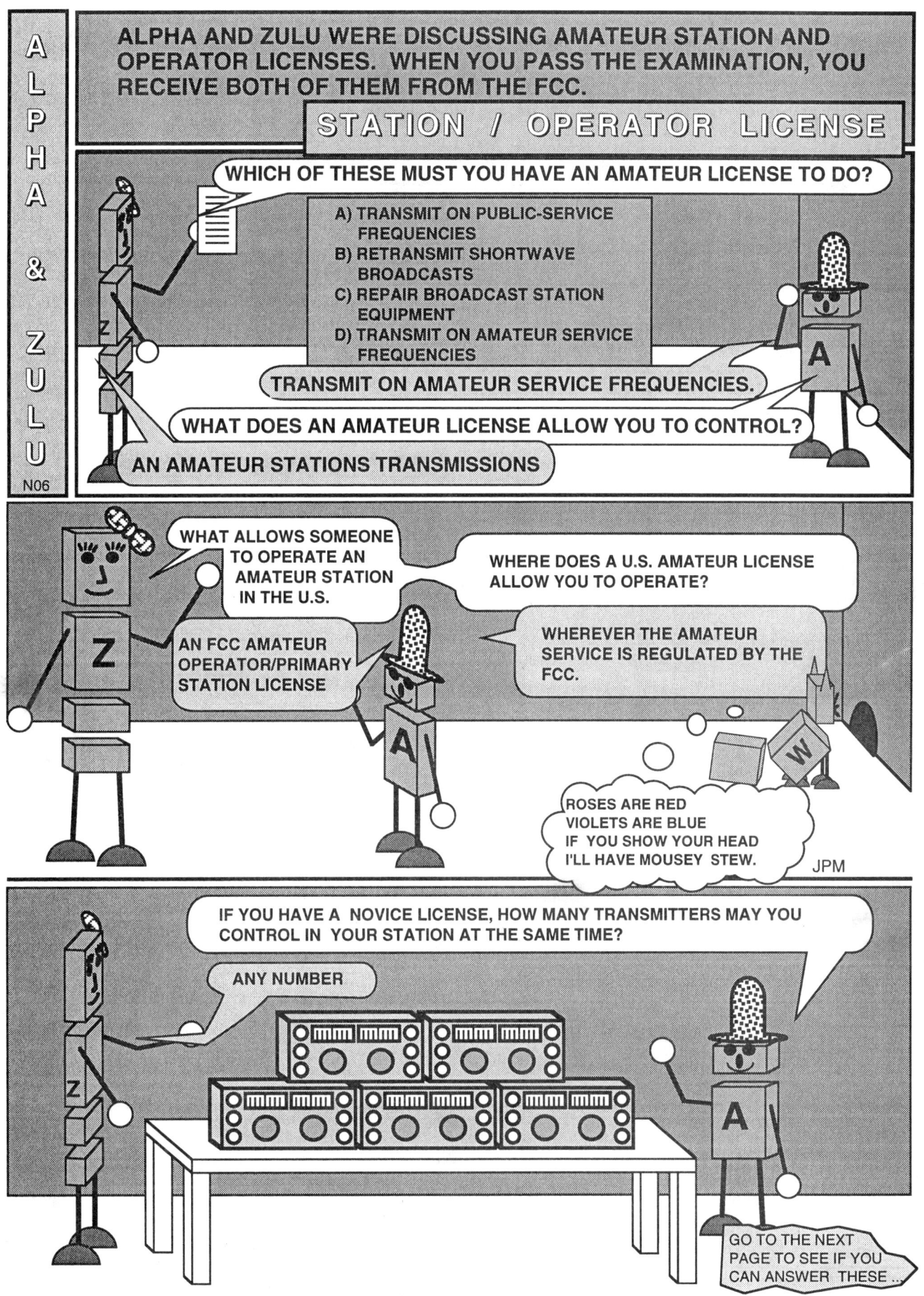

Riding the Airwaves with Alpha & Zulu

SEE IF YOU CAN ANSWER THESE QUESTIONS WITHOUT LOOKING BACK TO THE LAST 'TOON'. WRITE YOUR ANSWERS IN THE ANSWER BOX BELOW. GOOD LUCK.

PRACTICE TEST PAGE   N06

N1B01 [97.5a]
Which one of these must you have an amateur license to do?
A. Transmit on public-service frequencies
B. Retransmit shortwave broadcasts
C. Repair broadcast station equipment
D. Transmit on amateur service frequencies

N1B02 [97.5a]
What does an amateur license allow you to control?
A. A shortwave-broadcast station's transmissions
B. An amateur station's transmissions
C. Non-commercial FM broadcast transmissions
D. Any type of transmitter, as long as it is used for non-commercial transmissions

N1B03 [97.5a]
What allows someone to operate an amateur station in the US?
A. An FCC operator's training permit for a licensed radio station
B. An FCC Form 610 together with a license examination fee
C. An FCC amateur operator/primary station license
D. An FCC Certificate of Successful Completion of Amateur Training

N1B04 [97.5d]
Where does a US amateur license allow you to operate?
A. Anywhere in the world
B. Wherever the amateur service is regulated by the FCC
C. Within 50 km of your primary station location
D. Only at your primary station location

N1B05 [97.5e]
If you have a Novice license, how many transmitters may you control in your station at the same time?
A. Only one at a time
B. Only one at a time, except for emergency communications
C. Any number
D. Any number, as long as they are transmitting on different bands

ANSWERS TO PREVIOUS TEST
N1A07
B
N1A08
C
N1A09
D
N1A10
C
N1A11
C

YOUR ANSWERS TO THIS TEST
N1B01
N1B02
N1B03
N1B04
N1B05

artsci inc

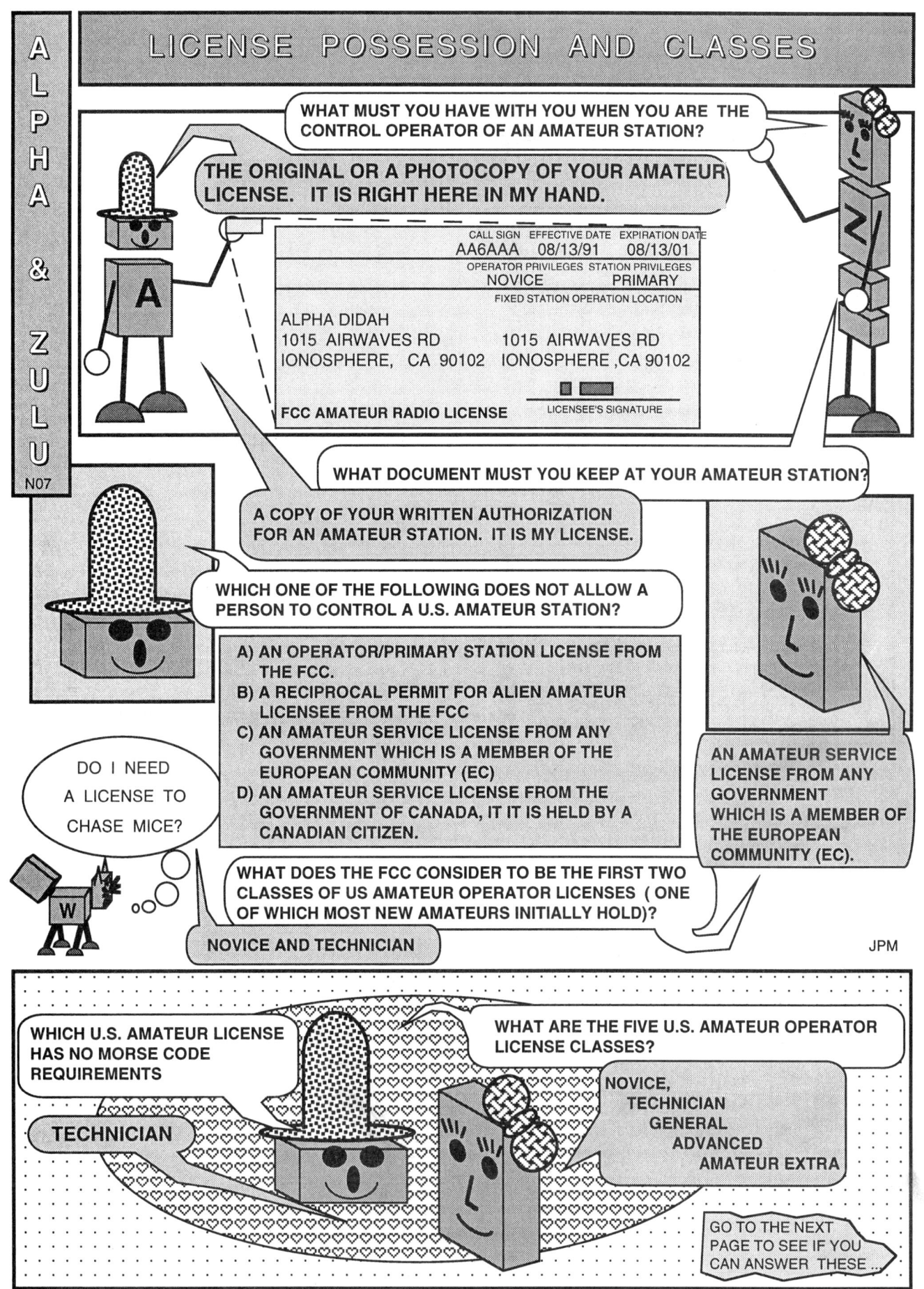

Riding the Airwaves with Alpha & Zulu

SEE IF YOU CAN ANSWER THESE QUESTIONS WITHOUT LOOKING BACK TO THE LAST 'TOON'. WRITE YOUR ANSWERS IN THE ANSWER BOX BELOW. GOOD LUCK.

PRACTICE TEST PAGE N07

N1B06 [97.5e]
What document must you keep at your amateur station?
- A. A copy of your written authorization for an amateur station
- B. A copy of the Rules and Regulations of the Amateur Service (Part 97)
- C. A copy of the Amateur Radio Handbook for instant reference
- D. A chart of the frequencies allowed for your class of license

N1B07) [97.7]
Which one of the following does not allow a person to control a US amateur station?
- A. An operator/primary station license from the FCC
- B. A reciprocal permit for alien amateur licensee from the FCC
- C. An amateur service license from any government which is a member of the European Community (EC)
- D. An amateur service license from the Government of Canada, if it is held by a Canadian citizen

N1B08 [97.9a]
What are the five US amateur operator License classes?
- A. Novice, Communicator, General, Advanced, Amateur Extra
- B. Novice, Technician, General, Advanced, Expert
- C. Novice, Communicator, General, Amateur, Extra
- D. Novice, Technician, General, Advanced, Amateur Extra

N1B09 [97.9]
What does the FCC consider to be the first two classes of US amateur operator licenses (one of which most new amateurs initially hold)?
- A. Novice and Technician
- B. CB and Communicator
- C. Novice and General
- D. CB and Novice

N1B10 [97.9]
What must you have with you when you are the control operator of an amateur station?
- A. A copy of the Rules and Regulations of the Amateur Service (Part 97)
- B. The original or a photocopy of your amateur license
- C. A list of countries which allow third-party communications from the US
- D. A chart of the frequencies allowed for your class of license

N1B11 [97.501d]
Which US amateur license has no Morse code requirements?
- A. Amateur Extra
- B. Advanced
- C. General
- D. Technician

ANSWERS TO PREVIOUS TEST

N1B01	D
N1B02	B
N1B03	C
N1B04	B
N1B05	C

YOUR ANSWERS TO THIS TEST

N1B06	
N1B07	
N1B08	
N1B09	
N1B10	
N1B11	

artsci inc

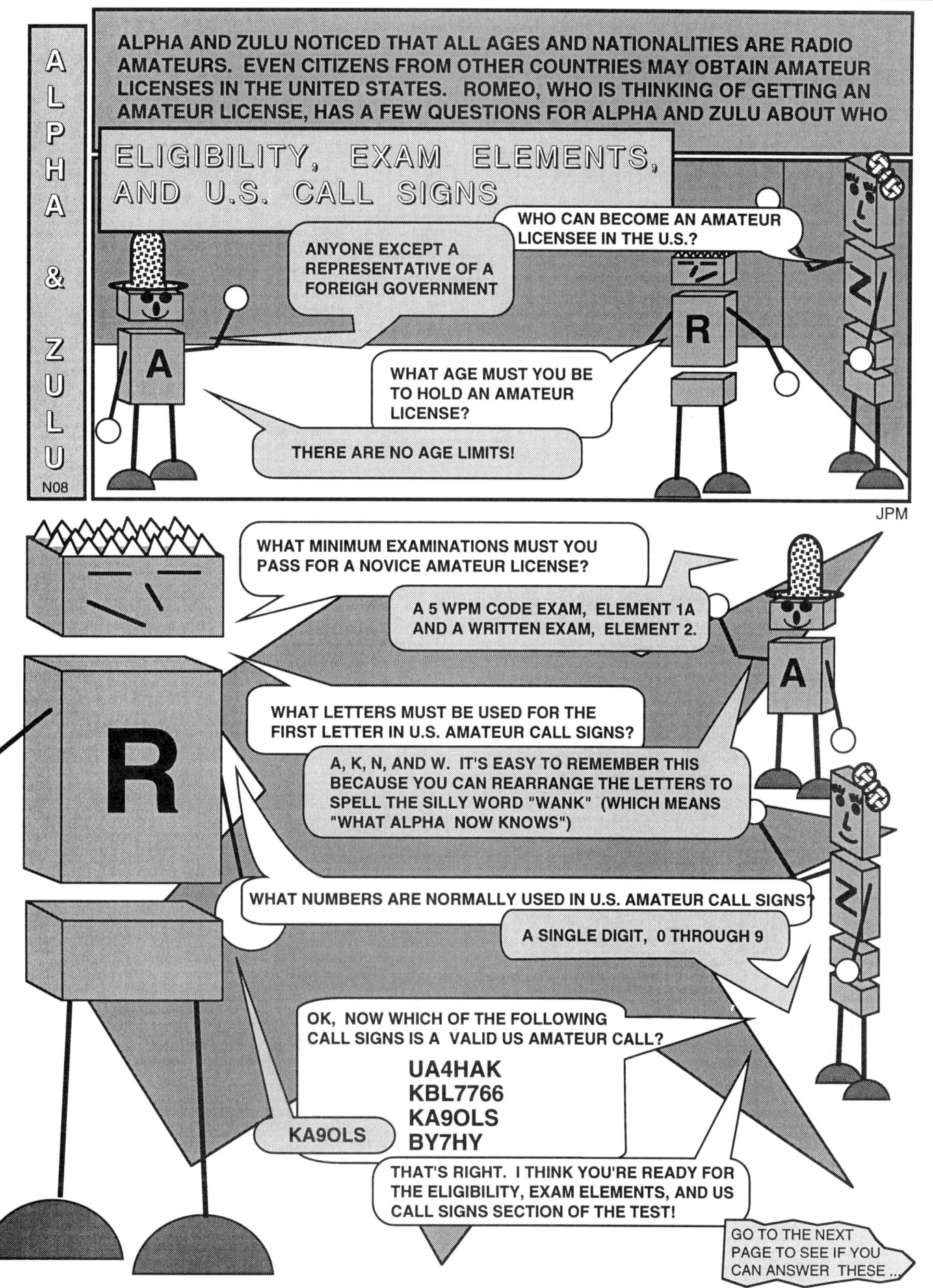

Riding the Airwaves with Alpha & Zulu

SEE IF YOU CAN ANSWER THESE QUESTIONS WITHOUT LOOKING BACK TO THE LAST 'TOON'. WRITE YOUR ANSWERS IN THE ANSWER BOX BELOW. GOOD LUCK.

PRACTICE TEST PAGE N08

N1D01 [97.5d1]
Who can become an amateur licensee in the US?
- A. Anyone except a representative of a foreign government
- B. Only a citizen of the United States
- C. Anyone except an employee of the US government
- D. Anyone

N1D03 [97.501e]
What minimum examinations must you pass for a Novice amateur license?
- A. A written exam, Element 1(A); and a 5 WPM code exam, Element 2(A)
- B. A 5 WPM code exam, Element 1(A); and a written exam, Element 3(A)
- C. A 5 WPM code exam, Element 1(A); and a written exam, Element 2
- D. A written exam, Element 2; and a 5 WPM code exam, Element 4

N1D02
What age must you be to hold an amateur license?
- A. 14 years or older
- B. 18 years or older
- C. 70 years or younger
- D. There are no age limits

N1D07
Which of the following call signs is a valid US amateur call?
- A. UA4HAK
- B. KBL7766
- C. KA9OLS
- D. BY7HY

N1D08
What letters must be used for the first letter in US amateur call signs?
- A. K, N, U and W
- B. A, K, N and W
- C. A, B, C and D
- D. A, N, V and W

N1D09
What numbers are normally used in US amateur call signs?
- A. Any two-digit number, 10 through 99
- B. Any two-digit number, 22 through 45
- C. A single digit, 1 though 9
- D. A single digit, 0 through 9

ANSWERS TO PREVIOUS TEST
N1B06
A
N1B07
C
N1B08
D
N1B09
A
N1B10
B
N1B11
D

YOUR ANSWERS TO THIS TEST
N1D01
N1D02
N1D03
N1D07
N1D08
N1D09

artsci inc

PRACTICE TEST PAGE N09

SEE IF YOU CAN ANSWER THESE QUESTIONS WITHOUT LOOKING BACK TO THE LAST 'TOON'. WRITE YOUR ANSWERS IN THE ANSWER BOX BELOW. GOOD LUCK.

N1D04 [97.21]
Why must an amateur operator have a current US Postal mailing address?
- A. So the FCC has a record of the location of each amateur station
- B. To follow the FCC rules and so the licensee can receive mail from the FCC
- C. So the FCC can send license-renewal notices
- D. So the FCC can publish a call-sign directory

N1D05 [97.27]
What must you do to replace your license if it is lost, mutilated or destroyed?
- A. Nothing; no replacement is needed
- B. Send a change of address to the FCC using a current FCC Form 610
- C. Retake all examination elements for your license
- D. Request a new one from the FCC, explaining what happened to the original

N1D06 [97.19]
What must you do to notify the FCC if your mailing address changes?
- A. Fill out an FCC Form 610 using your new address, attach a copy of your license, and mail it to your local FCC Field Office
- B. Fill out an FCC Form 610 using your new address, attach a copy of your license, and mail it to the FCC office in Gettysburg, PA
- C. Call your local FCC Field Office and give them your new address over the phone
- D. Call the FCC office in Gettysburg, PA, and give them your new address over the phone

N1D10 [97.23]
For how many years is an amateur license normally issued?
- A. 2
- B. 5
- C. 10
- D. 15

N1D11 [97.19c]
How soon before your license expires should you send the FCC a completed 610 for a renewal?
- A. 60 to 90 days
- B. within 21 days of the expiration date
- C. 6 to 9 months
- D. 6 months to a year

ANSWERS TO PREVIOUS TEST

N1D01	A
N1D02	D
N1D03	C
N1D07	C
N1D08	B
N1D09	D

YOUR ANSWERS TO THIS TEST

- N1D04 ____
- N1D06 ____
- N1D05 ____
- N1D10 ____
- N1D11 ____

Riding the Airwaves with Alpha & Zulu

SEE IF YOU CAN ANSWER THESE QUESTIONS WITHOUT LOOKING BACK TO THE LAST 'TOON'. WRITE YOUR ANSWERS IN THE ANSWER BOX BELOW. GOOD LUCK.

PRACTICE TEST PAGE N10

N2A04
If you are in contact with another station and you hear an emergency call for help on your frequency, what should you do?
  A. Tell the calling station that the frequency is in use
  B. Direct the calling station to the nearest emergency net frequency
  C. Call your local Civil Preparedness Office and inform them of the emergency
  D. Stop your QSO immediately and take the emergency call

N1J10 [97.403]
When may you use your amateur station to transmit an "SOS" or "MAYDAY"?
  A. Never
  B. Only at specific times (at 15 and 30 minutes after the hour)
  C. In a life or property threatening emergency
  D. When the National Weather Service has announced a severe weather watch

N1J08 [97.405a]
If you hear a voice distress signal on a frequency outside of your license privileges, what are you allowed to do to help the station in distress?
  A. You are NOT allowed to help because the frequency of the signal is outside your privileges
  B. You are allowed to help only if you keep your signals within the nearest frequency band of your privileges
  C. You are allowed to help on a frequency outside your privileges only if you use international Morse code
  D. You are allowed to help on a frequency outside your privileges in any way possible

N1J11 [97.405a]
When may you send a distress signal on any frequency?
  A. Never
  B. In a life or property threatening emergency
  C. Only at specific times (at 15 and 30 minutes after the hour)
  D. When the National Weather Service has announced a severe weather watch

ANSWERS TO PREVIOUS TEST
N1D04
B
N1D06
B
N1D05
D
N1D10
C
N1D11
A

YOUR ANSWERS TO THIS TEST
N2A04
N1J08
N1J10
N1J11

artsci inc

Riding the Airwaves with Alpha & Zulu

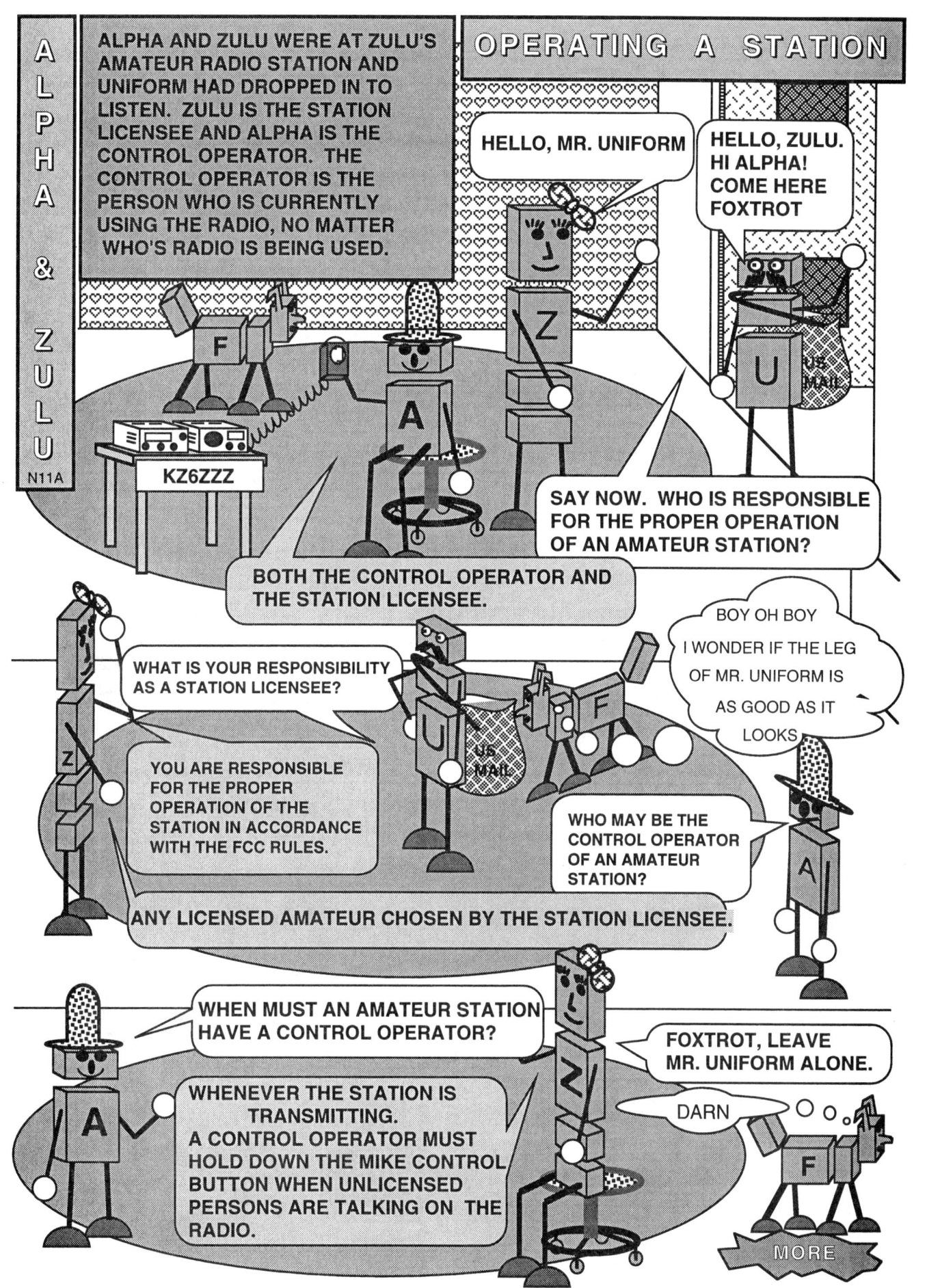

Riding the Airwaves with Alpha & Zulu

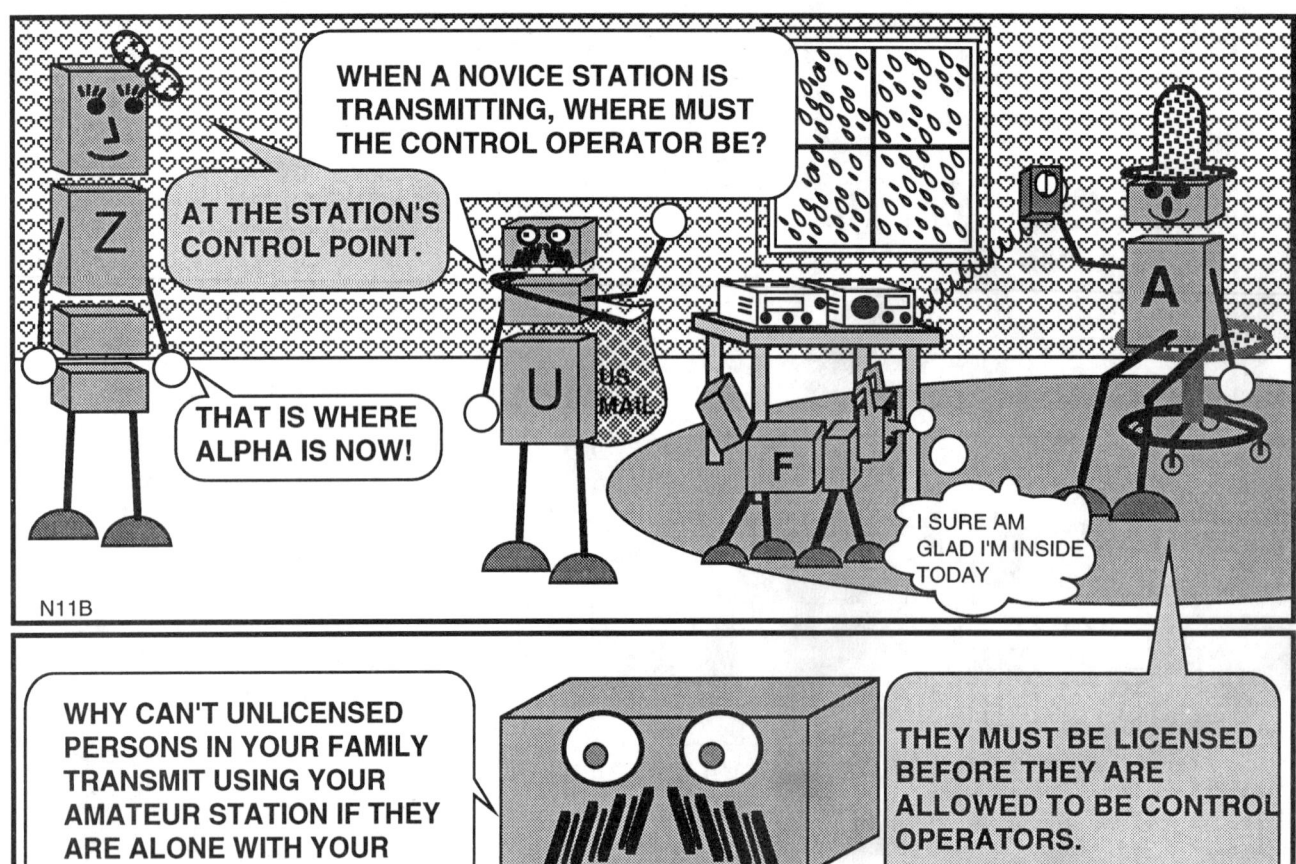

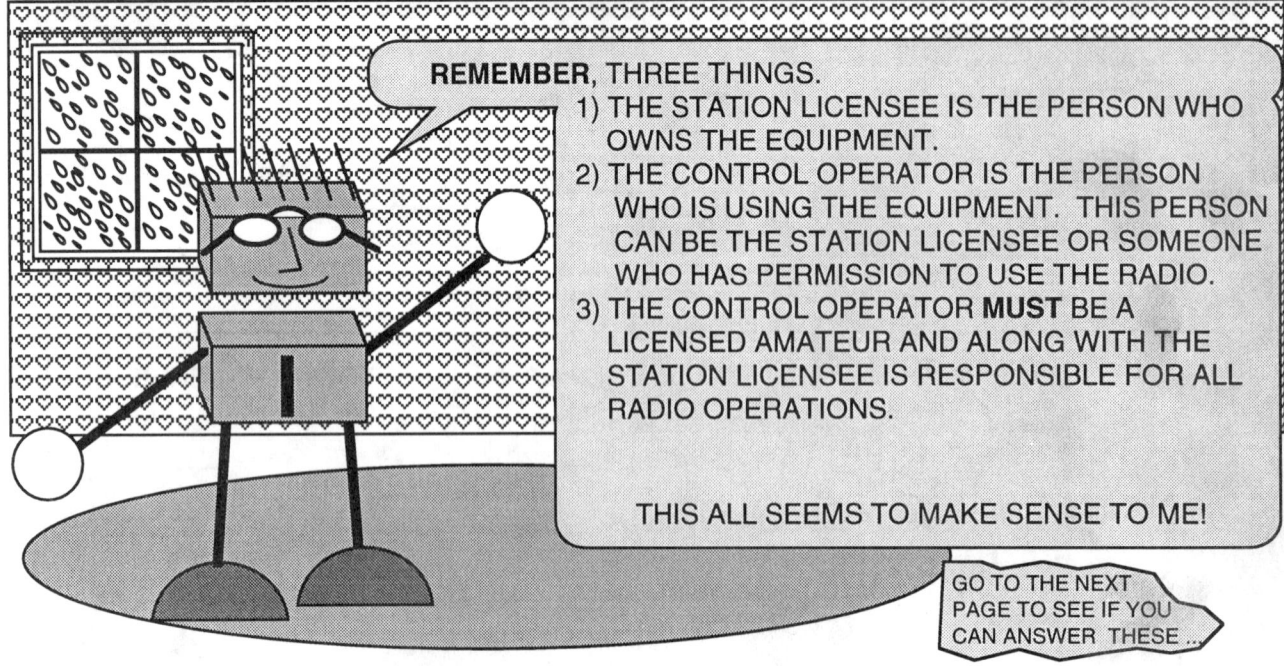

Riding the Airwaves with Alpha & Zulu

SEE IF YOU CAN ANSWER THESE QUESTIONS WITHOUT LOOKING BACK TO THE LAST 'TOON'. WRITE YOUR ANSWERS IN THE ANSWER BOX BELOW. GOOD LUCK.

PRACTICE TEST PAGE N11

N1G02 [97.103a]
Who is responsible for the proper operation of an amateur station?
A. Only the control operator
B. Only the station licensee
C. Both the control operator and the station licensee
D. The person who owns the station equipment

N1G04 [97.103a]
What is your responsibility as a station licensee?
A. You must allow another amateur to operate your station upon request
B. You must be present whenever the station is operated
C. You must notify the FCC if another amateur acts as the control operator
D. You are responsible for the proper operation of the station in accordance with the FCC rules

N1G05 [97.103b]
Who may be the control operator of an amateur station?
A. Any person over 21 years of age
B. Any person over 21 years of age with a General class license or higher
C. Any licensed amateur chosen by the station licensee
D. Any licensed amateur with a Technician class license or higher

N1G09 [97.7]
When must an amateur station have a control operator?
A. Only when training another amateur
B. Whenever the station receiver is operated
C. Whenever the station is transmitting
D. A control operator is not needed

N1G10 [97.109b]
When a Novice station is transmitting, where must its control operator be?
A. At the station's control point
B. Anywhere in the same building as the transmitter
C. At the station's entrance, to control entry to the room
D. Anywhere within 50 km of the station location

N1G11 [97.109b]
Why can't unlicensed persons in your family transmit using your amateur station if they are alone with your equipment?
A. They must not use your equipment without your permission
B. They must be licensed before they are allowed to be control operators
C. They must first know how to use the right abbreviations and Q signals
D. They must first know the right frequencies and emissions for transmitting

ANSWERS TO PREVIOUS TEST

N2A04	D
N1J08	D
N1J10	C
N1J11	B

YOUR ANSWERS TO THIS TEST

N1G02	
N1G04	
N1G05	
N1G09	
N1G10	
N1G11	

artsci inc

Riding the Airwaves with Alpha & Zulu

ALPHA & ZULU'S WORD SEARCH
ANTENNA TERMS

	1	2	3	4	5	6	7	8	9	10	11	12	13	14	15
A	N	A	J	V	H	F	M	M	E	T	E	R	A	R	I
B	B	A	S	E	E	R	A	O	X	D	N	U	O	R	G
C	Z	H	M	R	B	E	I	U	R	A	Q	C	H	I	A
D	W	A	T	T	B	W	L	N	B	Q	U	A	D	B	Y
E	H	O	R	I	Z	O	N	T	A	L	A	B	U	H	F
F	I	M	P	C	D	P	W	K	L	V	R	L	C	O	H
G	P	N	J	A	D	Q	H	F	U	F	T	E	K	C	E
H	L	I	O	L	I	O	C	O	N	N	E	C	T	O	R
I	E	Q	H	C	P	T	N	K	I	J	R	I	U	H	W
J	N	T	N	S	O	D	J	R	A	F	W	D	V	E	S
K	G	B	A	W	L	A	E	I	G	S	A	V	N	T	P
L	T	U	E	O	E	V	X	S	M	D	V	Z	R	A	D
M	H	R	G	K	H	Z	T	H	G	I	E	G	R	S	B

ZULU, CAN YOU SOLVE THIS PUZZLE?

OSCAR, I LOVE WORD SEARCH PUZZLES

BALUN	GROUND	QUAD
BAND	HAM	QUARTERWAVE
BASE	HF	RG EIGHT
BEAM	HORIZONTAL	SWITCH
BNC	KHZ	SWR
CABLE	LENGTH	TOWER
COAX	LOG	TVI
COIL	METER	UHF
CONNECTOR	MHZ	VERTICAL
DIPOLE	MOUNT	VHF
DUCK	OMNI	WATT
GAIN	POWER	WHIP
		YAGI

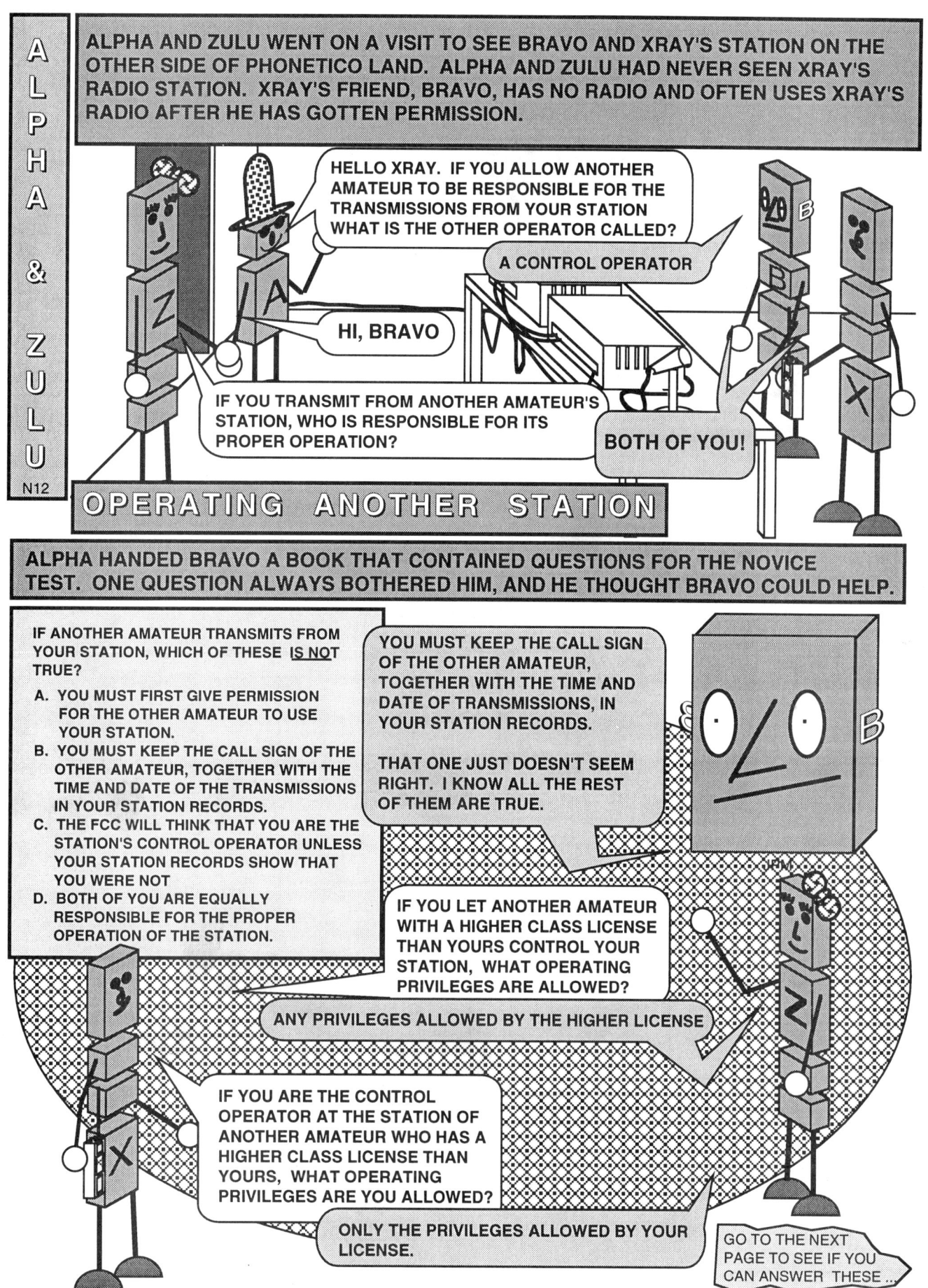

Riding the Airwaves with Alpha & Zulu

SEE IF YOU CAN ANSWER THESE QUESTIONS WITHOUT LOOKING BACK TO THE LAST 'TOON'. WRITE YOUR ANSWERS IN THE ANSWER BOX BELOW. GOOD LUCK.

PRACTICE TEST PAGE N12

N1G01 [97.3a11]
If you allow another amateur to be responsble for the transmissions from your station, what is the other operator called?
- A. An auxiliary operator
- B. The operations coordinator
- C. A third-party operator
- D. A control operator

N1G06 [97.103]
If another amateur transmits from your station, which of these is NOT true?
- A. You must first give permission for the other amateur to use your station
- B. You must keep the call sign of the other amateur, together with the time and date of transmissions, in your station records
- C. The FCC will think that you are the station's control operator unless your station records show that you were not
- D. Both of you are equally responsible for the proper operation of the station

N1G03 [97.103a]
If you transmit from another amateur's station, who is responsible for its proper operation?
- A. Both of you
- B. The other amateur (the station licensee)
- C. You, the control operator
- D. The station licensee, unless the station records show that you were the control operator at the time

N1G07 [97.105b]
If you let another amateur with a higher class license than yours control your station, what operating privileges are allowed?
- A. Any privileges allowed by the higher license
- B. Only the privileges allowed by your license
- C. All the emission privileges of the higher license, but only the frequency privileges of your license
- D. All the frequency privileges of the higher license, but only the emission privileges of your license

N1G08 [97.105b]
If you are the control operator at the station of another amateur who has a higher class license than yours, what operating privileges are you allowed?
- A. Any privileges allowed by the higher license
- B. Only the privileges allowed by your license
- C. All the emission privileges of the higher license, but only the frequency privileges of your license
- D. All the frequency privileges of the higher license, but only the emission privileges of your license

ANSWERS TO PREVIOUS TEST

N1G02	C
N1G04	D
N1G05	C
N1G09	C
N1G10	A
N1G11	B

YOUR ANSWERS TO THIS TEST

N1G01	
N1G03	
N1G06	
N1G07	
N1G08	

artsci inc

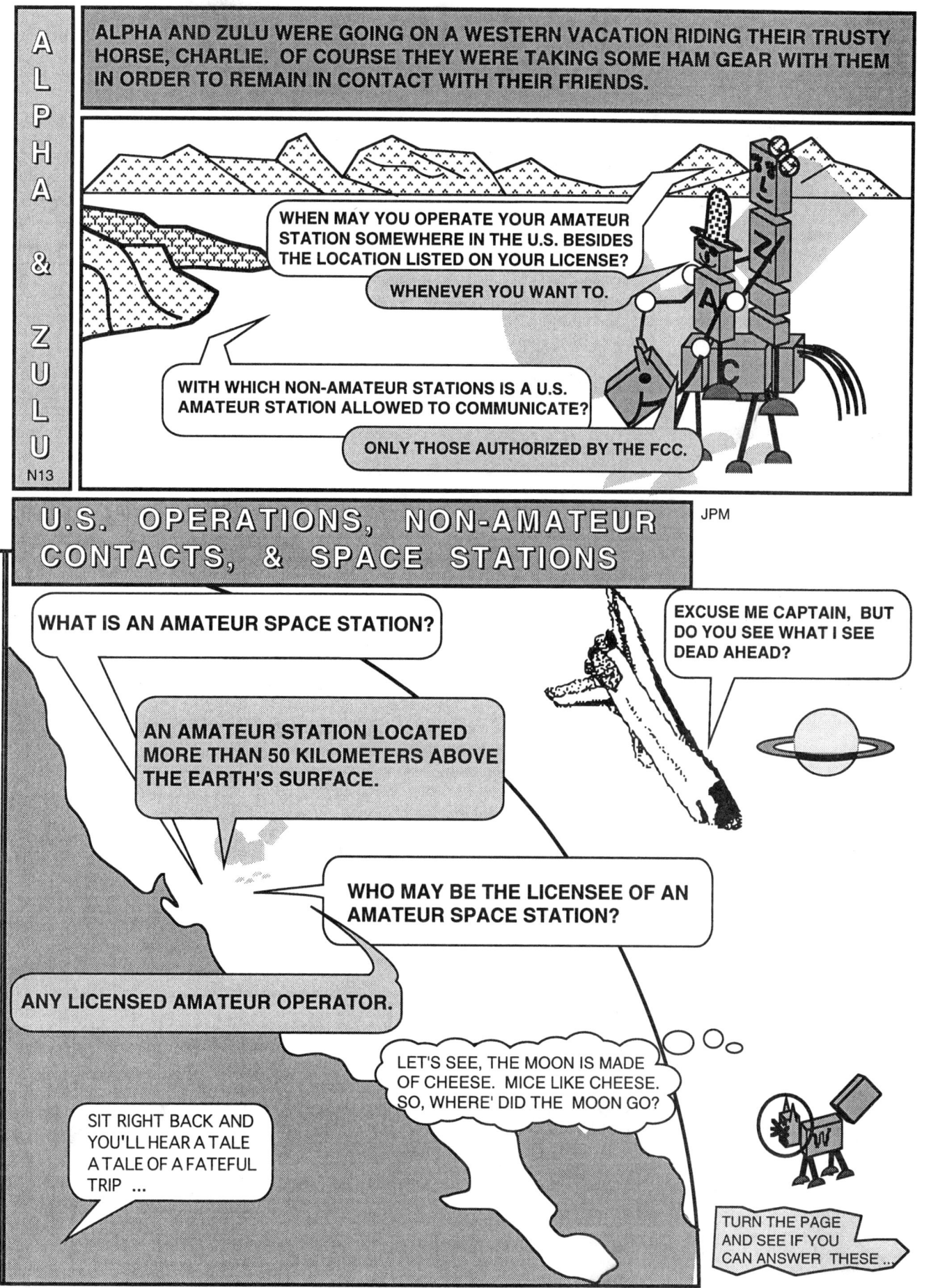

Riding the Airwaves with Alpha & Zulu

SEE IF YOU CAN ANSWER THESE QUESTIONS WITHOUT LOOKING BACK TO THE LAST 'TOON'. WRITE YOUR ANSWERS IN THE ANSWER BOX BELOW. GOOD LUCK.

PRACTICE TEST PAGE N13

N1H01 [97.5a]
When may you operate your amateur station somewhere in the US besides the location listed on your license?
  A. Only during times of emergency
  B. Only after giving proper notice to the FCC
  C. During an emergency or an FCC-approved emergency practice
  D. Whenever you want to

N1H02 [97.111]
With which non-amateur stations is a US amateur station allowed to communicate?
  A. No non-amateur stations
  B. All non-amateur stations
  C. Only those authorized by the FCC
  D. Only those who use international Morse code

N1I03 [97.3a36]
What is an amateur space station?
  A. An amateur station operated on an unused frequency
  B. An amateur station awaiting its new call letters from the FCC
  C. An amateur station located more than 50 kilometers above the Earth's surface
  D. An amateur station that communicates with Space Shuttles

N1I04 [New 97.207a per FCC 92-310]
Who may be the licensee of an amateur space station?
  A. An amateur holding an Amateur Extra class operator license
  B. Any licensed amateur operator
  C. Anyone designated by the commander of the spacecraft
  D. No one unless specifically authorized by the government

ANSWERS TO PREVIOUS TEST
N1G01
D
N1G03
A
N1G06
B
N1G07
A
N1G08
B

YOUR ANSWERS TO THIS TEST
N1H01
N1H02
N1I03
N1I04

artsci inc

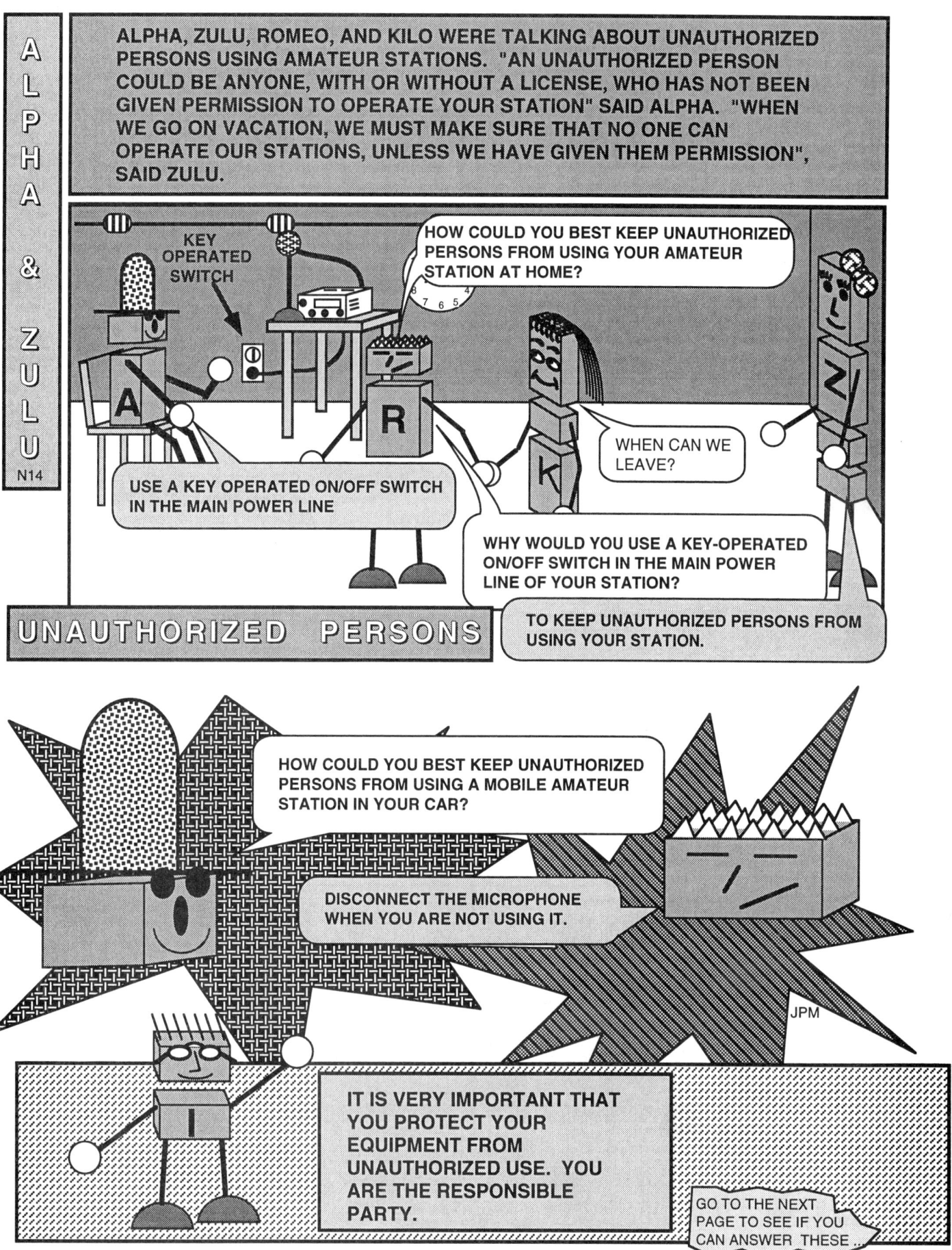

Riding the Airwaves with Alpha & Zulu

SEE IF YOU CAN ANSWER THESE QUESTIONS WITHOUT LOOKING BACK TO THE LAST 'TOON'. WRITE YOUR ANSWERS IN THE ANSWER BOX BELOW. GOOD LUCK.

PRACTICE TEST PAGE N14

N4A01
How could you best keep unauthorized persons from using your amateur station at home?
- A. Use a carrier-operated relay in the main power line
- B. Use a key-operated on/off switch in the main power line
- C. Put a "Danger - High Voltage" sign in the station
- D. Put fuses in the main power line

N4A02
How could you best keep unauthorized persons from using a mobile amateur station in your car?
- A. Disconnect the microphone when you are not using it
- B. Put a "do not touch" sign on the radio
- C. Turn the radio off when you are not using it
- D. Tune the radio to an unused frequency when you are done using it

N4A03
Why would you use a key-operated on/off switch in the main power line of your station?
- A. To keep unauthorized persons from using your station
- B. For safety, in case the main fuses fail
- C. To keep the power company from turning off your electricity during an emergency
- D. For safety, to turn off the station in the event of an emergency

ANSWERS TO PREVIOUS TEST

N1H01	D
N1H02	C
N1I03	C
N1I04	B

YOUR ANSWERS TO THIS TEST

N4A01	
N4A02	
N4A03	

artsci inc

Riding the Airwaves with Alpha & Zulu

SEE IF YOU CAN ANSWER THESE QUESTIONS WITHOUT LOOKING BACK TO THE LAST 'TOON'. WRITE YOUR ANSWERS IN THE ANSWER BOX BELOW. GOOD LUCK.

PRACTICE TEST PAGE N15

N1H03 [97.113a]
When are communications for business allowed in the amateur service?
- A. Only if they are for the safety of human life or immediate protection of property
- B. There are no rules against business communications
- C. No business communications are ever allowed
- D. Business communications are allowed between the hours of 9 AM to 5 PM, weekdays

N1H04 [97.113a]
Which of the following CANNOT be discussed on an amateur club net?
- A. Business planning
- B. Recreation planning
- C. Code practice planning
- D. Emergency planning

N1H05 [97.113a]
If you wanted to join a radio club, would you be allowed to send an message to them via amateur radio requesting an application?
- A. Yes, if the club is a not-for-profit organization
- B. No. This would facilitate the commercial affairs of the club
- C. Yes, but only during normal business hours, between 9 AM and 5 PM, weekdays
- D. Yes, since there are no rules against business communications in the amateur service

N1I05 [97.113b]
When may someone be paid to transmit messages from an amateur station?
- A. Only if he or she works for a public service agency such as the Red Cross
- B. Under no circumstances
- C. Only if he or she reports all such payments to the IRS
- D. Only if he or she works for a club station and special requirements are met

ANSWERS TO PREVIOUS TEST
N4A01
B
N4A02
A
N4A03
A

YOUR ANSWERS TO THIS TEST
N1H03
N1H04
N1H05
N1I05

artsci inc

Riding the Airwaves with Alpha & Zulu

Riding the Airwaves with Alpha & Zulu

PRACTICE TEST PAGE N16

N1I01 [97.3a39]
What is the definition of third-party communications?
A. A message sent between two amateur stations for someone else
B. Public service communications for a political party
C. Any messages sent by amateur stations
D. A three-minute transmission to another amateur

N1I09 [97.3a42]
What is a "third-party" in amateur communications?
A. An amateur station that breaks in to talk
B. A person who is sent a message by amateur communications other than a control operator who handles the message
C. A shortwave listener who monitors amateur communications
D. An unlicensed control operator

N1H10 [97.115c]
Besides normal identification, what else must a US station do when sending third-party communications internationally?
A. The US station must transmit its own call sign at the beginning of each communication, and at least every ten minutes after that
B. The US station must transmit both call signs at the end of each communication
C. The US station must transmit its own call sign at the beginning of each communication, and at least every five minutes after that
D. Each station must transmit its own call sign at the end of each communication, and at least every five minutes after that

N1I10 [97.115a2]
If you are allowing a non-amateur friend to use your station to talk to someone in the US, and a foreign station breaks in to talk to your friend, what should you do?
A. Have your friend wait until you find out if the US has a third-party agreement with the foreign station's government
B. Stop all discussions and quickly sign off
C. Since you can talk to any foreign amateurs, your friend may keep talking as long as you are the control operator
D. Report the incident to the foreign amateur's government

N1I02 [97.111a1]
When are you allowed to communicate with an amateur in a foreign country?
A. Only when the foreign amateur uses English
B. Only when you have permission from the FCC
C. Only when a third-party agreement exists between the US and the foreign country
D. At any time, unless it is not allowed by either government

N1I11 [97.115a2]
When are you allowed to transmit a message to a station in a foreign country for a third party?
A. Anytime
B. Never
C. Anytime, unless there is a third-party agreement between the US and the foreign government
D. If there is a third-party agreement with the US government, or if the third party could be the control operator

ANSWERS TO PREVIOUS TEST

N1H03	A
N1H04	A
N1H05	B
N1I05	D

YOUR ANSWERS TO THIS TEST

N1I01	
N1I09	
N1H10	
N1I10	
N1I02	
N1I11	

artsci inc

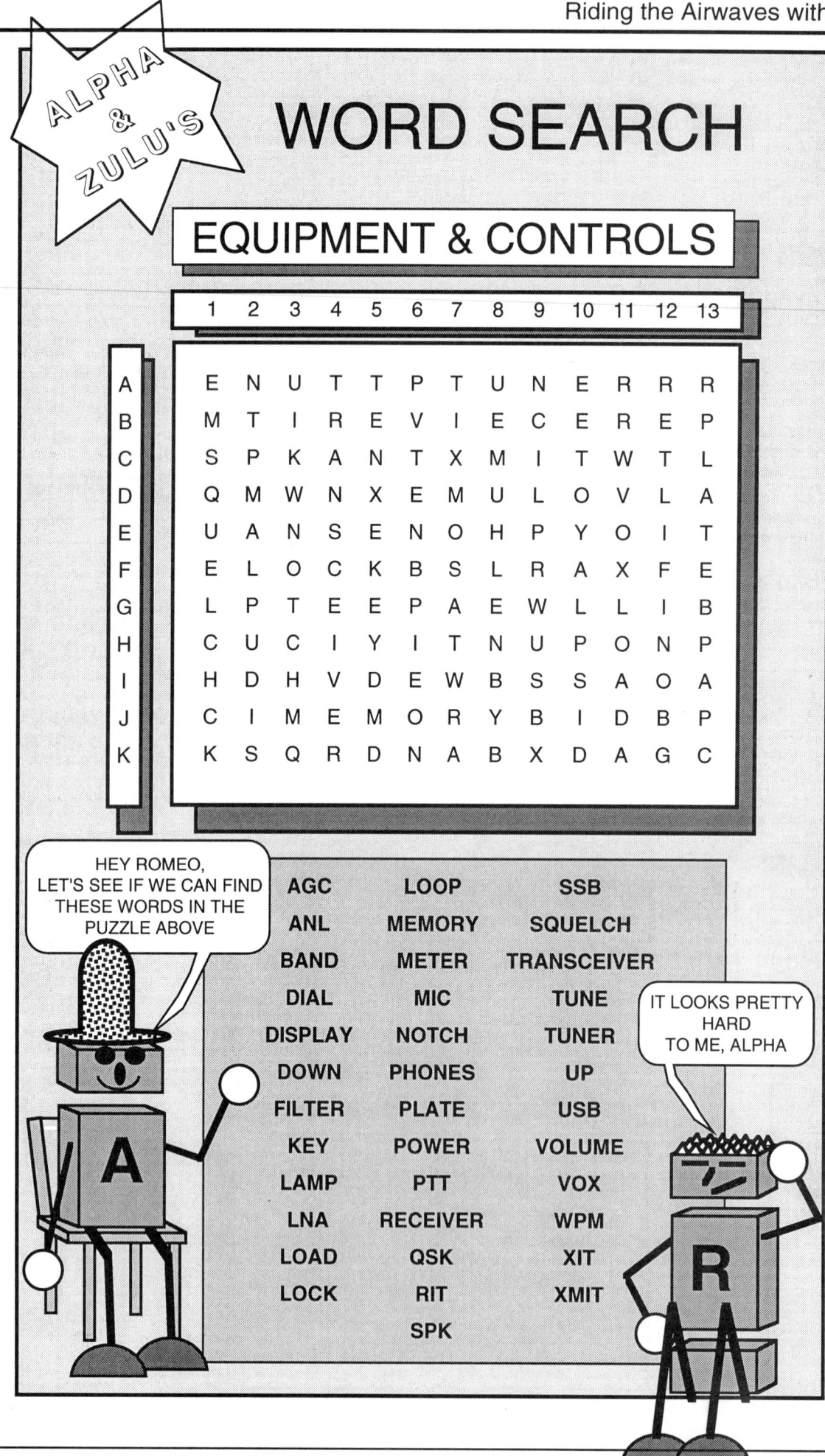

Riding the Airwaves with Alpha & Zulu

Riding the Airwaves with Alpha & Zulu

SEE IF YOU CAN ANSWER THESE QUESTIONS WITHOUT LOOKING BACK TO THE LAST 'TOON'. WRITE YOUR ANSWERS IN THE ANSWER BOX BELOW. GOOD LUCK.

PRACTICE TEST PAGE N17

N1I06 [97.113c]
When is an amateur allowed to broadcast information to the general public?
- A. Never
- B. Only when the operator is being paid
- C. Only when broadcasts last less than 1 hour
- D. Only when broadcasts last longer than 15 minutes

N1I07 [97.113d]
When is an amateur station permitted to transmit music?
- A. Never
- B. Only if the music played produces no spurious emissions
- C. Only if it is used to jam an illegal transmission
- D. Only if it is above 1280 MHz

N1I08 [97.113d]
When is the use of codes or ciphers allowed to hide the meaning of an amateur message?
- A. Only during contests
- B. Only during nationally declared emergencies
- C. Never, except when special requirements are met
- D. Only on frequencies above 1280 MHz

N1J01 [97.3a21]
What is a transmission called that disturbs other communications?
- A. Interrupted CW
- B. Harmful interference
- C. Transponder signals
- D. Unidentified transmissions

N1J02 [97.3a21]
Why is transmitting on a police frequency as a "joke" called harmful interference that deserves a large penalty?
- A. It annoys everyone who listens
- B. It blocks police calls which might be an emergency and interrupts police communications
- C. It is in bad taste to communicate with non-amateurs, even as a joke
- D. It is poor amateur practice to transmit outside the amateur bands

N1J03 [97.101d]
When may you deliberately interfere with another station's communications?
- A. Only if the station is operating illegally
- B. Only if the station begins transmitting on a frequency you are using
- C. Never
- D. You may expect, and cause, deliberate interference because it can't be helped during crowded band conditions

ANSWERS TO PREVIOUS TEST
N1I01
A
N1I09
B
N1H10
B
N1I10
A
N1I02
D
N1I11
D

YOUR ANSWERS TO THIS TEST
N1I06
N1I07
N1I08
N1J01
N1J02
N1J03

Riding the Airwaves with Alpha & Zulu

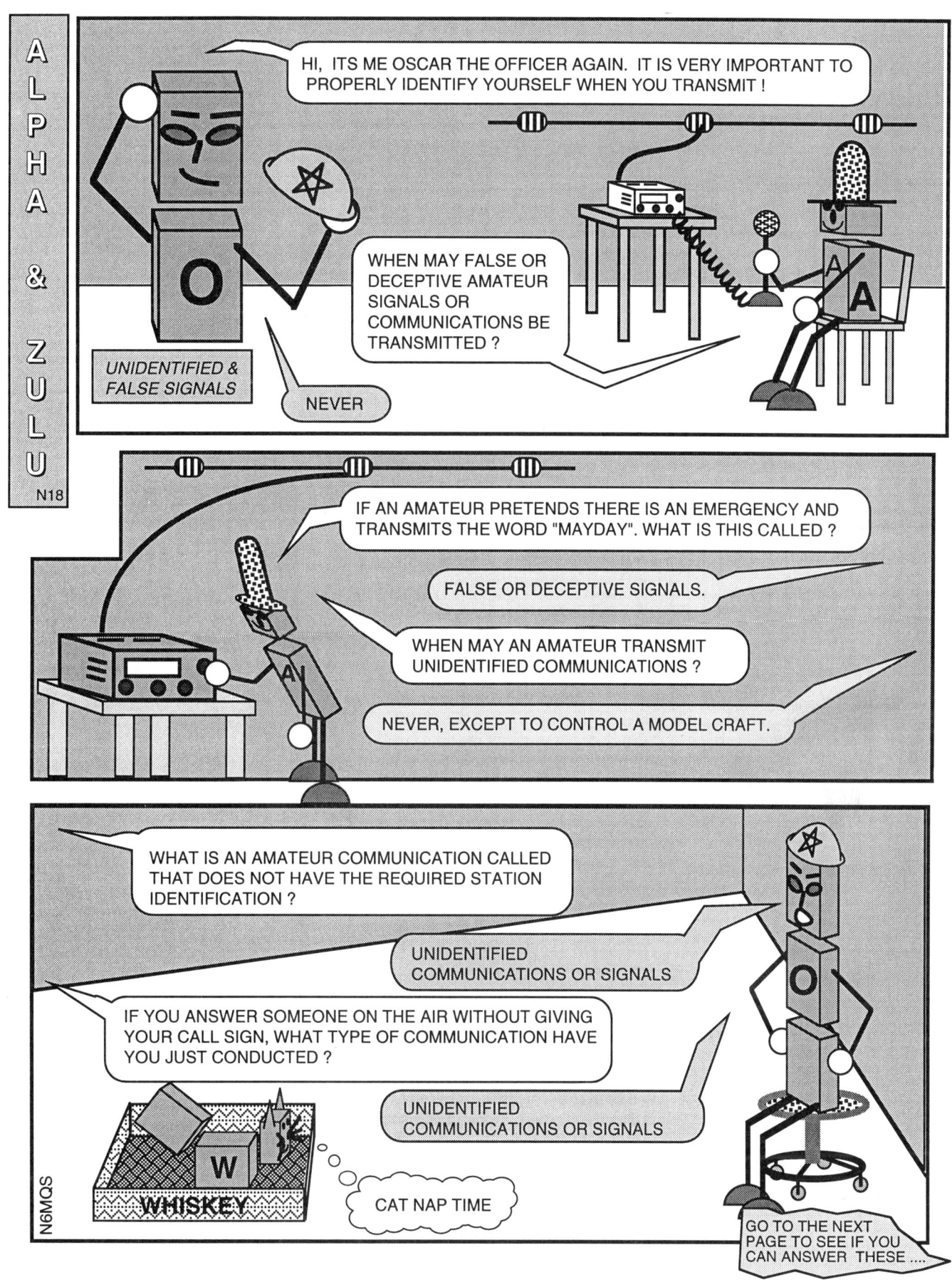

Riding the Airwaves with Alpha & Zulu

SEE IF YOU CAN ANSWER THESE QUESTIONS WITHOUT LOOKING BACK TO THE LAST 'TOON'. WRITE YOUR ANSWERS IN THE ANSWER BOX BELOW. GOOD LUCK.

PRACTICE TEST PAGE N18

N1J04 [97.113d]
When may false or deceptive amateur signals or communications be transmitted?
- A. Never
- B. When operating a beacon transmitter in a "fox hunt" exercise
- C. When playing a harmless "practical joke"
- D. When you need to hide the meaning of a message for secrecy

N1J05 [97.113d]
If an amateur pretends there is an emergency and transmits the word "MAYDAY," what is this called?
- A. A traditional greeting in May
- B. An emergency test transmission
- C. False or deceptive signals
- D. Nothing special; "MAYDAY" has no meaning in an emergency

N1J06 [97.119a]
When may an amateur transmit unidentified communications?
- A. Only for brief tests not meant as messages
- B. Only if it does not interfere with others
- C. Never, except to control a model craft
- D. Only for two-way or third-party communications

N1J07 [97.119a]
What is an amateur communication called that does not have the required station identification?
- A. Unidentified communications or signals
- B. Reluctance modulation
- C. Test emission
- D. Tactical communication

N1J09 [97.119a]
If you answer someone on the air without giving your call sign, what type of communication have you just conducted?
- A. Test transmission
- B. Tactical signal
- C. Packet communication
- D. Unidentified communication

ANSWERS TO PREVIOUS TEST
N1I06
A
N1I07
A
N1I08
C
N1J01
B
N1J02
B
N1J03
C

YOUR ANSWERS TO THIS TEST
N1J04
N1J05
N1J06
N1J07
N1J09

Riding the Airwaves with Alpha & Zulu

SEE IF YOU CAN ANSWER THESE QUESTIONS WITHOUT LOOKING BACK TO THE LAST 'TOON'. WRITE YOUR ANSWERS IN THE ANSWER BOX BELOW. GOOD LUCK.

PRACTICE TEST PAGE N19

N2A01
What should you do before you transmit on any frequency?
  A. Listen to make sure others are not using the frequency
  B. Listen to make sure that someone will be able to hear you
  C. Check your antenna for resonance at the selected frequency
  D. Make sure the SWR on your antenna feed line is high enough

N2A02
If you make contact with another station and your signal is extremely strong and perfectly readable, what adjustment might you make to your transmitter?
  A. Turn on your speech processor
  B. Reduce you SWR
  C. Continue with your contact, making no changes
  D. Turn down your power output to the minimum necessary

N2A03
What is one way to shorten transmitter tune-up time on the air to cut down on interference?
  A. Use a random wire antenna
  B. Tune up on 40 meters first, then switch to the desired band
  C. Tune the transmitter into a dummy load
  D. Use twin lead instead of coaxial-cable feed lines

ANSWERS TO PREVIOUS TEST
N1J04
A
N1J05
C
N1J06
C
N1J07
A
N1J09
D

YOUR ANSWERS TO THIS TEST
N2A01
N2A02
N2A03

Riding the Airwaves with Alpha & Zulu

SEE IF YOU CAN ANSWER THESE QUESTIONS WITHOUT LOOKING BACK TO THE LAST 'TOON'. WRITE YOUR ANSWERS IN THE ANSWER BOX ON THE NEXT PAGE. GOOD LUCK.

PRACTICE TEST PAGE N20A

N2B07
What is simplex operation?
- A. Transmitting and receiving on the same frequency
- B. Transmitting and receiving over a wide area
- C. Transmitting on one frequency and receiving on another
- D. Transmitting one-way communications

N2B08
When should you use simplex operation instead of a repeater?
- A. When the most reliable communications are needed
- B. When a contact is possible without using a repeater
- C. When an emergency telephone call is needed
- D. When you are traveling and need some local information

N2B09
What is a good way to make contact on a repeater?
- A. Say the call sign of the station you want to contact three times
- B. Say the other operator's name, then your call sign three times
- C. Say the call sign of the station you want to contact, then your call sign
- D. Say, "Breaker, breaker," then your call sign

N2B10
When using a repeater to communicate, what do you need to know about the repeater besides its output frequency?
- A. Its input frequency
- B. Its call sign
- C. Its power level
- D. Whether or not it has a phone patch

N2B11
What is the main purpose of a repeater?
- A. To make local information available 24 hours a day
- B. To link amateur stations with the telephone system
- C. To retransmit NOAA weather information during severe storm warnings
- D. To increase the range of portable and mobile stations

ANSWERS TO PREVIOUS TEST
N2A01
A
N2A02
D
N2A03
C

MORE TEST ON THE NEXT PAGE →

artsci inc

PRACTICE TEST PAGE PART 2
N20B

N2B12
What does it mean to say that a repeater has an input and an output frequency?
  A. The repeater receives on one frequency and transmits on another
  B. The repeater offers a choice of operating frequency, in case one is busy
  C. One frequency is used to control the repeater and another is used to retransmit received signals
  D. The repeater must receive an access code on one frequency before retransmitting received signals

N2B13
What is an autopatch?
  A. Something that automatically selects the strongest signal to be repeated
  B. A device which connects a mobile station to the next repeater if it moves out of range of the first
  C. A device that allows repeater users to make telephone calls from their stations
  D. A device which locks other stations out of a repeater when there is an important conversation in progress

N2B14
What is the purpose of a repeater time-out timer?
  A. It lets a repeater have a rest period after heavy use
  B. It logs repeater transmit time to predict when a repeater will fail
  C. It tells how long someone has been using a repeater
  D. It limits the amount of time someone can transmit on a repeater

N2B15
What is a CTCSS (or PL) tone?
  A. A special signal used for telecommand control of model craft
  B. A sub-audible tone added to a carrier which may cause a receiver to accept a signal
  C. A tone used by repeaters to mark the end of a transmission
  D. A special signal used for telemetry between amateur space stations and Earth stations

YOUR ANSWERS TO THIS TEST

N2B07

N2B08

N2B09

N2B10

N2B11

N2B12

N2B13

N2B14

N2B15

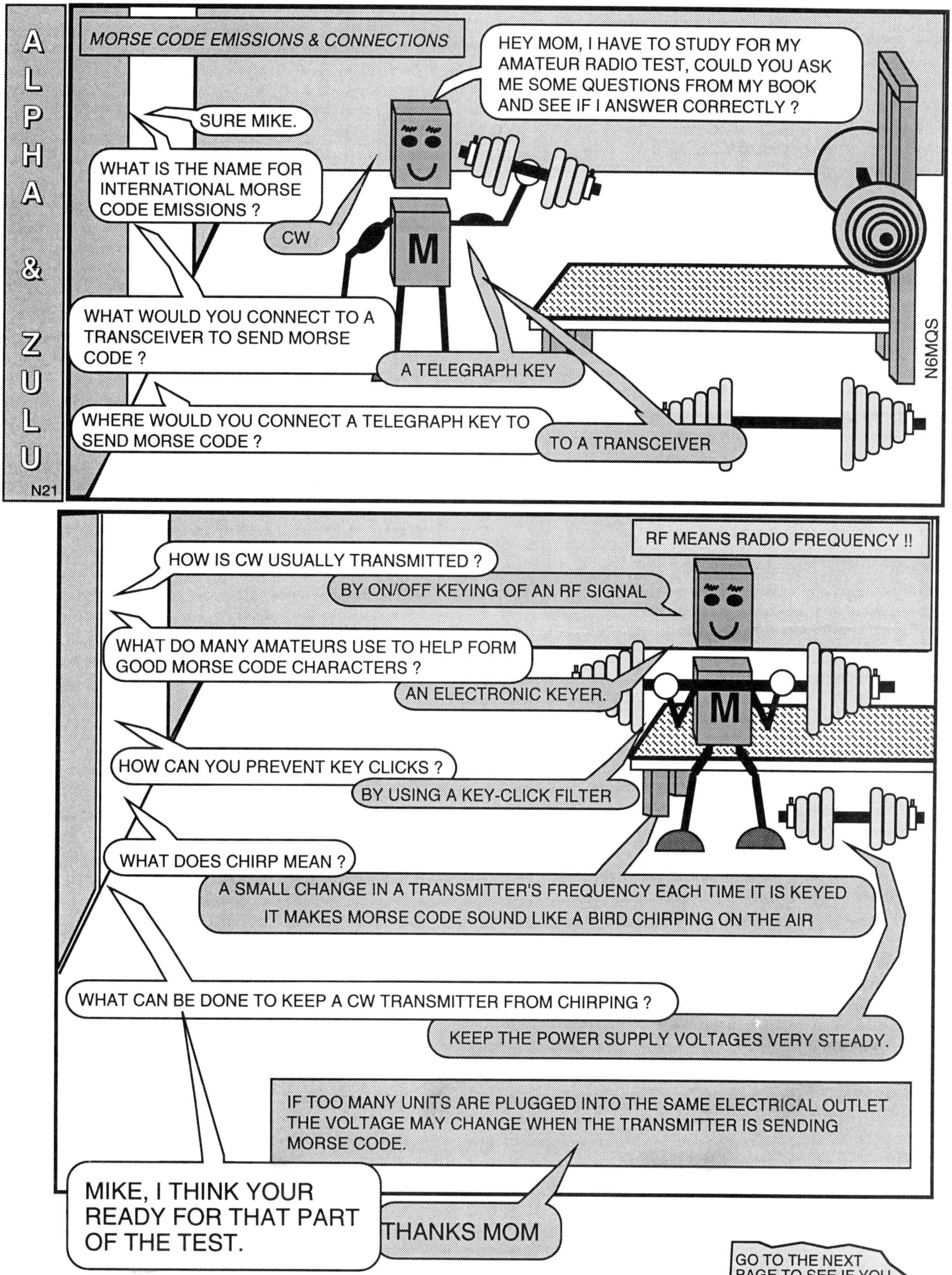

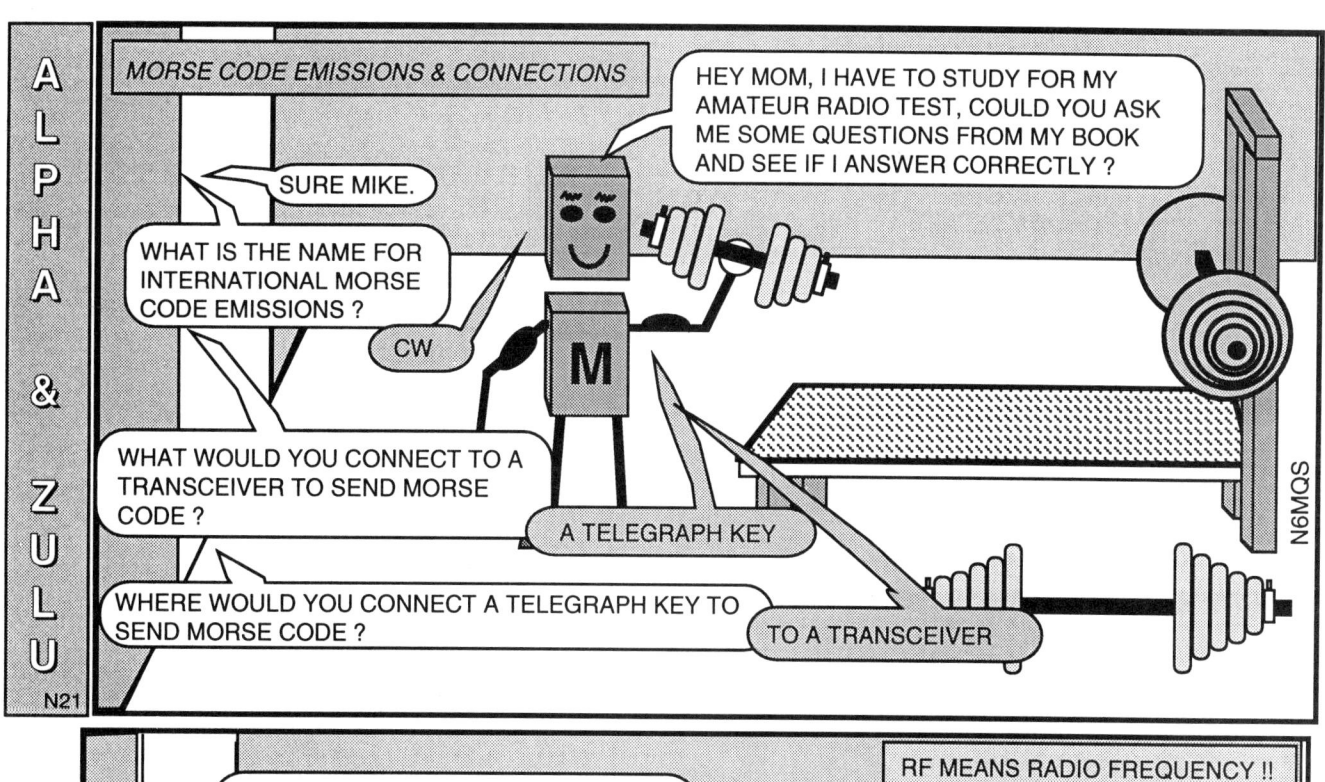

Riding the Airwaves with Alpha & Zulu

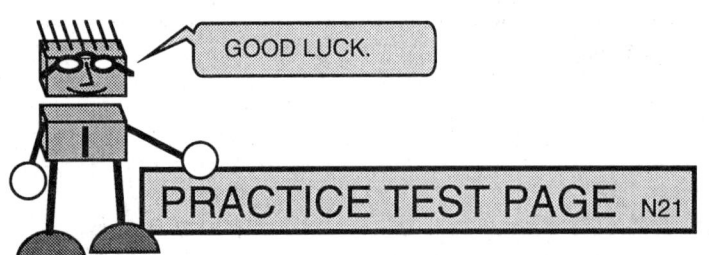

PRACTICE TEST PAGE N21

N8A07
How can you prevent key clicks?
  A. By sending CW more slowly
  B. By increasing power
  C. By using a better power supply
  D. By using a key-click filter

N8A01
How is CW usually transmitted?
  A. By frequency-shift keying an RF signal
  B. By on/off keying an RF signal
  C. By audio-frequency-shift keying an oscillator tone
  D. By on/off keying an audio-frequency signal

N8A08
What does chirp mean?
  A. An overload in a receiver's audio circuit whenever CW is received
  B. A high-pitched tone which is received along with a CW signal
  C. A small change in a transmitter's frequency each time it is keyed
  D. A slow change in transmitter frequency as the circuit warms up

N8A03
What is the name for international Morse code emissions?
  A. RTTY
  B. Data
  C. CW
  D. Phone

N8A09
What can be done to keep a CW transmitter from chirping?
  A. Add a low-pass filter
  B. Use an RF amplifier
  C. Keep the power supply current very steady
  D. Keep the power supply voltages very steady

N7B01
What would you connect to a transceiver to send Morse code?
  A. A terminal-node controller
  B. A telegraph key
  C. An SWR meter
  D. An antenna switch

N7B02
Where would you connect a telegraph key to send Morse code?
  A. To a power supply
  B. To an antenna switch
  C. To a transceiver
  D. To an antenna

N7B03
What do many amateurs use to help form good Morse code characters?
  A. A key-operated on/off switch
  B. An electronic keyer
  C. A key-click filter
  D. A DTMF keypad

ANSWERS TO PREVIOUS TEST	
N2B07	A
N2B08	B
N2B09	C
N2B10	A
N2B11	D
N2B12	A
N2B13	C
N2B14	D
N2B15	B

YOUR ANSWERS TO THIS TEST
N8A01
N8A03
N8A07
N8A08
N8A09
N7B01
N7B02
N7B03

Riding the Airwaves with Alpha & Zulu

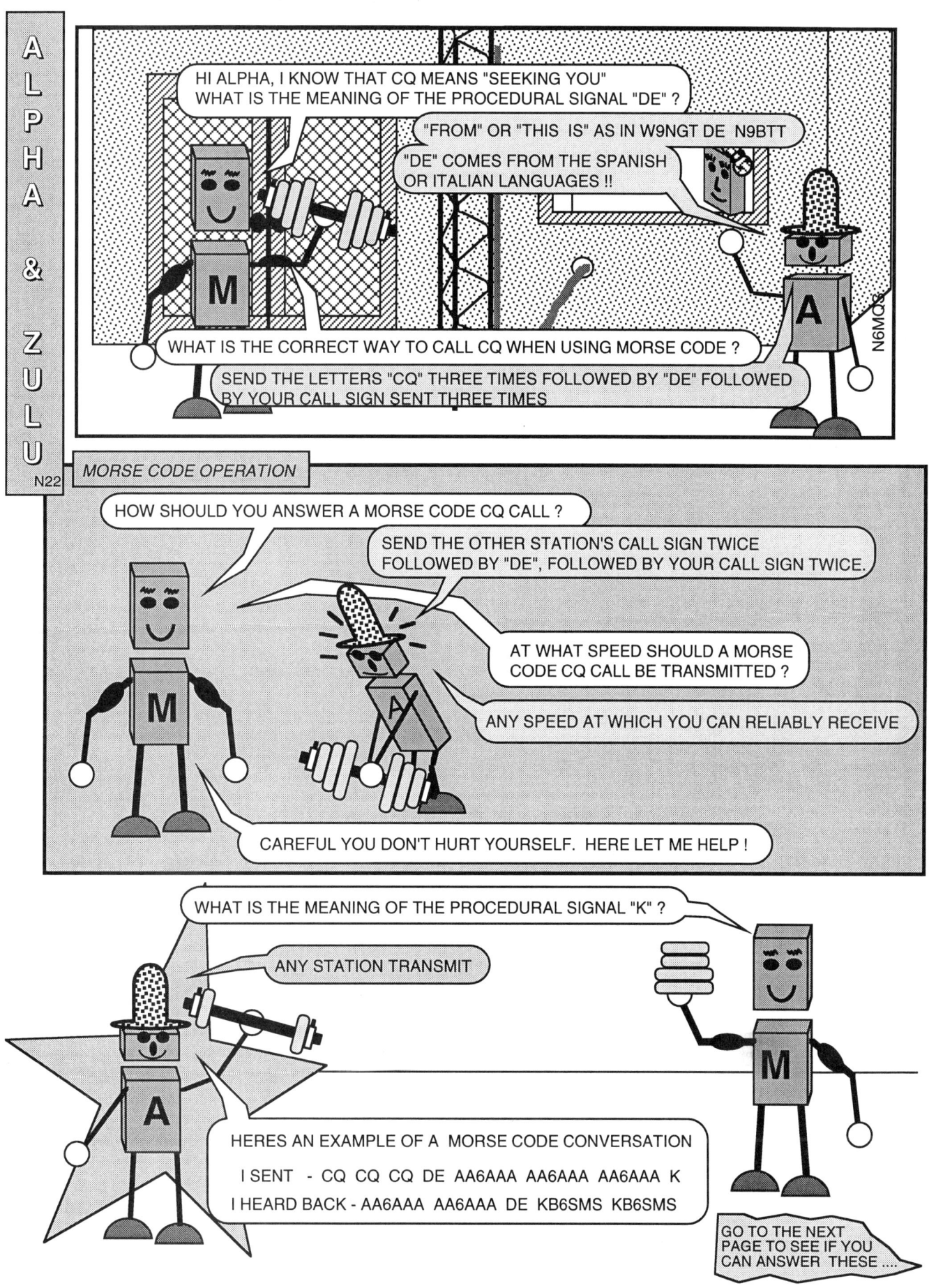

Riding the Airwaves with Alpha & Zulu

SEE IF YOU CAN ANSWER THESE QUESTIONS WITHOUT LOOKING BACK TO THE LAST 'TOON'. WRITE YOUR ANSWERS IN THE ANSWER BOX BELOW. GOOD LUCK.

PRACTICE TEST PAGE N22

N2A05
What is the correct way to call CQ when using Morse code?
- A Send the letters "CQ" three times, followed by "DE," followed by your call sign sent once
- B. Send the letters "CQ" three times, followed by "DE," followed by your call sign sent three times
- C. Send the letters "CQ" ten times, followed by "DE," followed by your call sign sent once
- D. Send the letters "CQ" over and over

N2A06
How should you answer a Morse code CQ call?
- A. Send your call sign four times
- B. Send the other station's call sign twice, followed by "DE," followed by your call sign twice
- C. Send the other station's call sign once, followed by "DE," followed by your call sign four times
- D. Send your call sign followed by your name, station location and a signal report

N2A07
At what speed should a Morse code CQ call be transmitted?
- A. Only speeds below five WPM
- B. The highest speed your keyer will operate
- C. Any speed at which you can reliably receive
- D. The highest speed at which you can control the keyerransmit"

N2A09
What is the meaning of the procedural signal "DE"?
- A. "From" or "this is," as in "W9NGT DE N9BTT"
- B. "Directional Emissions" from your antenna
- C. "Received all correctly"
- D. "Calling any station"

N2A10
What is the meaning of the procedural signal "K"?
- A. "Any station transmit"
- B. "All received correctly"
- C. "End of message"
- D. "Called station only transmit"

ANSWERS TO PREVIOUS TEST

N8A01	B
N8A03	C
N8A07	D
N8A08	C
N8A09	D
N7B01	B
N7B02	C
N7B03	B

YOUR ANSWERS TO THIS TEST

N2A05	
N2A06	
N2A07	
N2A09	
N2A10	

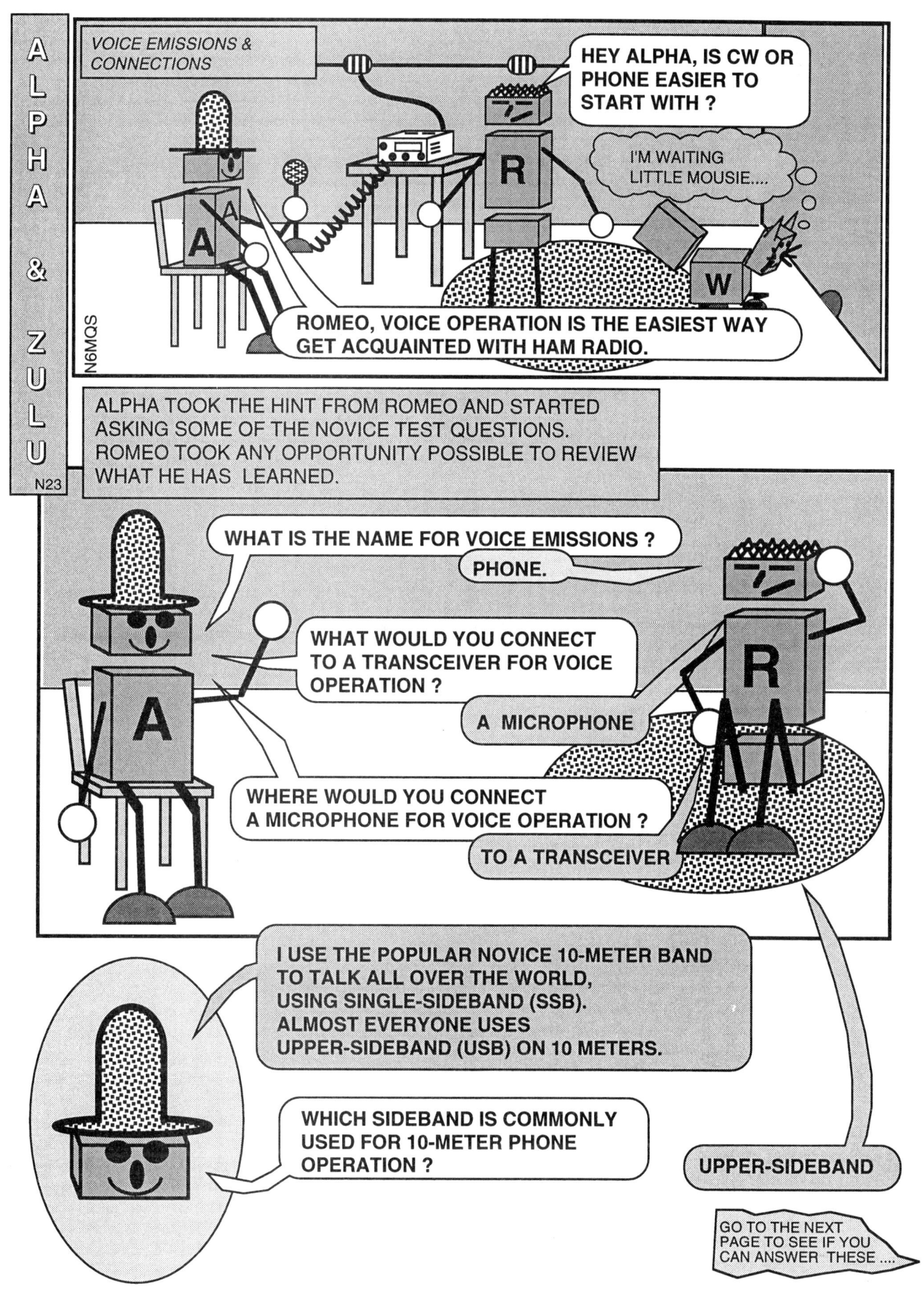

Riding the Airwaves with Alpha & Zulu

SEE IF YOU CAN ANSWER THESE QUESTIONS WITHOUT LOOKING BACK TO THE LAST 'TOON'. WRITE YOUR ANSWERS IN THE ANSWER BOX BELOW. GOOD LUCK.

PRACTICE TEST PAGE N23

N8A06
What is the name for voice emissions?
  A. RTTY
  B. Data
  C. CW
  D. Phone

N7B04
Where would you connect a microphone for voice operation?
  A. To a power supply
  B. To an antenna switch
  C. To a transceiver
  D. To an antenna

N7B05
What would you connect to a transceiver for voice operation?
  A. A splatter filter
  B. A terminal-voice controller
  C. A receiver audio filter
  D. A microphone

N8A11
Which sideband is commonly used for 10-meter phone operation?
  A. Upper-sideband
  B. Lower-sideband
  C. Amplitude-compandored sideband
  D. Double-sideband

ANSWERS TO PREVIOUS TEST
N2A05
B
N2A06
B
N2A07
C
N2A09
A
N2A10
A

YOUR ANSWERS TO THIS TEST
N8A06
N7B04
N7B05
N8A11

artsci inc

Riding the Airwaves with Alpha & Zulu

SEE IF YOU CAN ANSWER THESE QUESTIONS WITHOUT LOOKING BACK TO THE LAST 'TOON'. WRITE YOUR ANSWERS IN THE ANSWER BOX BELOW. GOOD LUCK.

PRACTICE TEST PAGE N24

N8A02
How is RTTY usually transmitted?
- A. By frequency-shift keying an RF signal
- B. By on/off keying an RF signal
- C. By digital pulse-code keying of an unmodulated carrier
- D. By on/off keying an audio-frequency signal

N8A04
What is the name for narrow-band direct-printing telegraphy emissions?
- A. RTTY
- B. Data
- C. CW
- D. Phone

N7B06
What would you connect to a transceiver for RTTY operation?
- A. A modem and a teleprinter or computer system
- B. A computer, a printer and a RTTY refresh unit
- C. A terminal voice controller
- D. A modem, a monitor and a DTMF keypad

N7B07
What would you connect between a transceiver and a computer system or teleprinter for RTTY operation?
- A. An RS-232 interface
- B. A DTMF keypad
- C. A modem
- D. A terminal-network controller

N7B10
In RTTY operation, what equipment connects to a modem?
- A. A DTMF keypad, a monitor and a transceiver
- B. A DTMF microphone, a monitor and a transceiver
- C. A transceiver and a terminal-network controller
- D. A transceiver and a teleprinter or computer system

ANSWERS TO PREVIOUS TEST
N8A06
D
N7B04
C
N7B05
D
N8A11
A

YOUR ANSWERS TO THIS TEST
N8A02
N8A04
N7B06
N7B07
N7B10

artsci inc

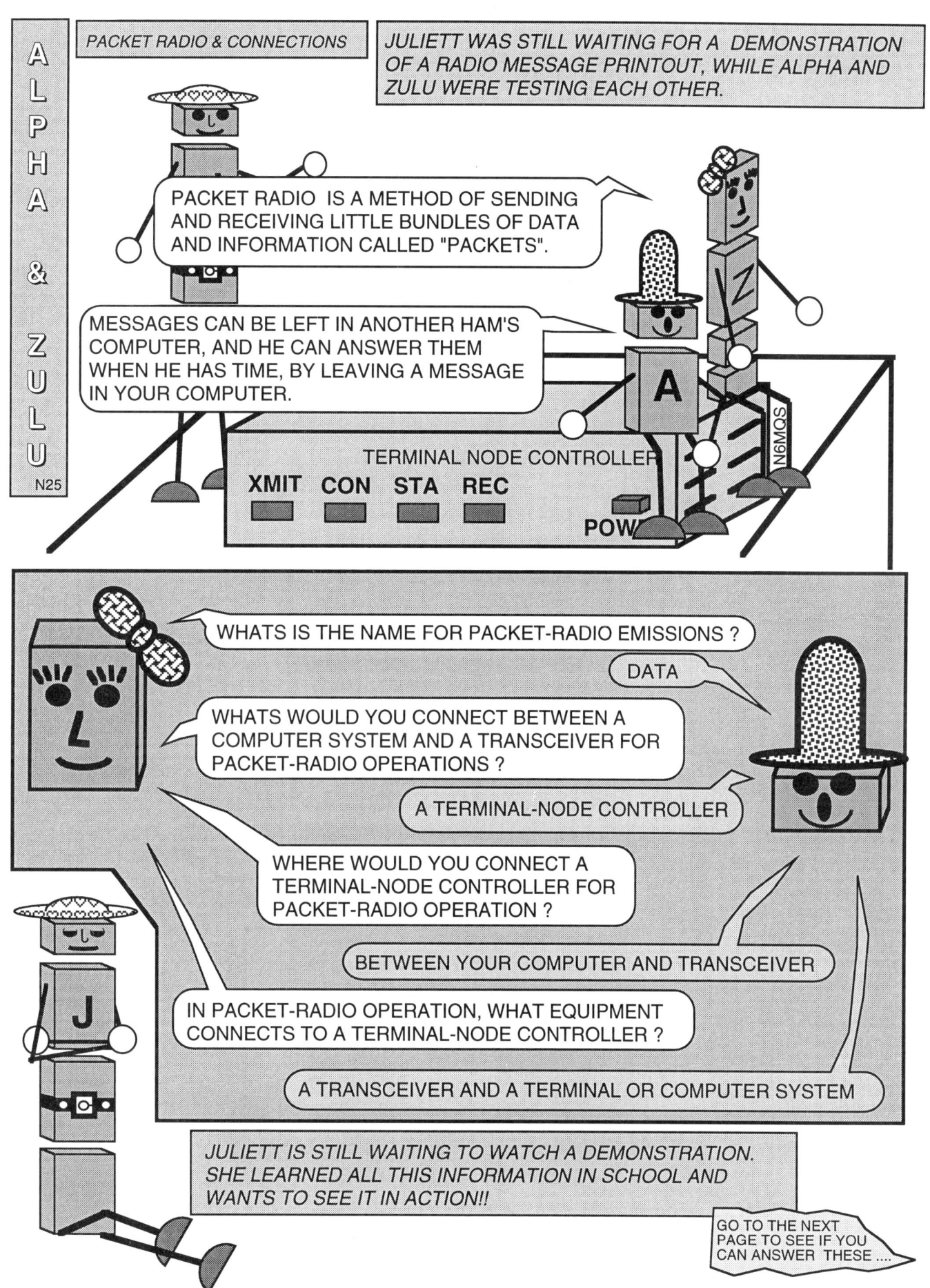

Riding the Airwaves with Alpha & Zulu

SEE IF YOU CAN ANSWER THESE QUESTIONS WITHOUT LOOKING BACK TO THE LAST 'TOON'. WRITE YOUR ANSWERS IN THE ANSWER BOX BELOW. GOOD LUCK.

PRACTICE TEST PAGE N25

N8A05
What is the name for packet-radio emissions?
- A. RTTY
- B. Data
- C. CW
- D. Phone

N7B08
What would you connect between a computer system and a transceiver for packet-radio operation?
- A. A terminal-node controller
- B. A DTMF keypad
- C. An SWR bridge
- D. An antenna tuner

N7B09
Where would you connect a terminal-node controller for packet-radio operation?
- A. Between your antenna and transceiver
- B. Between your computer and monitor
- C. Between your computer and transceiver
- D. Between your keyboard and computer

N7B11
In packet-radio operation, what equipment connects to a terminal-node controller?
- A. A transceiver and a modem
- B. A transceiver and a terminal or computer system
- C. A DTMF keypad, a monitor and a transceiver
- D. A DTMF microphone, a monitor and a transceiver

ANSWERS TO PREVIOUS TEST

N8A02	A
N8A04	A
N7B06	A
N7B07	C
N7B10	D

YOUR ANSWERS TO THIS TEST

N8A05 _____
N7B08 _____
N7B09 _____
N7B11 _____

58 artsci inc

Riding the Airwaves with Alpha & Zulu

SEE IF YOU CAN ANSWER THESE QUESTIONS WITHOUT LOOKING BACK TO THE LAST 'TOON'. WRITE YOUR ANSWERS IN THE ANSWER BOX BELOW. GOOD LUCK.

PRACTICE TEST PAGE N26

N2B01
What is the correct way to call CQ when using RTTY?
A. Send the letters "CQ" three times, followed by "DE," followed by your call sign sent once
B. Send the letters "CQ" three to six times, followed by "DE," followed by your call sign sent three times
C. Send the letters "CQ" ten times, followed by the procedural signal "DE," followed by your call sent one time
D. Send the letters "CQ" over and over

N2B02
What speed should you use when answering a CQ call using RTTY?
A. Half the speed of the received signal
B. The same speed as the received signal
C. Twice the speed of the received signal
D. Any speed, since RTTY systems adjust to any signal speed

N2B05
What is a digipeater?
A. A packet-radio station that retransmits only data that is marked to be retransmitted
B. A packet-radio station that retransmits any data that it receives
C. A repeater that changes audio signals to digital data
D. A repeater built using only digital electronics parts

N2B03
What does "connected" mean in a packet-radio link?
A. A telephone link is working between two stations
B. A message has reached an amateur station for local delivery
C. A transmitting station is sending data to only one receiving station; it replies that the data is being received correctly
D. A transmitting and receiving station are using a digipeater, so no other contacts can take place until they are finished

N2B04
What does "monitoring" mean on a packet-radio frequency?
A. The FCC is copying all messages
B. A member of the Amateur Auxiliary to the FCC's Field Operations Bureau is copying all messages
C. A receiving station is displaying all messages sent to it, and replying that the messages are being received correctly
D. A receiving station is displaying messages that may not be sent to it, and is not replying to any message

N2B06
What does "network" mean in packet radio?
A. A way of connecting terminal-node controllers by telephone so data can be sent over long distances
B. A way of connecting packet-radio stations so data can be sent over long distances
C. The wiring connections on a terminal-node controller board
D. The programming in a terminal-node controller that rejects other callers if a station is already connected

ANSWERS TO PREVIOUS TEST
N8A05
B
N7B08
A
N7B09
C
N7B11
B

YOUR ANSWERS TO THIS TEST
N2B01
N2B02
N2B05
N2B03
N2B04
N2B06

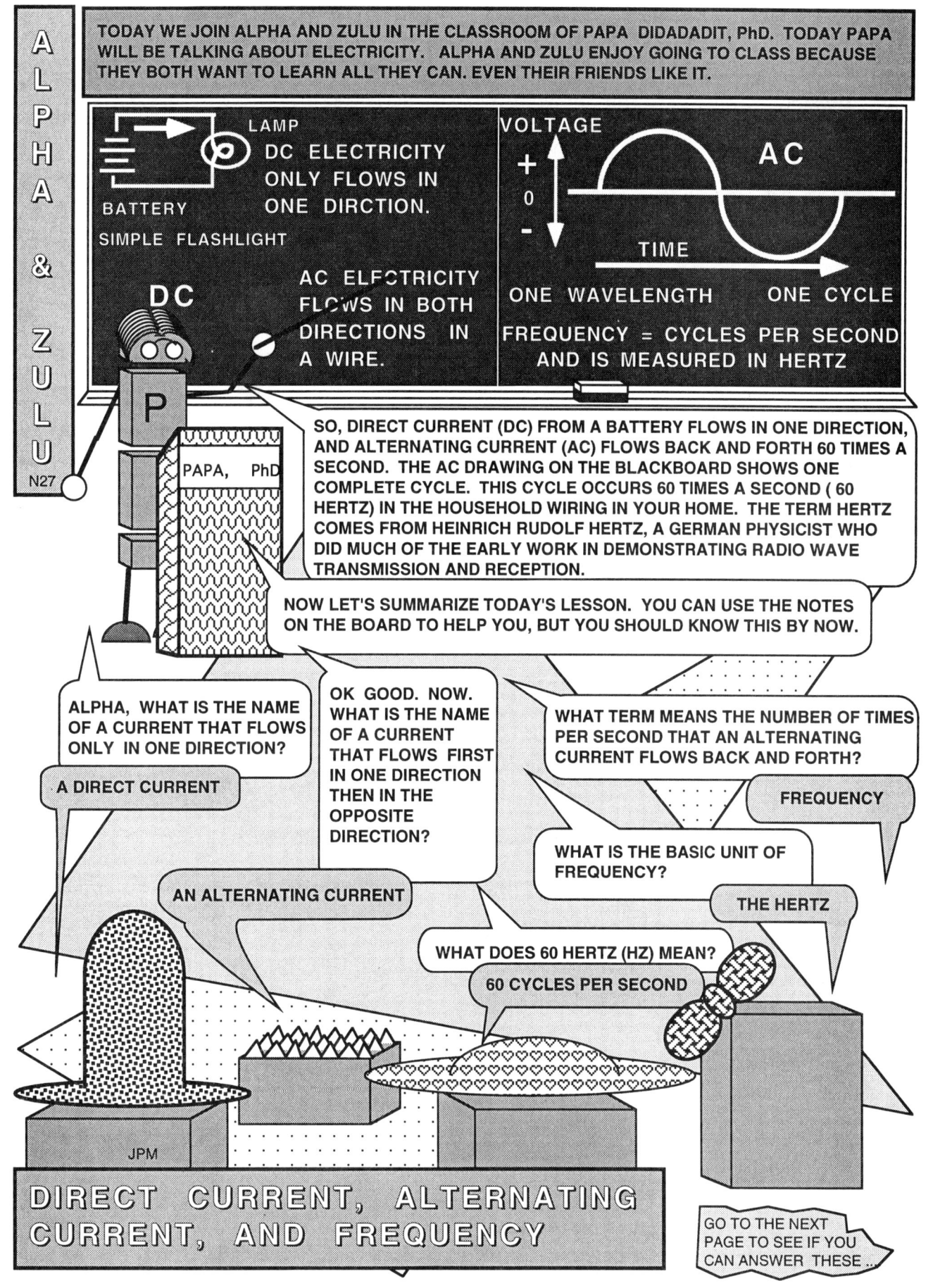

Riding the Airwaves with Alpha & Zulu

SEE IF YOU CAN ANSWER THESE QUESTIONS WITHOUT LOOKING BACK TO THE LAST 'TOON'. WRITE YOUR ANSWERS IN THE ANSWER BOX BELOW. GOOD LUCK.

PRACTICE TEST PAGE N27

N5C10
What is the name of a current that flows only in one direction?
- A. An alternating current
- B. A direct current
- C. A normal current
- D. A smooth current

N5C11
What is the name of a current that flows back and forth, first in one direction, then in the opposite direction?
- A. An alternating current
- B. A direct current
- C. A rough current
- D. A reversing current

N5D01
What term means the number of times per second that an alternating current flows back and forth?
- A. Pulse rate
- B. Speed
- C. Wavelength
- D. Frequency

N5D02
What is the basic unit of frequency?
- A. The hertz
- B. The watt
- C. The ampere
- D. The ohm

N5D11
What does 60 hertz (Hz) mean?
- A. 6000 cycles per second
- B. 60 cycles per second
- C. 6000 meters per second
- D. 60 meters per second

ANSWERS TO PREVIOUS TEST

N2B01	B
N2B02	B
N2B05	A
N2B03	C
N2B04	D
N2B06	B

YOUR ANSWERS TO THIS TEST

N5C10	
N5C11	
N5D01	
N5D02	
N5D11	

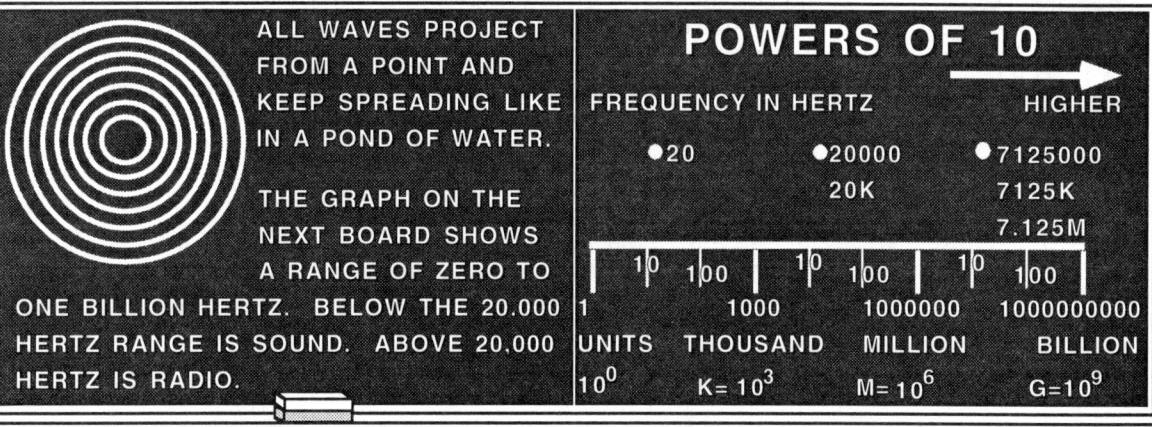

Riding the Airwaves with Alpha & Zulu

SEE IF YOU CAN ANSWER THESE QUESTIONS WITHOUT LOOKING BACK TO THE LAST 'TOON'. WRITE YOUR ANSWERS IN THE ANSWER BOX BELOW. GOOD LUCK.

PRACTICE TEST PAGE N28

N5D04
Why do we call signals in the range 20 Hz to 20,000 Hz audio frequencies?
- A. Because the human ear cannot sense anything in this range
- B. Because the human ear can sense sounds in this range
- C. Because this range is too low for radio energy
- D. Because the human ear can sense radio waves in this range

N5D03
What frequency can humans hear?
- A. 0 - 20 Hz
- B. 20 - 20,000 Hz
- C. 200 - 200,000 Hz
- D. 10,000 - 30,000 Hz

N5D05
What is the lowest frequency of electrical energy that is usually known as a radio frequency?
- A. 20 Hz
- B. 2,000 Hz
- C. 20,000 Hz
- D. 1,000,000 Hz

N5D07
If a radio wave makes 3,725,000 cycles in one second, what does this mean?
- A. The radio wave's voltage is 3,725 kilovolts
- B. The radio wave's wavelength is 3,725 kilometers
- C. The radio wave's frequency is 3,725 kilohertz
- D. The radio wave's speed is 3,725 kilometers per second

N5D06
Electrical energy at a frequency of 7125 kHz is in what frequency range?
- A. Audio
- B. Radio
- C. Hyper
- D. Super-high

ANSWERS TO PREVIOUS TEST
N5C10
B
N5C11
A
N5D01
D
N5D02
A
N5D11
B

YOUR ANSWERS TO THIS TEST
N5D04
_____
N5D03
_____
N5D05
_____
N5D07
_____
N5D06
_____

artsci inc

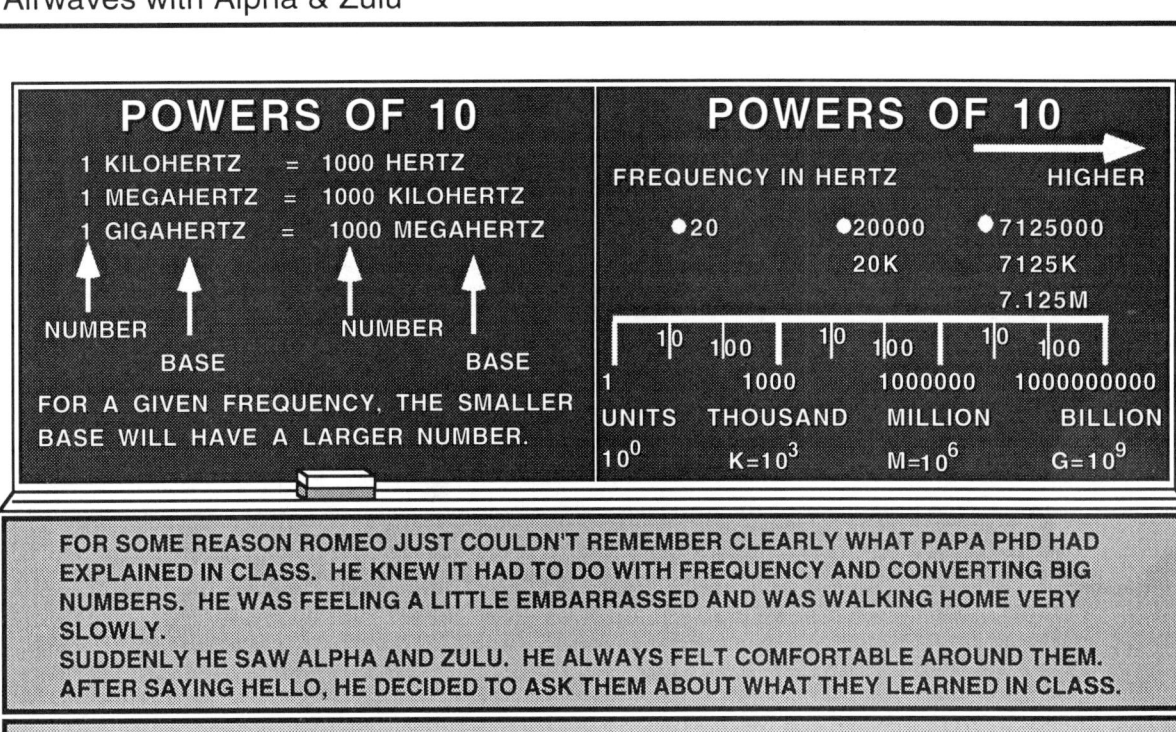

SEE IF YOU CAN ANSWER THESE QUESTIONS WITHOUT LOOKING BACK TO THE LAST 'TOON'. WRITE YOUR ANSWERS IN THE ANSWER BOX BELOW. GOOD LUCK.

PRACTICE TEST PAGE N29

N5A09
How many hertz are in a kilohertz?
  A. 10
  B. 100
  C. 1000
  D. 1000000

N5A10
How many kilohertz are in a megahertz?
  A. 10
  B. 100
  C. 1000
  D. 1000000

N5A01
If a dial marked in kilohertz shows a reading of 7125 kHz, what would it show if it were marked in megahertz?
  A. 0.007125 MHz
  B. 7.125 MHz
  C. 71.25 MHz
  D. 7,125,000 MHz

N5A03
If a dial marked in kilohertz shows a reading of 3725 kHz, what would it show if it were marked in hertz?
  A. 3,725 Hz
  B. 37.25 Hz
  C. 3,725 Hz
  D. 3,725,000 Hz

N5A02
If a dial marked in megahertz shows a reading of 3.525 MHz, what would it show if it were marked in kilohertz?
  A. 0.003525 kHz
  B. 35.25 kHz
  C. 3525 kHz
  D. 3,525,000 kHz

ANSWERS TO PREVIOUS TEST
N5D04
B
N5D03
B
N5D05
C
N5D07
C
N5D06
B

YOUR ANSWERS TO THIS TEST
N5A09
N5A10
N5A01
N5A03
N5A02

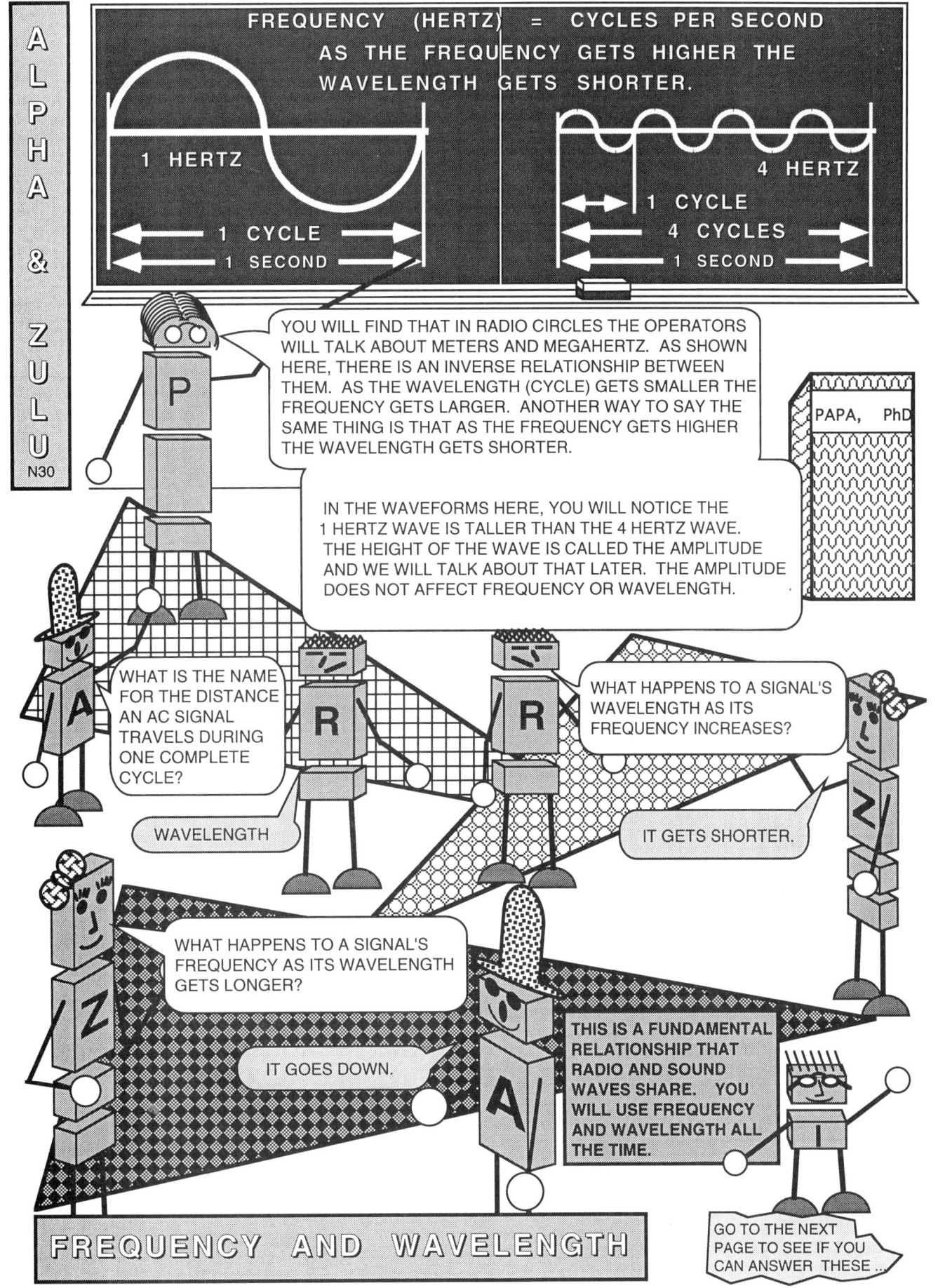

Riding the Airwaves with Alpha & Zulu

SEE IF YOU CAN ANSWER THESE QUESTIONS WITHOUT LOOKING BACK TO THE LAST 'TOON'. WRITE YOUR ANSWERS IN THE ANSWER BOX BELOW. GOOD LUCK.

PRACTICE TEST PAGE N30

N5D08
What is the name for the distance an AC signal travels during one complete cycle?
A. Wave speed
B. Waveform
C. Wavelength
D. Wave spread

N5D09
What happens to a signal's wavelength as its frequency increases?
A. It gets shorter
B. It gets longer
C. It stays the same
D. It disappears

N5D10
What happens to a signal's frequency as its wavelength gets longer?
A. It goes down
B. It goes up
C. It stays the same
D. It disappears

ANSWERS TO PREVIOUS TEST

N5A09	C
N5A10	C
N5A01	B
N5A03	D
N5A02	C

YOUR ANSWERS TO THIS TEST

N5D08	___
N5D09	___
N5D10	___

Riding the Airwaves with Alpha & Zulu

ALPHA AND ZULU WERE TALKING ABOUT THE NOVICE AMATEUR FREQUENCY "BANDS", AND WHERE THEY ARE IN THE RADIO WORLD!

AS THE FREQUENCIES GET HIGHER, THE WAVELENGTHS BECOME SHORTER.

80 METER	40 METER	15 METER	10 METER	1.25 METER	23 CENTIMETER
3675 KHZ TO 3725 KHZ	7100 KHZ TO 7150 KHZ	21.100 MHZ TO 21.200 MHZ	28.100 MHZ TO 28.500 MHZ	222.1 MHZ TO 223.91 MHZ	1270 MHZ TO 1295 MHZ

THESE RADIO BANDS (AS DEFINED BY THE WAVELENGTH, IE. 80 METER BAND) ARE THE MOST COMMONLY USED BY THE HAM RADIO OPERATOR.

NOVICE HAM OPERATORS ARE ALLOWED TO USE MORSE CODE (CW) IN THE LOWER FREQUENCIES, WHILE THE HIGHER FREQUENCIES CAN BE USED FOR VOICE, CW, DATA, AND RTTY.

WHAT ARE THE FREQUENCY LIMITS OF THE 80-METER NOVICE BAND?

3675 - 3725 KHZ

IF YOU ARE OPERATING ON 3700 KHZ, IN WHAT AMATEUR BAND ARE YOU OPERATING?

80 METERS.

CW ONLY FREQUENCIES

WHAT ARE THE FREQUENCY LIMITS OF THE 40-METER NOVICE BAND (ITU REGION 2)?

7100-7150 KHZ

ITU REGION 2 IS NORTH AMERICA

IF YOU ARE OPERATING ON 7125 KHZ, IN WHAT AMATEUR BAND ARE YOU OPERATING?

40 METERS

WHAT ARE THE FREQUENCY LIMITS OF THE 15-METER NOVICE BAND?

21.100 - 21.200 MHZ

IF YOU ARE OPERATING ON 21.150 MHZ, IN WHAT AMATEUR BAND ARE YOU OPERATING?

15 METERS

GO TO THE NEXT PAGE TO SEE IF YOU CAN ANSWER THESE ...

artsci inc

Riding the Airwaves with Alpha & Zulu

SEE IF YOU CAN ANSWER THESE QUESTIONS WITHOUT LOOKING BACK TO THE LAST 'TOON'. WRITE YOUR ANSWERS IN THE ANSWER BOX BELOW. GOOD LUCK.

PRACTICE TEST PAGE N31

N1C01 [97.301e]
What are the frequency limits of the 80-meter Novice band?
- A. 3500 - 4000 kHz
- B. 3675 - 3725 kHz
- C. 7100 - 7150 kHz
- D. 7000 - 7300 kHz

N1C03 [97.301e]
What are the frequency limits of the 15-meter Novice band?
- A. 21.100 - 21.200 MHz
- B. 21.000 - 21.450 MHz
- C. 28.000 - 29.700 MHz
- D. 28.100 - 28.200 MHz

N1C02 [97.301e]
What are the frequency limits of the 40-meter Novice band (ITU Region 2)?
- A. 3500 - 4000 kHz
- B. 3700 - 3750 kHz
- C. 7100 - 7150 kHz
- D. 7000 - 7300 kHz

N1C07 [97.301e]
If you are operating on 3700 kHz, in what amateur band are you operating?
- A. 80 meters
- B. 40 meters
- C. 15 meters
- D. 10 meters

N1C08 [97.301e]
If you are operating on 7125 kHz, in what amateur band are you operating?
- A. 80 meters
- B. 40 meters
- C. 15 meters
- D. 10 meters

N1C09 [97.301e]
If you are operating on 21.150 MHz, in what amateur band are you operating?
- A. 80 meters
- B. 40 meters
- C. 15 meters
- D. 10 meters

ANSWERS TO PREVIOUS TEST

N5D08 — C
N5D09 — A
N5D10 — A

YOUR ANSWERS TO THIS TEST

N1C01
N1C02
N1C03
N1C07
N1C08
N1C09

Riding the Airwaves with Alpha & Zulu

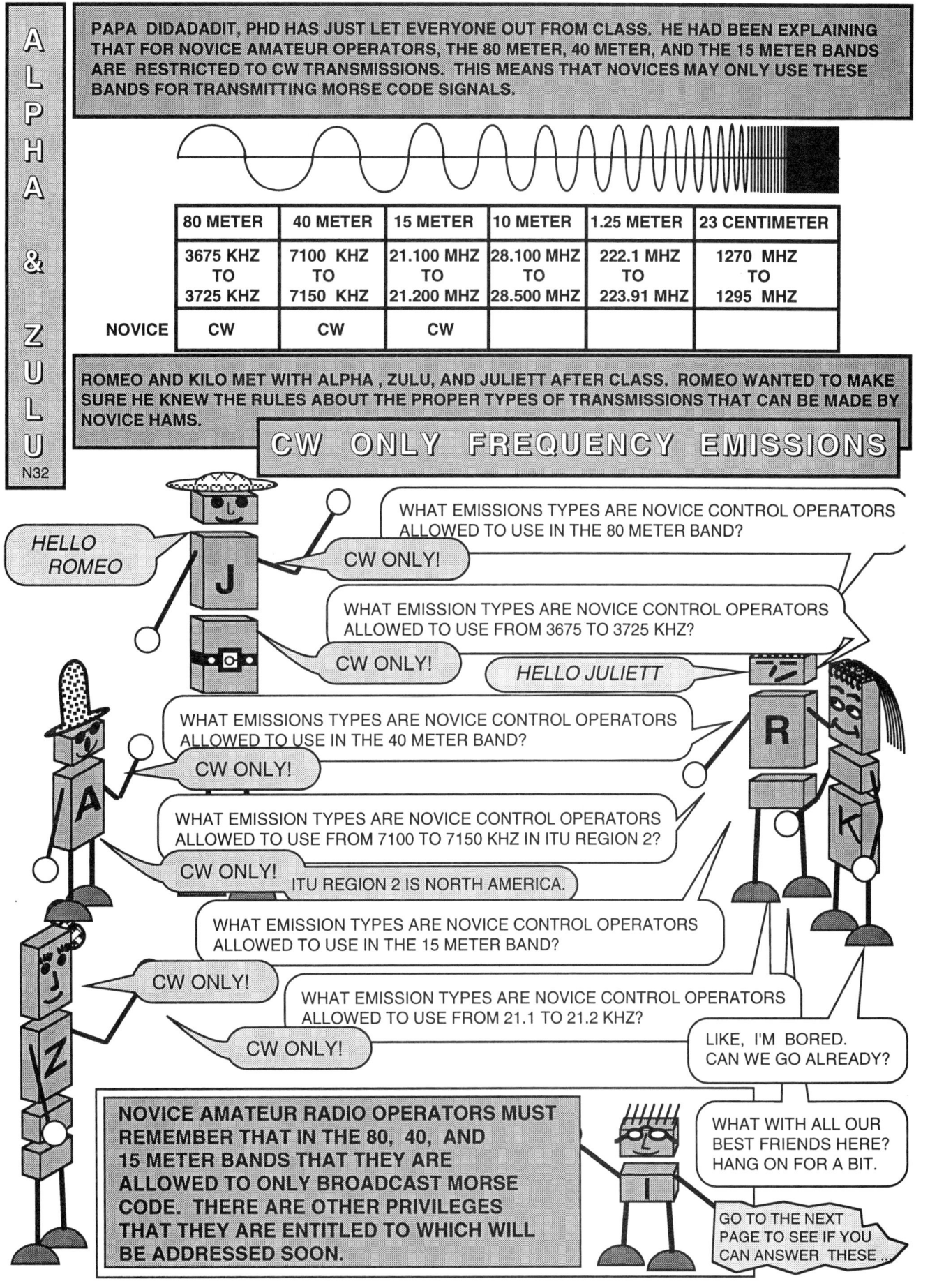

PRACTICE TEST PAGE N32

SEE IF YOU CAN ANSWER THESE QUESTIONS WITHOUT LOOKING BACK TO THE LAST 'TOON'. WRITE YOUR ANSWERS IN THE ANSWER BOX BELOW. GOOD LUCK.

N1E01 [97.305/.307f9]
What emission types are Novice control operators allowed to use in the 80-meter band?
- A. CW only
- B. Data only
- C. RTTY only
- D. Phone only

N1E02 ([97.305/307f9]
What emission types are Novice control operators allowed to use in the 40-meter band?
- A. CW only
- B. Data only
- C. RTTY only
- D. Phone only

N1E03 [97.305/307f9]
What emission types are Novice control operators allowed to use in the 15-meter band?
- A. CW only
- B. Data only
- C. RTTY only
- D. Phone only

N1E04 [97.305/307f9]
What emission types are Novice control operators allowed to use from 3675 to 3725 kHz?
- A. Phone only
- B. Image only
- C. Data only
- D. CW only

N1E05 [97.305/307f9]
What emission types are Novice control operators allowed to use from 7100 to 7150 kHz in ITU Region 2?
- A. CW and data
- B. Phone
- C. Data only
- D. CW only

N1E06 [97.305/307f9]
What emission types are Novice control operators allowed to use on frequencies from 21.1 to 21.2 MHz?
- A. CW and data
- B. CW and phone
- C. Data only
- D. CW only

ANSWERS TO PREVIOUS TEST	
N1C01	B
N1C02	C
N1C03	A
N1C07	A
N1C08	B
N1C09	C

YOUR ANSWERS TO THIS TEST	
N1E01	
N1E02	
N1E03	
N1E04	
N1E05	
N1E06	

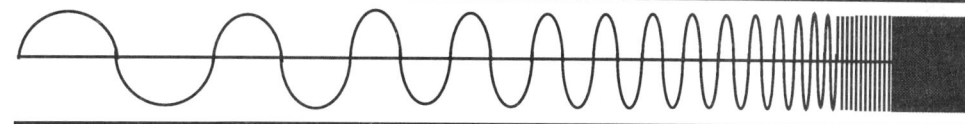

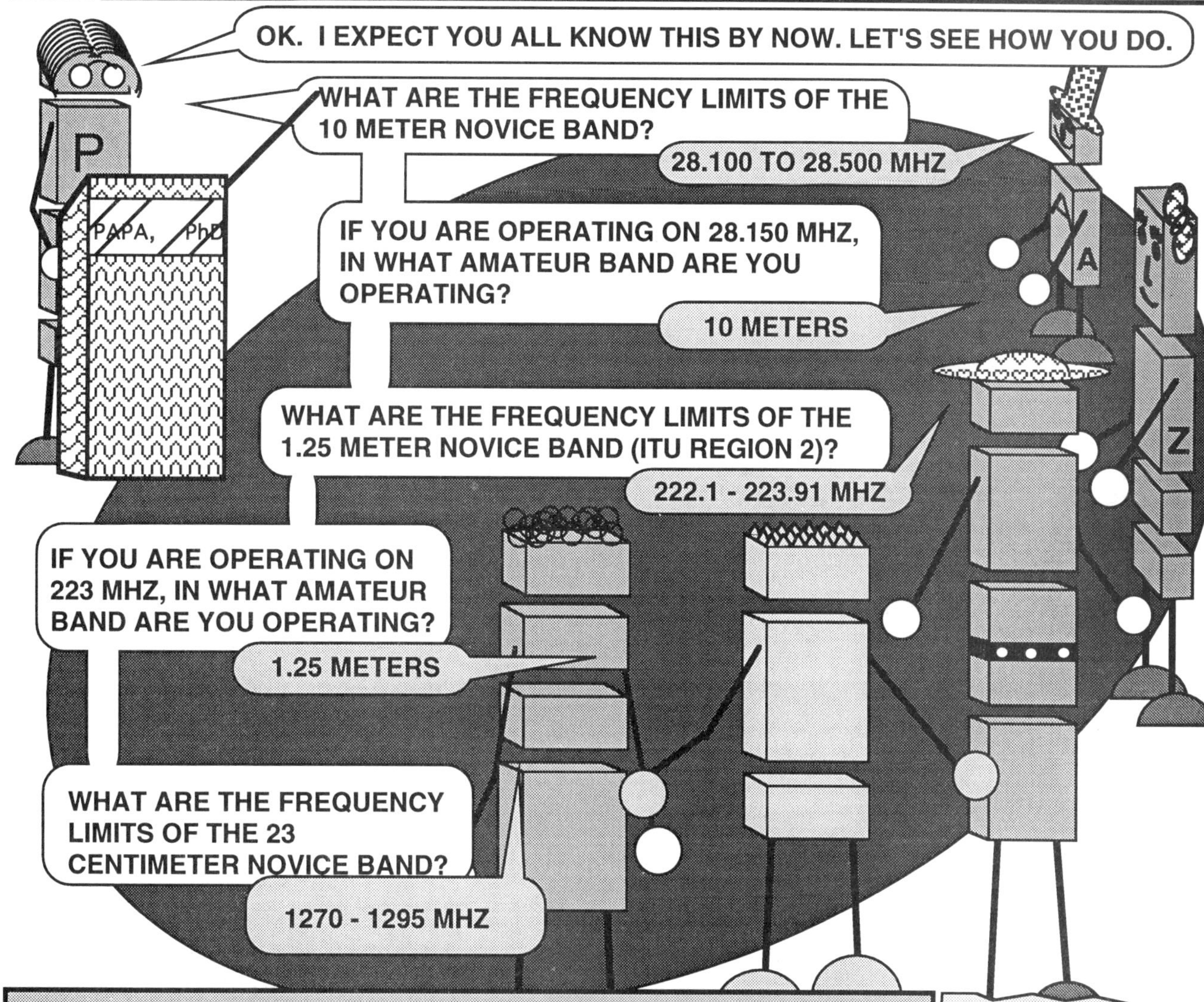

Riding the Airwaves with Alpha & Zulu

SEE IF YOU CAN ANSWER THESE QUESTIONS WITHOUT LOOKING BACK TO THE LAST 'TOON'. WRITE YOUR ANSWERS IN THE ANSWER BOX BELOW. GOOD LUCK.

PRACTICE TEST PAGE N33

N1C04 [97.301e]
What are the frequency limits of the 10-meter Novice band?
  A. 28.000 - 28.500 MHz
  B. 28.100 - 29.500 MHz
  C. 28.100 - 28.500 MHz
  D. 29.100 - 29.500 MHz

N1C05 [97.301f]
What are the frequency limits of the 1.25-meter Novice band (ITU Region 2)?
  A. 225.0 - 230.5 MHz
  B. 222.1 - 223.91 MHz
  C. 224.1 - 225.1 MHz
  D. 222 - 225 MHzHz

N1C06 [97.301f]
What are the frequency limits of the 23-centimeter Novice band?
  A. 1260 - 1270 MHz
  B. 1240 - 1300 MHz
  C. 1270 - 1295 MHz
  D. 1240 - 1246 MHz

N1C10 [97.301e]
If you are operating on 28.150 MHz, in what amateur band are you operating?
  A. 80 meters
  B. 40 meters
  C. 15 meters
  D. 10 meters

N1C11 [97.301f]
If you are operating on 223 MHz, in what amateur band are you operating?
  A. 15 meters
  B. 10 meters
  C. 2 meters
  D. 1.25 meters

ANSWERS TO PREVIOUS TEST
N1E01 A
N1E02 A
N1E03 A
N1E04 D
N1E05 D
N1E06 D

YOUR ANSWERS TO THIS TEST
N1C04
N1C05
N1C06
N1C10
N1C11

Riding the Airwaves with Alpha & Zulu

A L P H A & Z U L U
N34

THE 10 METER, 1.25 METER, AND 23 CENTIMETER BANDS ALLOW THE NOVICE AMATEUR RADIO OPERATOR MORE FLEXIBILITY. THERE ARE STILL SOME RESTRICTIONS BUT VOICE, DATA, AND RTTY TRANSMISSIONS ARE ALLOWED ALONG WITH CW.

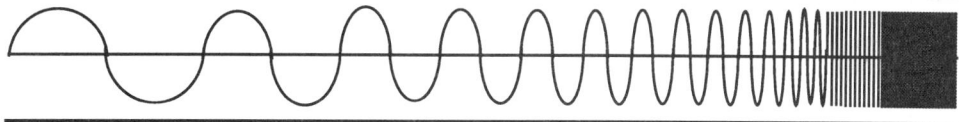

	80 METER	40 METER	15 METER	10 METER	1.25 METER	23 CENTIMETER
	3675 KHZ TO 3725 KHZ	7100 KHZ TO 7150 KHZ	21.100 MHZ TO 21.200 MHZ	28.100 MHZ TO 28.500 MHZ	222.1 MHZ TO 223.91 MHZ	1270 MHZ TO 1295 MHZ
NOVICE	CW	CW	CW	CW VOICE DATA	CW VOICE DATA	CW VOICE DATA

EVERYONE WAS GETTING VERY EXCITED NOW THAT THEY WERE ABLE TO FIND OUT THE FREQUENCIES THEY WOULD BE ALLOWED TO SPEAK ON. MORSE CODE IS FUN, ITS LIKE A GAME, BUT TALKING IS SO MUCH EASIER.

CW, VOICE AND DATA FREQUENCY EMISSIONS

A: WHAT EMISSION TYPES ARE NOVICE CONTROL OPERATORS ALLOWED TO USE ON FREQUENCIES FROM 28.1 TO 28.3 MHZ?

R: CW, RTTY, AND DATA

J: ON WHAT FREQUENCIES IN THE 10 METER BAND MAY NOVICE CONTROL OPERATORS USE RTTY?

Z: 28.1 - 28.3 MHZ

A: ON WHAT FREQUENCIES IN THE 10 METER BAND MAY NOVICE CONTROL OPERATORS USE DATA EMISSIONS?

Z: 28.1 - 28.3 MHZ

A: BY THE WAY. HAVE YOU NOTICED THAT JULIETT LIKES ROMEO? I DON'T THINK HE EVEN KNOWS.

Z: ROMEO ALREADY HAS A GIRLFRIEND. PLEASE DON'T GET INVOLVED IN THIS.

A: I KNOW. BUT HIS GIRLFRIEND, KILO, DOESN'T SEEM FRIENDLY, AND SHE NEVER JOINS IN. I LIKE JULIETT BETTER.

MORE ON THE NEXT PAGE

Riding the Airwaves with Alpha & Zulu

Riding the Airwaves with Alpha & Zulu

SEE IF YOU CAN ANSWER THESE QUESTIONS WITHOUT LOOKING BACK TO THE LAST 'TOON'. WRITE YOUR ANSWERS IN THE ANSWER BOX BELOW. GOOD LUCK.

PRACTICE TEST PAGE N34

N1E07 [97.305]
What emission types are Novice control operators allowed to use on frequencies from 28.1 to 28.3 MHz?
- A. All authorized amateur emission privileges
- B. Data or phone
- C. CW, RTTY and data
- D. CW and phone

N1E08 [97.305/307f10]
What emission types are Novice control operators allowed to use on frequencies from 28.3 to 28.5 MHz?
- A. All authorized amateur emission privileges
- B. CW and data
- C. CW and single-sideband phone
- D. Data and phone

N1E09 [97.305]
What emission types are Novice control operators allowed to use on the amateur 1.25-meter band in ITU Region 2?
- A. CW and phone
- B. CW and data
- C. Data and phone
- D. All amateur emission privileges authorized for use on the band

N1E10 [97.305]
What emission types are Novice control operators allowed to use on the amateur 23-centimeter band?
- A. Data and phone
- B. CW and data
- C. CW and phone
- D. All amateur emission privileges authorized for use on the band

N1E11 [97.305/.307f10]
On what HF frequencies may Novice control operators use single-sideband (SSB) phone?
- A. 3700 - 3750 kHz
- B. 7100 - 7150 kHz
- C. 21100 - 21200 kHz
- D. 28300 - 28500 kHz

N1E12 [97.305]
On what frequencies in ITU Region 2 may Novice control operators use FM phone?
- A. 28.3 - 28.5 MHz
- B. 144.0 - 148.0 MHz
- C. 222.1 - 223.91 MHz
- D. 1240 - 1270 MHz

N1E13 [97.301e/.305]
On what frequencies in the 10-meter band may Novice control operators use RTTY?
- A. 28.0 - 28.3 MHz
- B. 28.1 - 28.3 MHz
- C. 28.0 - 29.3 MHz
- D. 29.1 - 29.3 MHz

N1E14 [97.301e/.305]
On what frequencies in the 10-meter band may Novice control operators use data emissions?
- A. 28.0 - 28.3 MHz
- B. 28.1 - 28.3 MHz
- C. 28.0 - 29.3 MHz
- D. 29.1 - 29.3 MHz

ANSWERS TO PREVIOUS TEST

N1C04	C
N1C05	B
N1C06	C
N1C10	D
N1C11	D

YOUR ANSWERS TO THIS TEST

N1E07	
N1E08	
N1E09	
N1E10	
N1E11	
N1E12	
N1E13	
N1E14	

artsci inc

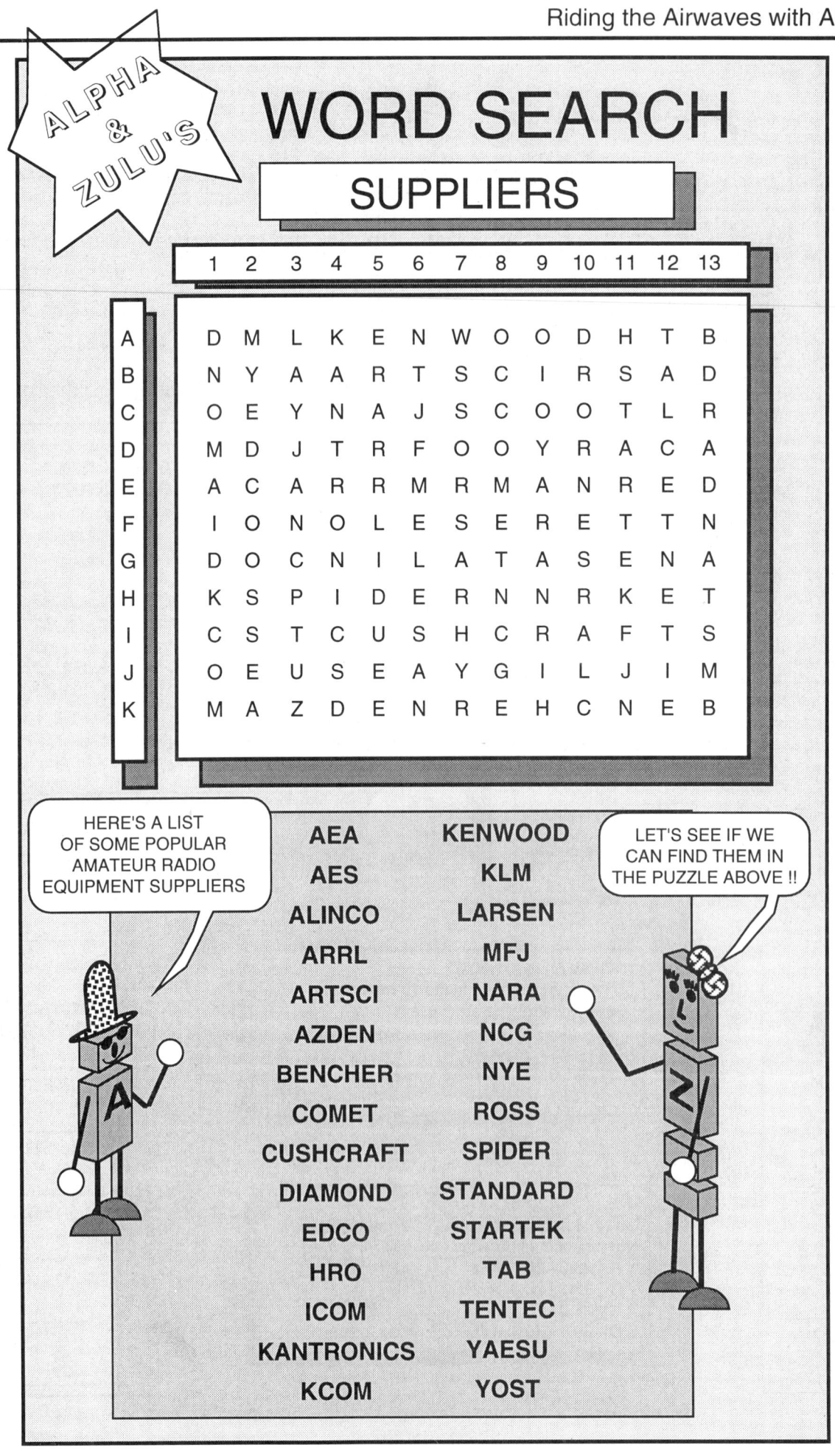

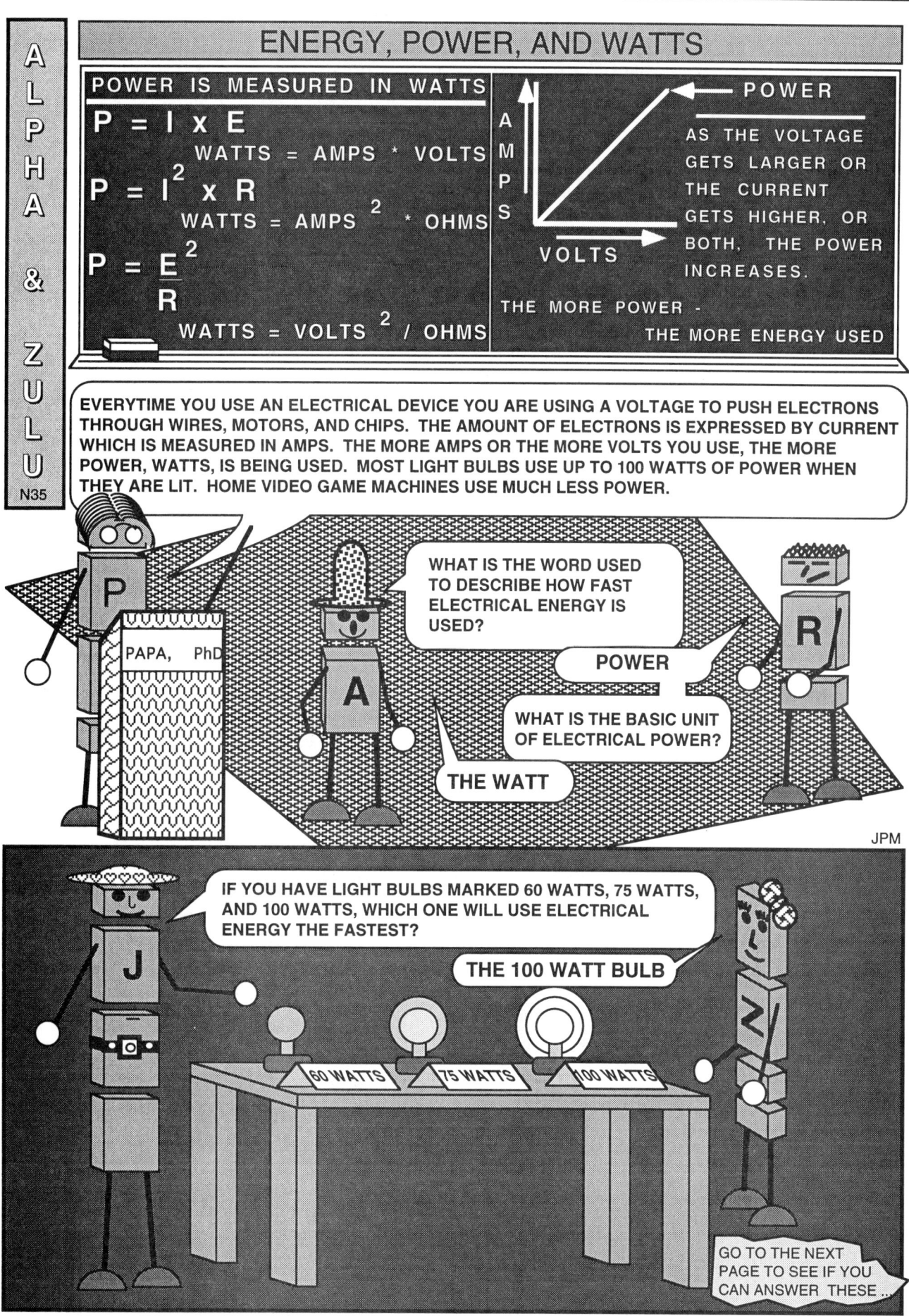

Riding the Airwaves with Alpha & Zulu

SEE IF YOU CAN ANSWER THESE QUESTIONS WITHOUT LOOKING BACK TO THE LAST 'TOON'. WRITE YOUR ANSWERS IN THE ANSWER BOX BELOW. GOOD LUCK.

PRACTICE TEST PAGE N35

N5C05
What is the word used to describe how fast electrical energy is used?
 A. Resistance
 B. Current
 C. Power
 D. Voltage

N5C07
What is the basic unit of electrical power?
 A. The ohm
 B. The watt
 C. The volt
 D. The amperesame

N5C06
If you have light bulbs marked 60 watts, 75 watts and 100 watts, which one will use electrical energy the fastest?
 A. The 60 watt bulb
 B. The 75 watt bulb
 C. The 100 watt bulb
 D. They will all be the same

ANSWERS TO PREVIOUS TEST

N1E07	C
N1E08	C
N1E09	D
N1E10	D
N1E11	D
N1E12	C
N1E13	B
N1E14	B

YOUR ANSWERS TO THIS TEST

N5C05	
N5C07	
N5C06	

artsci inc

HIGH FREQUENCY POWER

PEAK ENVELOPE POWER
PEP

THE AVERAGE POWER OF A SIGNAL AT ITS LARGEST AMPLITUDE PEAK.

WHEN TRANSMITTING AN AMATEUR SHOULD USE THE MINIMUM POWER NECESSARY IN ORDER TO COMMUNICATE. IT IS NOT NECESSARY TO USE FULL POWER TO TALK TO ANOTHER HAM IN TOWN.

HIGH FREQUENCY BANDS

THE 80, 40, 15, AND 10 METER BANDS HAVE A 200 WATT PEP OUTPUT POWER LIMIT FOR NOVICE AMATEURS.

IT IS SOMETIMES EASIER TO REMEMBER THAT THE MAXIMUM PEP POWER OUTPUT FOR FREQUENCIES BELOW 30 MHZ IS 200 WATTS.

P: THIS SHOULD BE AN EASY LESSON FOR YOU. WE ARE CONCERNED WITH THE BANDS IN THE HIGH FREQUENCY RANGE.

P: ALPHA, WHAT AMOUNT OF TRANSMITTER POWER MUST AMATEUR STATIONS USE AT ALL TIMES?

A: THE MIMIMUN LEGAL POWER NECESSARY TO COMMUNICATE.

P: RIGHT. NOW, ON WHICH BANDS MAY A NOVICE STATION USE UP TO 200 WATTS PEP OUTPUT POWER?

A: 80, 40, 15, AND 10 METERS

P: WHAT IS THE MOST TRANSMITTER POWER AN AMATEUR STATION MAY USE ON 3700 KHZ?

Z: 200 WATTS PEP OUTPUT

P: WHAT IS THE MOST TRANSMITTER POWER AN AMATEUR STATION MAY USE ON 7125 KHZ?

J: 200 WATTS PEP OUTPUT

P: WHAT IS THE MOST TRANSMITTER POWER AN AMATEUR STATION MAY USE ON 21.125 MHZ?

J: 200 WATTS PEP OUTPUT

P: WHAT IS THE MOST TRANSMITTER POWER AN AMATEUR STATION MAY USE ON 28.125 MHZ?

R: 200 WATTS PEP OUTPUT

P: WHAT IS THE MOST TRANSMITTER POWER A NOVICE STATION MAY USE ON THE 10 METER BAND?

R: 200 WATTS PEP OUTPUT

GO TO THE NEXT PAGE TO SEE IF YOU CAN ANSWER THESE...

Riding the Airwaves with Alpha & Zulu

SEE IF YOU CAN ANSWER THESE QUESTIONS WITHOUT LOOKING BACK TO THE LAST 'TOON'. WRITE YOUR ANSWERS IN THE ANSWER BOX BELOW. GOOD LUCK.

PRACTICE TEST PAGE N36

N1F01 [97.313a]
What amount of transmitter power must amateur stations use at all times?
A. 25 watts PEP output
B. 250 watts PEP output
C. 1500 watts PEP output
D. The minimum legal power necessary to communicate

N1F02 [97.313c1]
What is the most transmitter power an amateur station may use on 3700 kHz?
A. 5 watts PEP output
B. 25 watts PEP output
C. 200 watts PEP output
D. 1500 watts PEP output

N1F03 [97.313c1]
What is the most transmitter power an amateur station may use on 7125 kHz?
A. 5 watts PEP output
B. 25 watts PEP output
C. 200 watts PEP output
D. 1500 watts PEP output

N1F04 [97.313c1]
What is the most transmitter power an amateur station may use on 21.125 MHz?
A. 5 watts PEP output
B. 25 watts PEP output
C. 200 watts PEP output
D. 1500 watts PEP output

N1F05 [97.313c2]
What is the most transmitter power a Novice station may use on 28.125 MHz?
A. 5 watts PEP output
B. 25 watts PEP output
C. 200 watts PEP output
D. 1500 watts PEP output

N1F06 [97.313c2]
What is the most transmitter power a Novice station may use on the 10-meter band?
A. 5 watts PEP output
B. 25 watts PEP output
C. 200 watts PEP output
D. 1500 watts PEP output

N1F09 [97.313c]
On which bands may a Novice station use up to 200 watts PEP output power?
A. 80, 40, 15, and 10 meters
B. 80, 40, 20, and 10 meters
C. 1.25 meters
D. 23 centimeters

ANSWERS TO PREVIOUS TEST

N5C05
C

N5C07
B

N5C06
C

YOUR ANSWERS TO THIS TEST

N1F01 ____

N1F02 ____

N1F03 ____

N1F04 ____

N1F05 ____

N1F06 ____

N1F09 ____

artsci inc

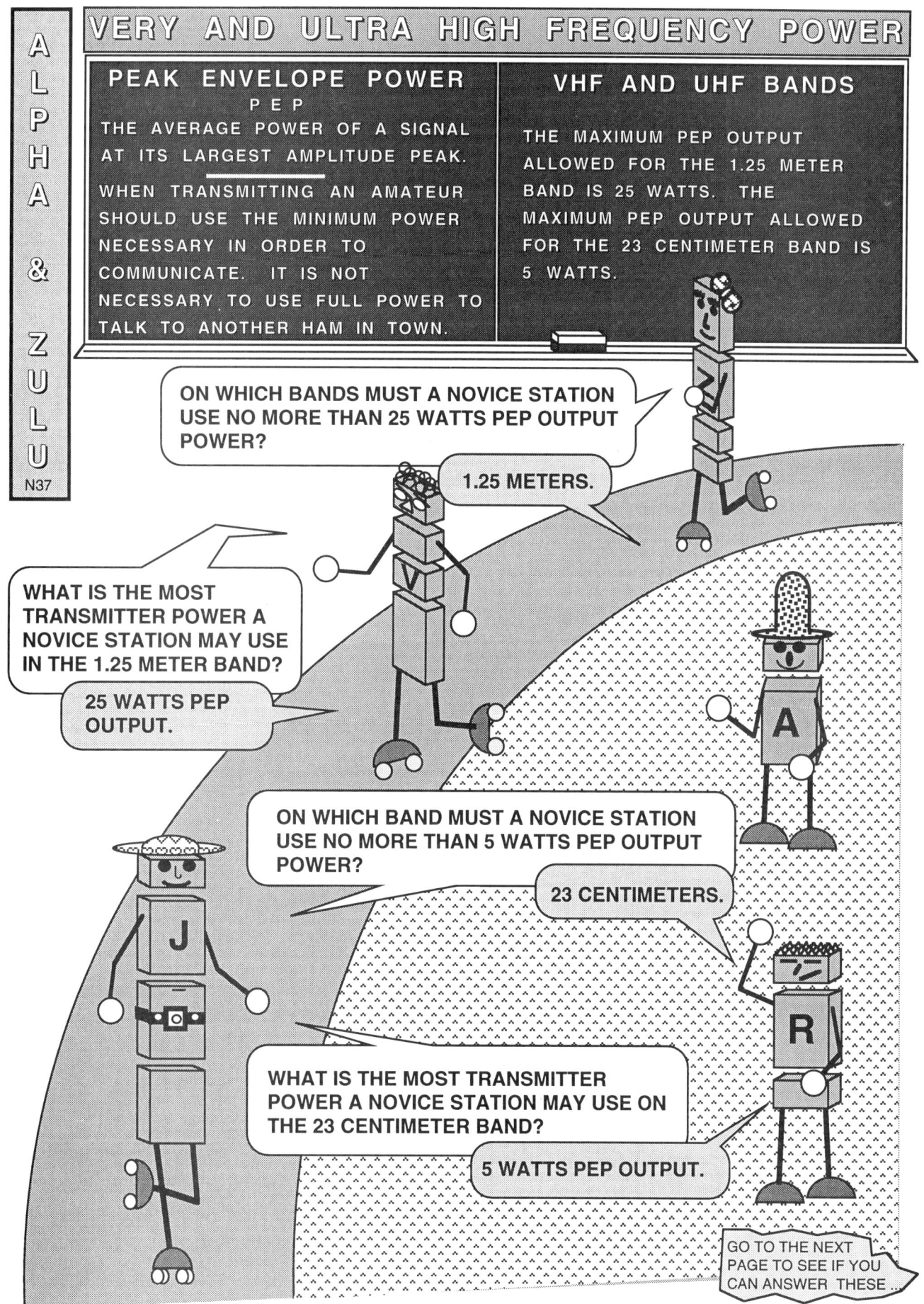

Riding the Airwaves with Alpha & Zulu

SEE IF YOU CAN ANSWER THESE QUESTIONS WITHOUT LOOKING BACK TO THE LAST 'TOON'. WRITE YOUR ANSWERS IN THE ANSWER BOX BELOW. GOOD LUCK.

PRACTICE TEST PAGE N37

N1F07 [97.313d]
What is the most transmitter power a Novice station may use on the 1.25-meter band?
  A. 5 watts PEP output
  B. 25 watts PEP output
  C. 200 watts PEP output
  D. 1500 watts PEP output

N1F08 [97.313e]
What is the most transmitter power a Novice station may use on the 23-centimeter band?
  A. 5 watts PEP output
  B. 25 watts PEP output
  C. 200 watts PEP output
  D. 1500 watts PEP output

N1F10 [97.313d]
On which bands must a Novice station use no more than 25 watts PEP output power?
  A. 80, 40, 15, and 10 meters
  B. 80, 40, 20, and 10 meters
  C. 1.25 meters
  D. 23 centimeters

N1F11 [97.313e]
On which bands must a Novice station use no more than 5 watts PEP output power?
  A. 80, 40, 15, and 10 meters
  B. 80, 40, 20, and 10 meters
  C. 1.25 meters
  D. 23 centimeters

ANSWERS TO PREVIOUS TEST

N1F01	D
N1F02	C
N1F03	C
N1F04	C
N1F05	C
N1F06	C
N1F09	A

YOUR ANSWERS TO THIS TEST

N1F07	
N1F08	
N1F10	
N1F11	

artsci inc

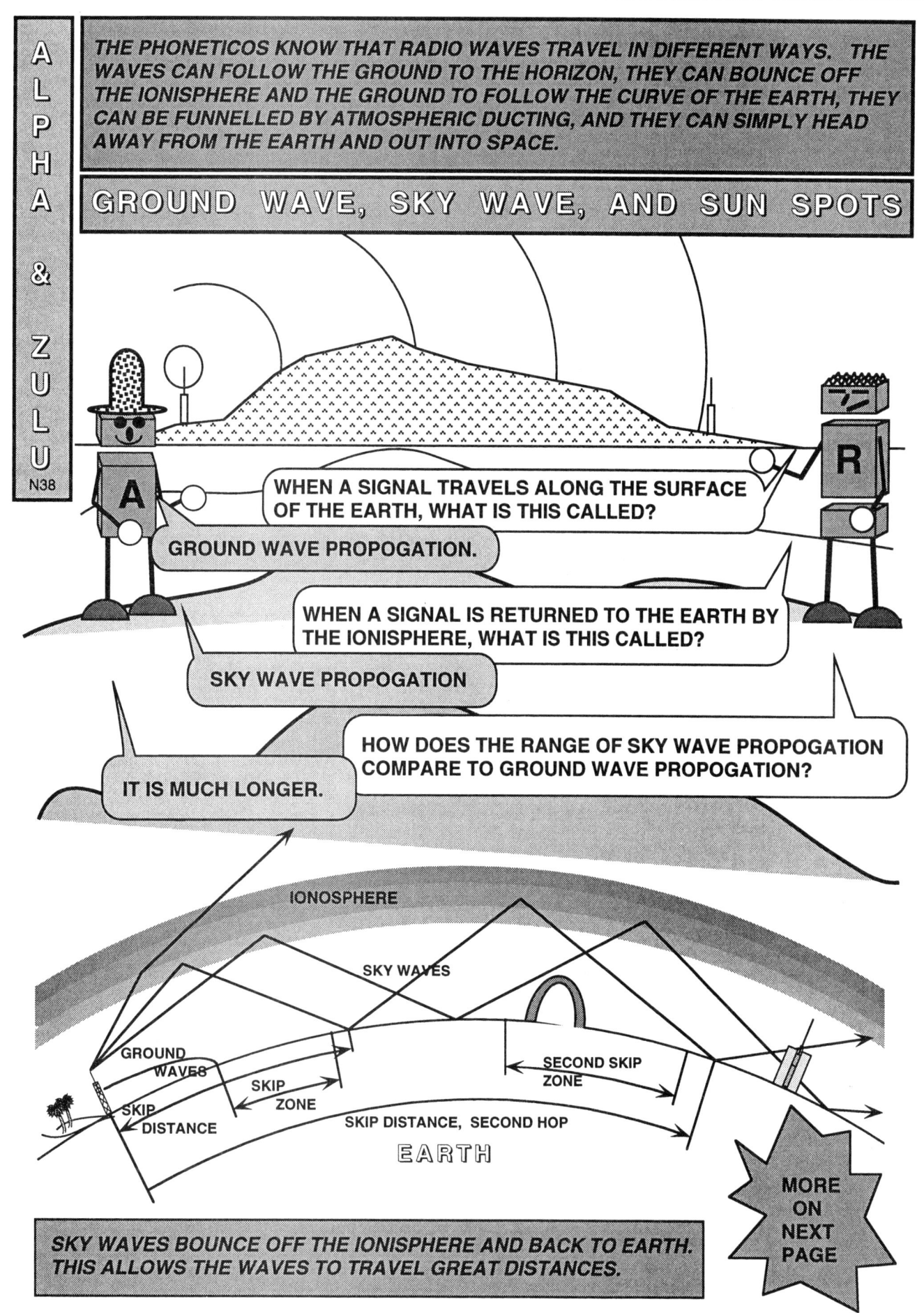

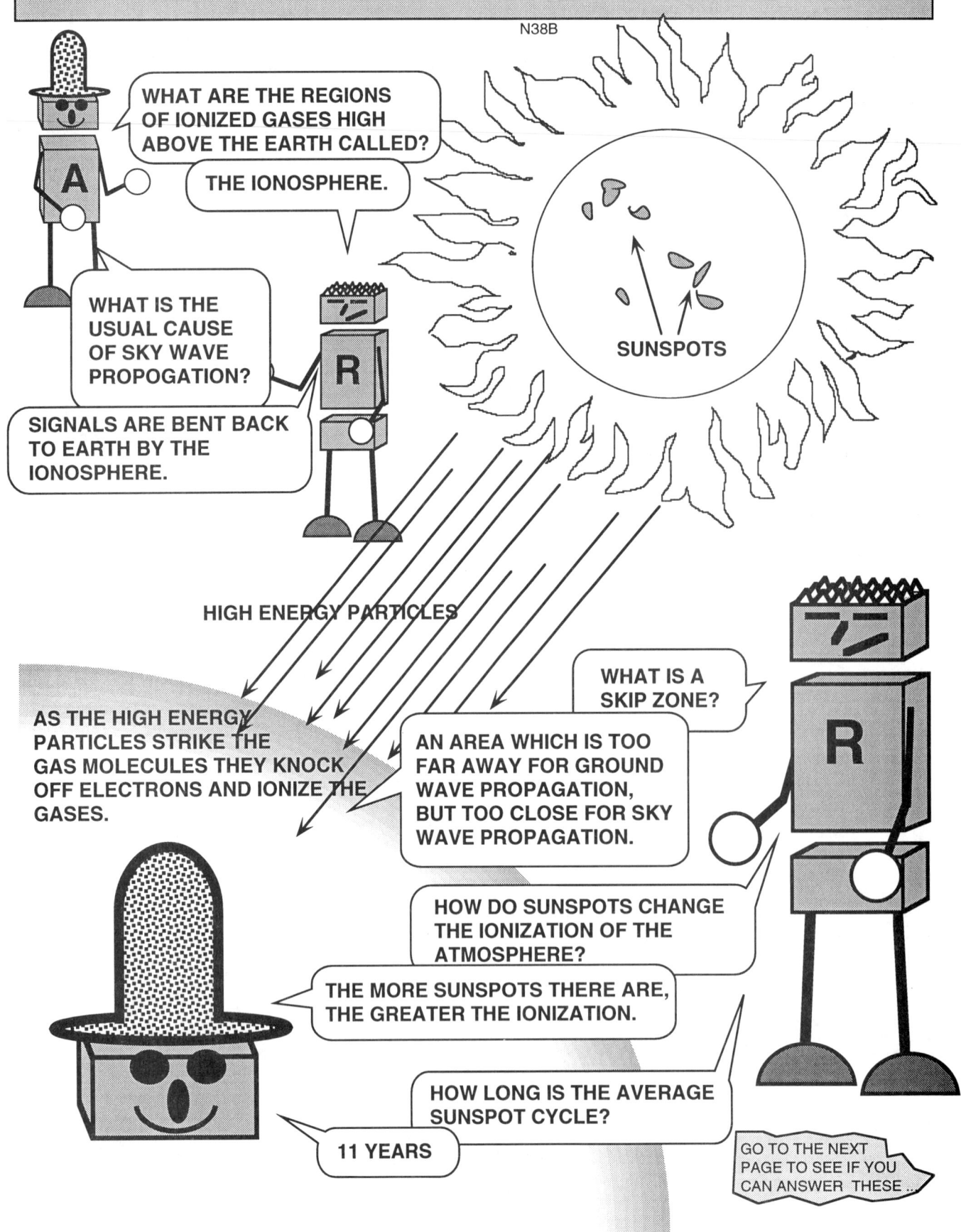

Riding the Airwaves with Alpha & Zulu

PRACTICE TEST PAGE N38

N3A05
When a signal travels along the surface of the Earth, what is this called?
A. Sky-wave propagation
B. Knife-edge diffraction
C. E-region propagation
D. Ground-wave propagation

N3A06
How does the range of sky-wave propagation compare to ground-wave propagation?
A. It is much shorter
B. It is much longer
C. It is about the same
D. It depends on the weather

N3A07
When a signal is returned to earth by the ionosphere, what is this called?
A. Sky-wave propagation
B. Earth-moon-earth propagation
C. Ground-wave propagation
D. Tropospheric propagation

N3A08
What is the usual cause of sky-wave propagation?
A. Signals are reflected by a mountain
B. Signals are reflected by the moon
C. Signals are bent back to earth by the ionosphere
D. Signals are repeated by a repeater

N3A09
What is a skip zone?
A. An area covered by ground-wave propagation
B. An area covered by sky-wave propagation
C. An area which is too far away for ground-wave propagation, but too close for sky-wave propagation
D. An area which is too far away for ground-wave or sky-wave propagation

N3A10
What are the regions of ionized gases high above the earth called?
A. The ionosphere
B. The troposphere
C. The gas region
D. The ion zone

N3A11
How do sunspots change the ionization of the atmosphere?
A. The more sunspots there are, the greater the ionization
B. The more sunspots there are, the less the ionization
C. Unless there are sunspots, the ionization is zero
D. They have no effect

N3A12
How long is an average sunspot cycle?
A. 2 years
B. 5 years
C. 11 years
D. 17 years

ANSWERS TO PREVIOUS TEST
N1F07
B
N1F08
A
N1F10
C
N1F11
D

YOUR ANSWERS TO THIS TEST
N3A05
N3A06
N3A07
N3A08
N3A09
N3A10
N3A11
N3A12

artsci inc

PHONETICO COLORING PAGE

AS WELL AS COLORING IN THIS PAGE, MAKE UP A SMALL CONVERSATION OR COMMENTS AND FILL IN THE BLANK TALK BALLOONS IN THE PICTURE. YOU CAN MAKE IT SERIOUS OR FUNNY, LONG OR SHORT. YOU CAN WRITE ANYTHING YOU WANT. SO HAVE FUN AND USE YOUR IMAGINATION.

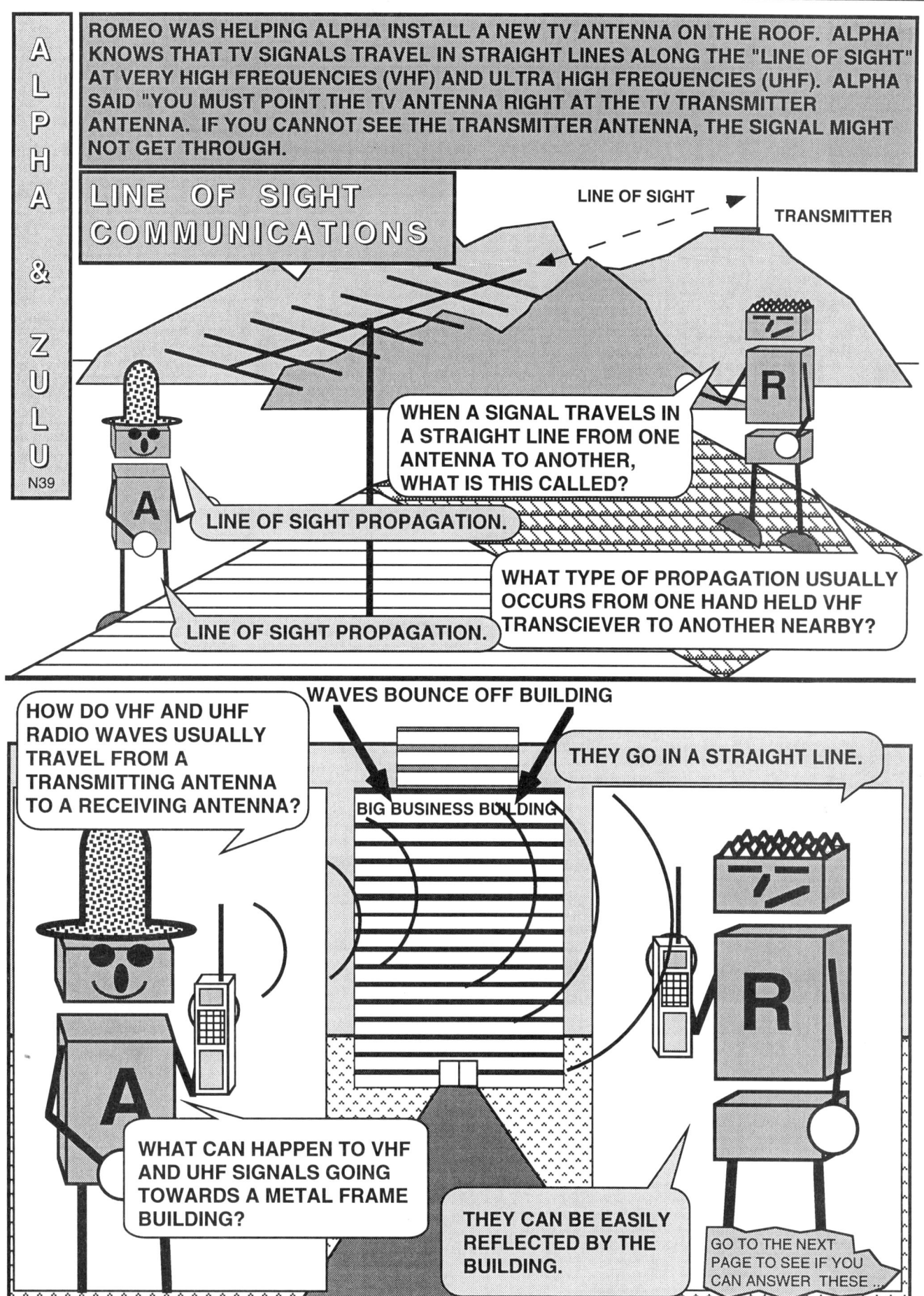

Riding the Airwaves with Alpha & Zulu

SEE IF YOU CAN ANSWER THESE QUESTIONS WITHOUT LOOKING BACK TO THE LAST 'TOON'. WRITE YOUR ANSWERS IN THE ANSWER BOX BELOW. GOOD LUCK.

PRACTICE TEST PAGE N39

N3A01
When a signal travels in a straight line from one antenna to another, what is this called?
- A. Line-of-sight propagation
- B. Straight-line propagation
- C. Knife-edge diffraction
- D. Tunnel propagation

N3A02
What type of propagation usually occurs from one hand held VHF transceiver to another nearby?
- A. Tunnel propagation
- B. Sky-wave propagation
- C. Line-of-sight propagation
- D. Auroral propagation

N3A03
How do VHF and UHF radio waves usually travel from a transmitting antenna to a receiving antenna?
- A. They bend through the ionosphere
- B. They go in a straight line
- C. They wander in any direction
- D. They move in a circle going either east or west from the transmitter

N3A04
What can happen to VHF or UHF signals going towards a metal-framed building?
- A. They will go around the building
- B. They can be bent by the ionosphere
- C. They can be easily reflected by the building
- D. They are sometimes scattered in the ectosphere

ANSWERS TO PREVIOUS TEST

N3A05	D
N3A06	B
N3A07	A
N3A08	C
N3A09	C
N3A10	A
N3A11	A
N3A12	C

YOUR ANSWERS TO THIS TEST

N3A01 ___

N3A02 ___

N3A03 ___

N3A04 ___

artsci inc

Riding the Airwaves with Alpha & Zulu

SEE IF YOU CAN ANSWER THESE QUESTIONS WITHOUT LOOKING BACK TO THE LAST 'TOON'. WRITE YOUR ANSWERS IN THE ANSWER BOX BELOW. GOOD LUCK.

PRACTICE TEST PAGE N40

N5B03
What is the pressure that forces electrons to flow through a circuit?
 A. Magnetomotive force, or inductance
 B. Electromotive force, or voltage
 C. Farad force, or capacitance
 D. Thermal force, or heat

N5B04
What is the basic unit of voltage?
 A. The volt
 B. The watt
 C. The ampere
 D. The ohm

N5B05
How much voltage does an automobile battery usually supply?
 A. About 12 volts
 B. About 30 volts
 C. About 120 volts
 D. About 240 volts

N5B06
How much voltage does a wall outlet usually supply (in the US)?
 A. About 12 volts
 B. About 30 volts
 C. About 120 volts
 D. About 480 volts

ANSWERS TO PREVIOUS TEST

N3A01 — A
N3A02 — C
N3A03 — B
N3A04 — C

YOUR ANSWERS TO THIS TEST

N5B03 ____
N5B04 ____
N5B05 ____
N5B06 ____

artsci inc

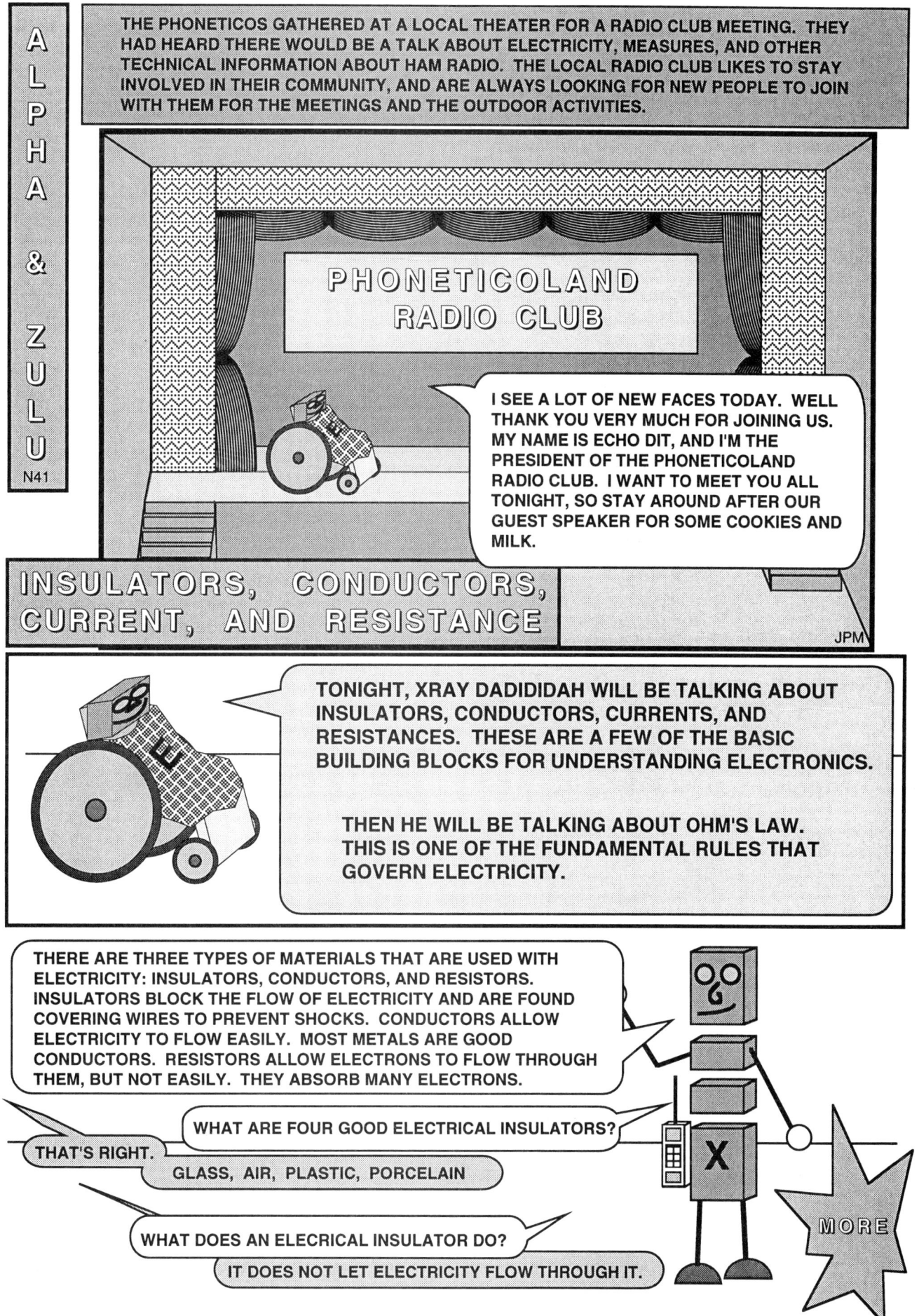

Riding the Airwaves with Alpha & Zulu

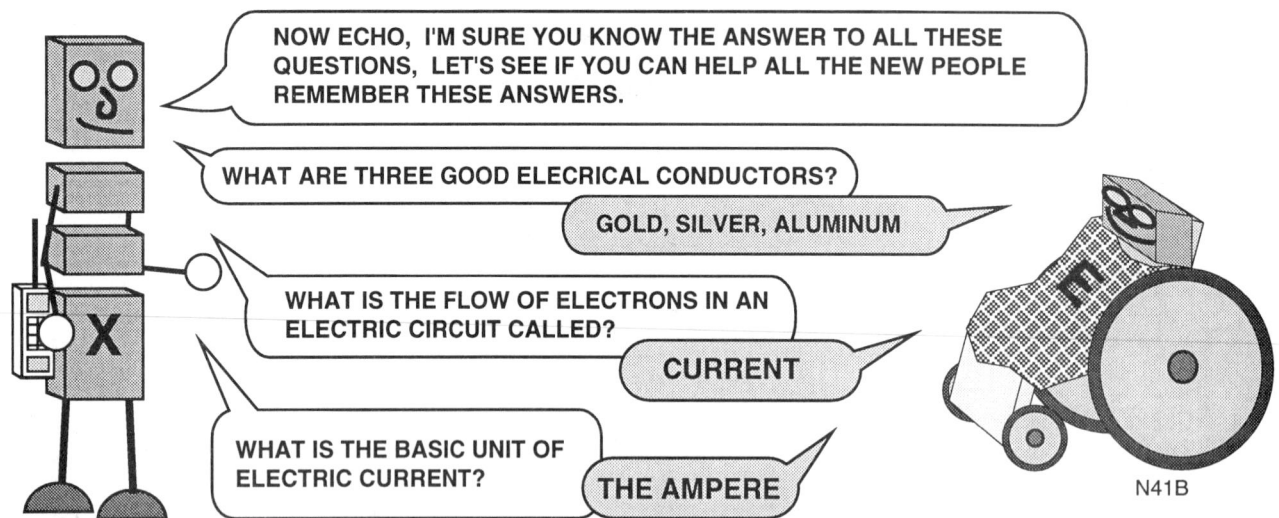

ECHO KNOWS ELECTRICITY VERY WELL. HE DECIDES TO ASK XRAY A FEW QUESTIONS AS WELL. FOR MANY OF THE PHONETICOS THAT ARE AT THIS MEETING FOR THE FIRST TIME, MUCH OF THIS TALK ABOUT ELECTRICIY IS NEW. NOT ONLY WILL THEY NEED IT TO PASS THEIR TEST, THE STUFF ON ELECTRICITY WILL BE USEFUL IN MANY OTHER WAYS AS WELL.

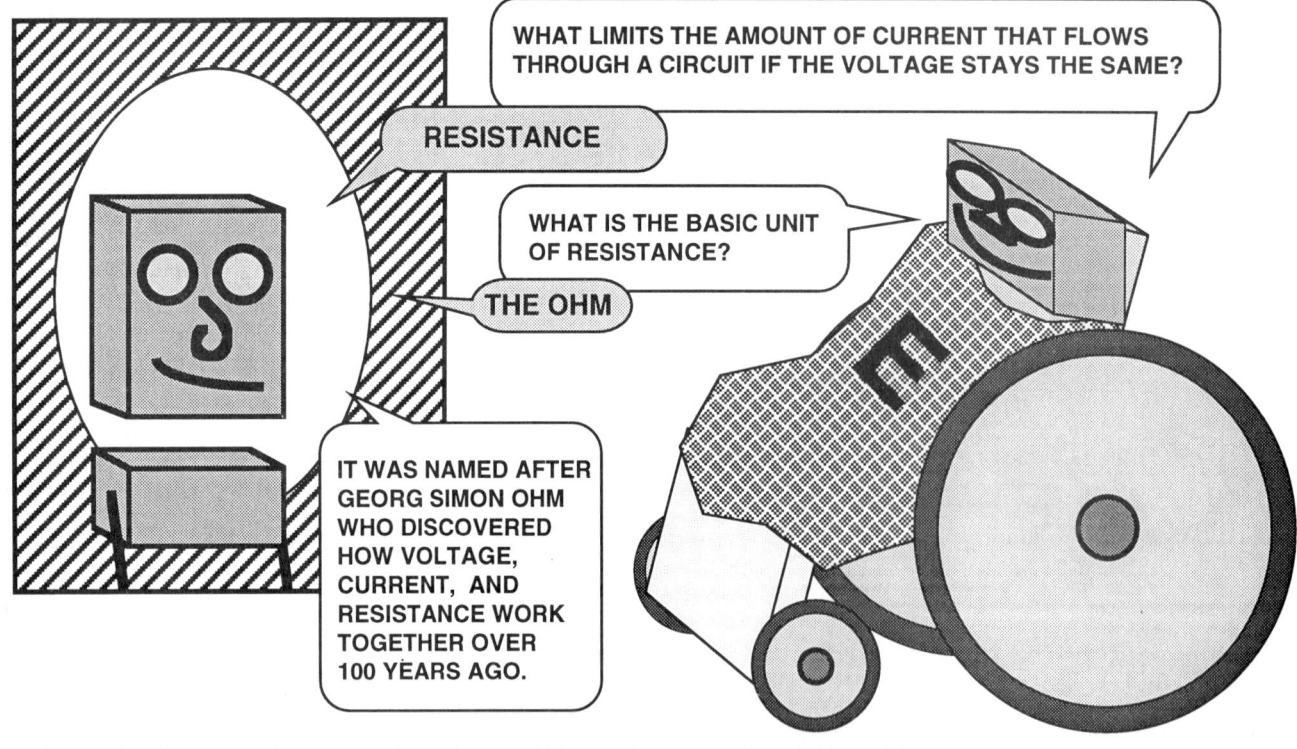

THE WAY IT REALLY WORKS IS THAT ALL MATERIALS ARE RESISTORS. THE ONES WITH VERY LOW RESISTANCE ARE CONDUCTORS. THE ONES WITH VERY HIGH RESISTANCES ARE INSULATORS. AS WE SAW EARLIER, VOLTAGE PUSHES ELECTRONS THROUGH WIRES. THE FLOW OF ELECTRONS IS CALLED THE CURRENT. RESISTANCE LIMITS THE CURRENT FLOWING THROUGH THE WIRE. GEORG SIMON OHM FOUND HOW THESE THREE PARTS OF ELECTRICITY, VOLTAGE, CURRENT, AND RESISTANCE, ACT TOGETHER. XRAY WILL BE COVERING THIS NEXT.

Riding the Airwaves with Alpha & Zulu

SEE IF YOU CAN ANSWER THESE QUESTIONS WITHOUT LOOKING BACK TO THE LAST 'TOON'. WRITE YOUR ANSWERS IN THE ANSWER BOX BELOW. GOOD LUCK.

PRACTICE TEST PAGE N41

N5B01
What is the flow of electrons in an electric circuit called?
- A. Voltage
- B. Resistance
- C. Capacitance
- D. Current

N5B02
What is the basic unit of electric current?
- A. The volt
- B. The watt
- C. The ampere
- D. The ohm

N5B07
What are three good electrical conductors?
- A. Copper, gold, mica
- B. Gold, silver, wood
- C. Gold, silver, aluminum
- D. Copper, aluminum, paper

N5B08
What are four good electrical insulators?
- A. Glass, air, plastic, porcelain
- B. Glass, wood, copper, porcelain
- C. Paper, glass, air, aluminum
- D. Plastic, rubber, wood, carbon

N5B09
What does an electrical insulator do?
- A. It lets electricity flow through it in one direction
- B. It does not let electricity flow through it
- C. It lets electricity flow through it when light shines on it
- D. It lets electricity flow through it

N5B10
What limits the amount of current that flows through a circuit if the voltage stays the same?
- A. Reliance
- B. Reactance
- C. Saturation
- D. Resistance

N5B11
What is the basic unit of resistance?
- A. The volt
- B. The watt
- C. The ampere
- D. The ohm

ANSWERS TO PREVIOUS TEST
N5B03
B
N5B04
A
N5B05
A
N5B06
C

YOUR ANSWERS TO THIS TEST
N5B01
N5B02
N5B07
N5B08
N5B09
N5B10
N5B11

artsci inc

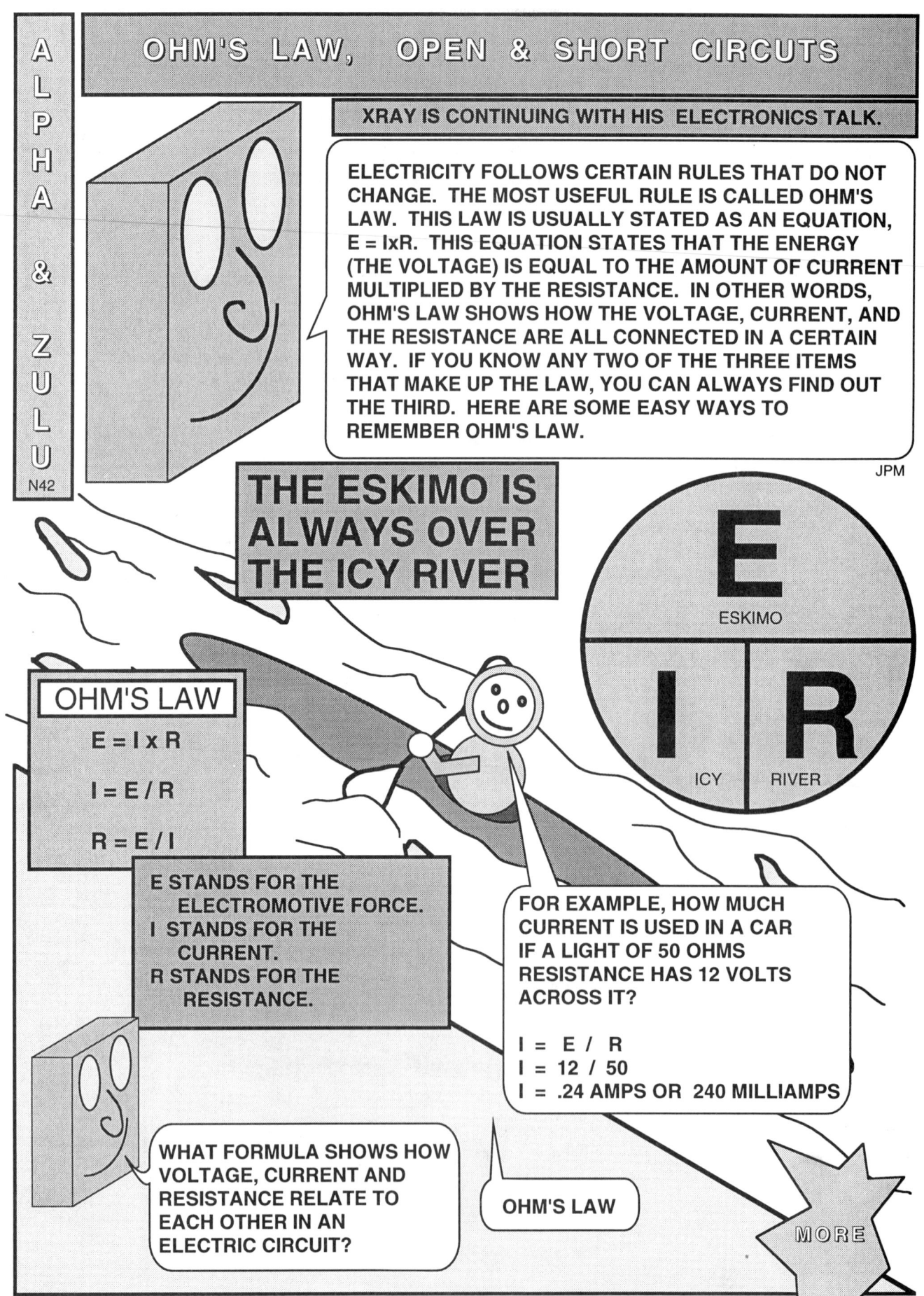

PRACTICE TEST PAGE N42

SEE IF YOU CAN ANSWER THESE QUESTIONS WITHOUT LOOKING BACK TO THE LAST 'TOON'. WRITE YOUR ANSWERS IN THE ANSWER BOX BELOW. GOOD LUCK.

N5C01
What formula shows how voltage, current and resistance relate to each other in an electric circuit?
A. Ohm's Law
B. Kirchhoff's Law
C. Ampere's Law
D. Tesla's Law

N5C02
If a current of 2 amperes flows through a 50-ohm resistor, what is the voltage across the resistor?
A. 25 volts
B. 52 volts
C. 100 volts
D. 200 volts

N5C03
If a 100-ohm resistor is connected to 200 volts, what is the current through the resistor?
A. 1/2 ampere
B. 2 amperes
C. 300 amperes
D. 20000 amperes

N5C04
If a current of 3 amperes flows through a resistor connected to 90 volts, what is the resistance?
A. 30 ohms
B. 93 ohms
C. 270 ohms
D. 1/30 ohm

N5C08
Which electrical circuit can have no current?
A. A closed circuit
B. A short circuit
C. An open circuit
D. A complete circuit

N5C09
Which electrical circuit uses too much current?
A. An open circuit
B. A dead circuit
C. A closed circuit
D. A short circuit

ANSWERS TO PREVIOUS TEST

N5B01	D
N5B02	C
N5B07	C
N5B08	A
N5B09	B
N5B10	D
N5B11	D

YOUR ANSWERS TO THIS TEST

N5C01 ___
N5C02 ___
N5C03 ___
N5C04 ___
N5C08 ___
N5C09 ___

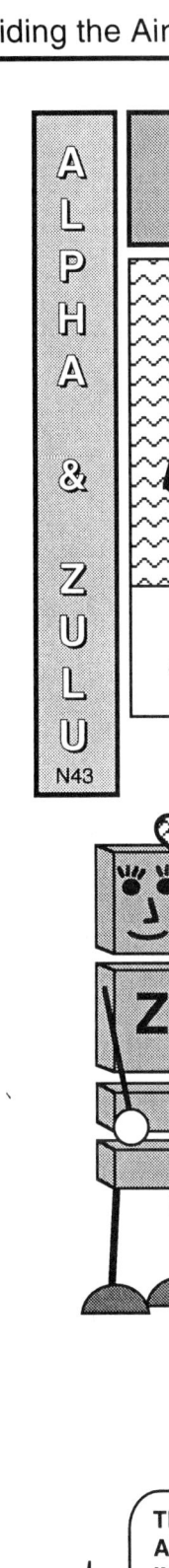

MEASURES
METERS, AMPERES, VOLTS, FARADS, WATTS

HEY ALPHA, DO YOU REMEMBER WHEN PAPA DIDADADIT WAS TALKING ABOUT THE POWERS OF TEN? THEY ALSO WORK WITH NUMBERS SMALLER THAN ONE.

THAT'S RIGHT ROMEO. MOST FORMS OF MEASUREMENTS USE THE SAME PREFIXES TO REPRESENT POWERS OF TEN. ZULU AND JULIETT HAVE A LIST OF THEM AND SOME COMMON USAGES ALL SET UP. LET'S SEE HOW THEY WORK.

JULIETT AND I DREW THIS BIG POSTER AS A SPECIAL PROJECT FOR OUR RADIO CLASS. IT SHOWS THE POWERS OF TEN FROM ONE ALL THE WAY DOWN TO ONE TRILLIONTH. THAT'S PRETTY SMALL.

1	ONE	1000 MILLI	10^{0}	ONE
.1	100 m	100 MILLI	10^{-1}	
.01	10 m	10 MILLI	10^{-2}	ALSO 1 CENTI
.001	1 m	1 MILLI	10^{-3}	1 / 1000
.0001	100 u	100 MICRO	10^{-4}	
.00001	10 u	10 MICRO	10^{-5}	
.000001	1 u	1 MICRO	10^{-6}	1/1,000,000
.0000001	100 n	100 NANO	10^{-7}	
.00000001	10 n	10 NANO	10^{-8}	
.000000001	1 n	1 NANO	10^{-9}	1/1,000,000,000
.0000000001	100 p	100 PICO	10^{-10}	
.00000000001	10 p	10 PICO	10^{-11}	
.000000000001	1 p	1 PICO	10^{-12}	1/1,000,000,000,000

THESE CAN BE USED WITH A LOT OF DIFFERENT TYPES OF MEASURES. FOR EXAMPLE, THERE ARE 100 CENTIMETERS IN ONE METER. THERE ARE 1000 MILLIAMPERES IN ONE AMPERE. THESE POWERS OF TEN ARE USED OFTEN IN RADIO AND YOU WILL QUICKLY LEARN TO USE THEM AS YOU PROGRESS.

THAT'S RIGHT. FOR EXAMPLE, THERE ARE 1000 MILLIVOLTS IN ONE VOLT, THERE ARE 1,000,000 MICROFARADS IN A FARAD, AND THERE ARE 1,000,000 PICOFARADS IN ONE MICROFARAD. THERE ARE 1000 MILLIWATTS IN ONE WATT. SEE HOW EASY IT IS TO CONVERT FROM ONE POWER OF TEN TO ANOTHER? IT ONLY TAKES A LITTLE PRACTICE AND YOU ARE READY.

I KNOW WE KNOW THIS. DO YOU THINK ALPHA AND ROMEO KNOW IT?

I'M NOT SURE. LET'S FIND OUT BY ASKING THEM, OK?

MORE

Riding the Airwaves with Alpha & Zulu

SEE IF YOU CAN ANSWER THESE QUESTIONS WITHOUT LOOKING BACK TO THE LAST 'TOON'. WRITE YOUR ANSWERS IN THE ANSWER BOX BELOW. GOOD LUCK.

PRACTICE TEST PAGE N43

N5A04
How long is an antenna that is 400 centimeters long?
 A. 0.0004 meters
 B. 4 meters
 C. 40 meters
 D. 40,000 meters

N5A05
If an ammeter marked in amperes is used to measure a 3000-milliampere current, what reading would it show?
 A. 0.003 amperes
 B. 0.3 amperes
 C. 3 amperes
 D. 3,000,000 amperes

N5A06
If a voltmeter marked in volts is used to measure a 3500-millivolt potential, what reading would it show?
 A. 0.35 volts
 B. 3.5 volts
 C. 35 volts
 D. 350 volts

N5A07
How many farads is 500,000 microfarads?
 A. 0.0005 farads
 B. 0.5 farads
 C. 500 farads
 D. 500,000,000 farads

N5A08
How many microfarads is 1,000,000 picofarads?
 A. 0.001 microfarads
 B. 1 microfarad
 C. 1,000 microfarads
 D. 1,000,000,000 microfarads

N5A11
If you have a hand held transceiver which puts out 500 milliwatts, how many watts would this be?
 A. 0.02
 B. 0.5
 C. 5
 D. 50

ANSWERS TO PREVIOUS TEST

N5C01	A
N5C02	C
N5C03	B
N5C04	A
N5C08	C
N5C09	D

YOUR ANSWERS TO THIS TEST

N5A04	
N5A05	
N5A06	
N5A07	
N5A08	
N5A11	

artsci inc

MORSE CODE PRACTICE

MORSE CODE USAGE DATES BACK OVER ONE HUNDRED YEARS TO THE TIME OF THE INVENTION OF THE TELEGRAPH. ALTHOUGH ITS USE HAS BEEN REPLACED OVER THE YEARS WITH VOICE, PICTURES, AND DATA, IT IS STILL A PART OF AMATEUR RADIO. THERE IS A GREAT DEGREE OF PRIDE THAT YOU WILL FEEL WHEN YOU HAVE MASTERED MORSE CODE. ALTHOUGH IT MAY SEEM DIFFICULT, WITH PRACTICE IT IS EASY TO LEARN. FOR MORE FUN, LEARN WITH A FRIEND. THE IMPORTANT THING IS THAT YOU LEARN SOMETHING NEW.

WHEN LEARNING MORSE CODE, USE THE TERMS Dit AND Dah. SAY THEM OUT LOUD TO YOURSELF AND YOUR FRIENDS TO GET USED TO THE SOUND. THIS MAKES LEARNING EASIER AND FASTER. TRY THE MESSAGE BELOW. WRITE WHAT YOU THINK THEY ARE SAYING IN THE BOXES. THE ENTIRE ALPHABET IS BELOW TO HELP YOU. DON'T FORGET THAT PUNCTUATION IS ALSO USED.

A	DiDah	M	DaDah	Y	DaDiDaDah
B	DaDiDiDit	N	DaDit	Z	DaDaDiDit
C	DaDiDaDit	O	DaDaDah	1	DiDaDaDaDah
D	DaDiDit	P	DiDaDaDit	2	DiDiDaDaDah
E	Dit	Q	DaDaDiDah	3	DiDiDiDaDah
F	DiDiDaDit	R	DiDaDit	4	DiDiDiDiDah
G	DaDaDit	S	DiDiDit	5	DiDiDiDiDit
H	DiDiDiDit	T	Dah	6	DaDiDiDiDit
I	DiDit	U	DiDiDah	7	DaDaDiDiDit
J	DiDaDaDah	V	DiDiDiDah	8	DaDaDaDiDit
K	DaDiDah	W	DiDaDah	9	DaDaDaDaDit
L	DiDaDiDit	X	DaDiDiDah	0	DaDaDaDaDah

PERIOD. DiDaDiDaDiDah QUESTION MARK? DiDiDaDaDiDit COMMA, DaDaDiDiDaDa

SLANT BAR / DaDiDiDaDit END OF TRANSMISSION AR DiDaDiDaDit

PAUSE BT DaDiDiDiDah END OF QSD SK DiDiDiDaDiDah

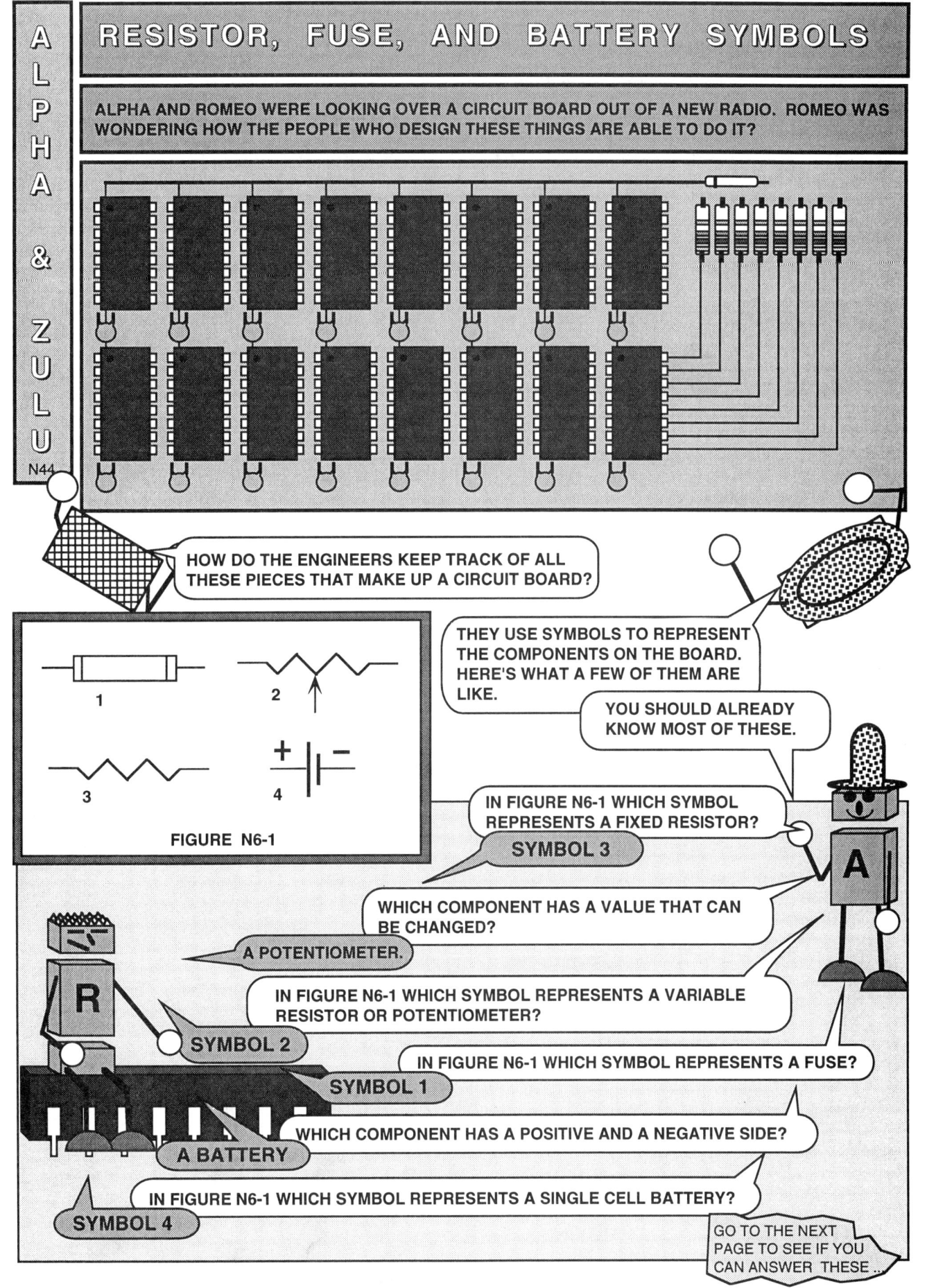

Riding the Airwaves with Alpha & Zulu

SEE IF YOU CAN ANSWER THESE QUESTIONS WITHOUT LOOKING BACK TO THE LAST 'TOON'. WRITE YOUR ANSWERS IN THE ANSWER BOX BELOW. GOOD LUCK.

PRACTICE TEST PAGE N44

N6A03
Which component has a positive and a negative side?
- A. A battery
- B. A potentiometer
- C. A fuse
- D. A resistor

N6A05
In Figure N6-1 which symbol represents a variable resistor or potentiometer?
- A. Symbol 1
- B. Symbol 2
- C. Symbol 3
- D. Symbol 4

N6A04
Which component has a value that can be changed?
- A. A single-cell battery
- B. A potentiometer
- C. A fuse
- D. A resistor

N6A06
In Figure N6-1 which symbol represents a fixed resistor?
- A. Symbol 1
- B. Symbol 2
- C. Symbol 3
- D. Symbol 4

N6A07
In Figure N6-1 which symbol represents a fuse?
- A. Symbol 1
- B. Symbol 2
- C. Symbol 3
- D. Symbol 4

N6A08
In Figure N6-1 which symbol represents a single-cell battery?
- A. Symbol 1
- B. Symbol 2
- C. Symbol 3
- D. Symbol 4

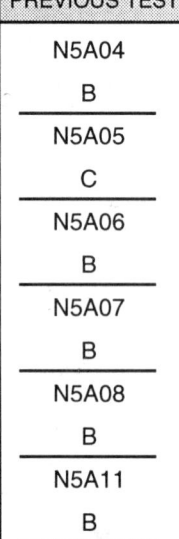

ANSWERS TO PREVIOUS TEST
N5A04
B
N5A05
C
N5A06
B
N5A07
B
N5A08
B
N5A11
B

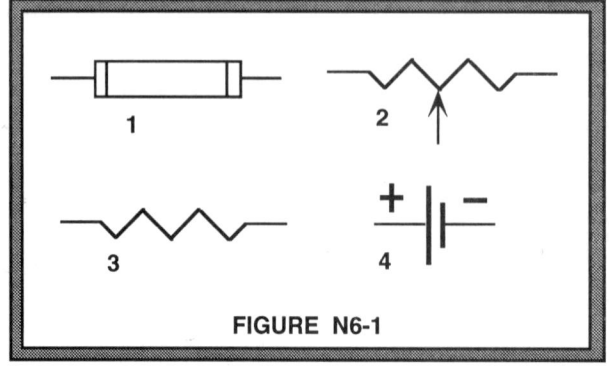

FIGURE N6-1

YOUR ANSWERS TO THIS TEST
N6A03
N6A04
N6A05
N6A06
N6A07
N6A08

Riding the Airwaves with Alpha & Zulu

SEE IF YOU CAN ANSWER THESE QUESTIONS WITHOUT LOOKING BACK TO THE LAST 'TOON'. WRITE YOUR ANSWERS IN THE ANSWER BOX BELOW. GOOD LUCK.

PRACTICE TEST PAGE N45

N6A01
What can a single-pole, double-throw switch do?
A. It can switch one input to one output
B. It can switch one input to either of two outputs
C. It can switch two inputs at the same time, one input to either of two outputs, and the other input to either of two outputs
D. It can switch two inputs at the same time, one input to one output, and the other input to another output

N6A02
What can a double-pole, single-throw switch do?
A. It can switch one input to one output
B. It can switch one input to either of two outputs
C. It can switch two inputs at the same time, one input to either of two outputs, and the other input to either of two outputs
D. It can switch two inputs at the same time, one input to one output, and the other input to the other output

N6A09
In Figure N6-2 which symbol represents a single-pole, single-throw switch?
A. Symbol 1
B. Symbol 2
C. Symbol 3
D. Symbol 4

N6A10
In Figure N6-2 which symbol represents a single-pole, double-throw switch?
A. Symbol 1
B. Symbol 2
C. Symbol 3
D. Symbol 4

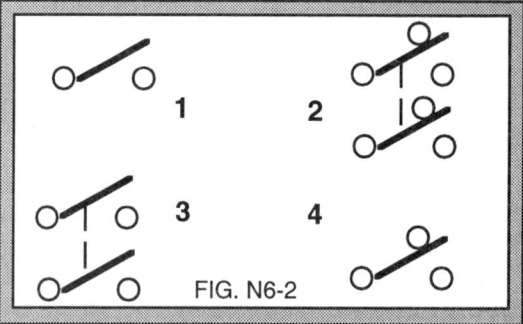
FIG. N6-2

N6A11
In Figure N6-2 which symbol represents a double-pole, single-throw switch?
A. Symbol 1
B. Symbol 2
C. Symbol 3
D. Symbol 4

N6A12
In Figure N6-2 which symbol represents a double-pole, double-throw switch?
A. Symbol 1
B. Symbol 2
C. Symbol 3
D. Symbol 4

ANSWERS TO PREVIOUS TEST
N6A03
A
N6A04
B
N6A05
B
N6A06
C
N6A07
A
N6A08
D

YOUR ANSWERS TO THIS TEST
N6A01
N6A02
N6A09
N6A10
N6A11
N6A12

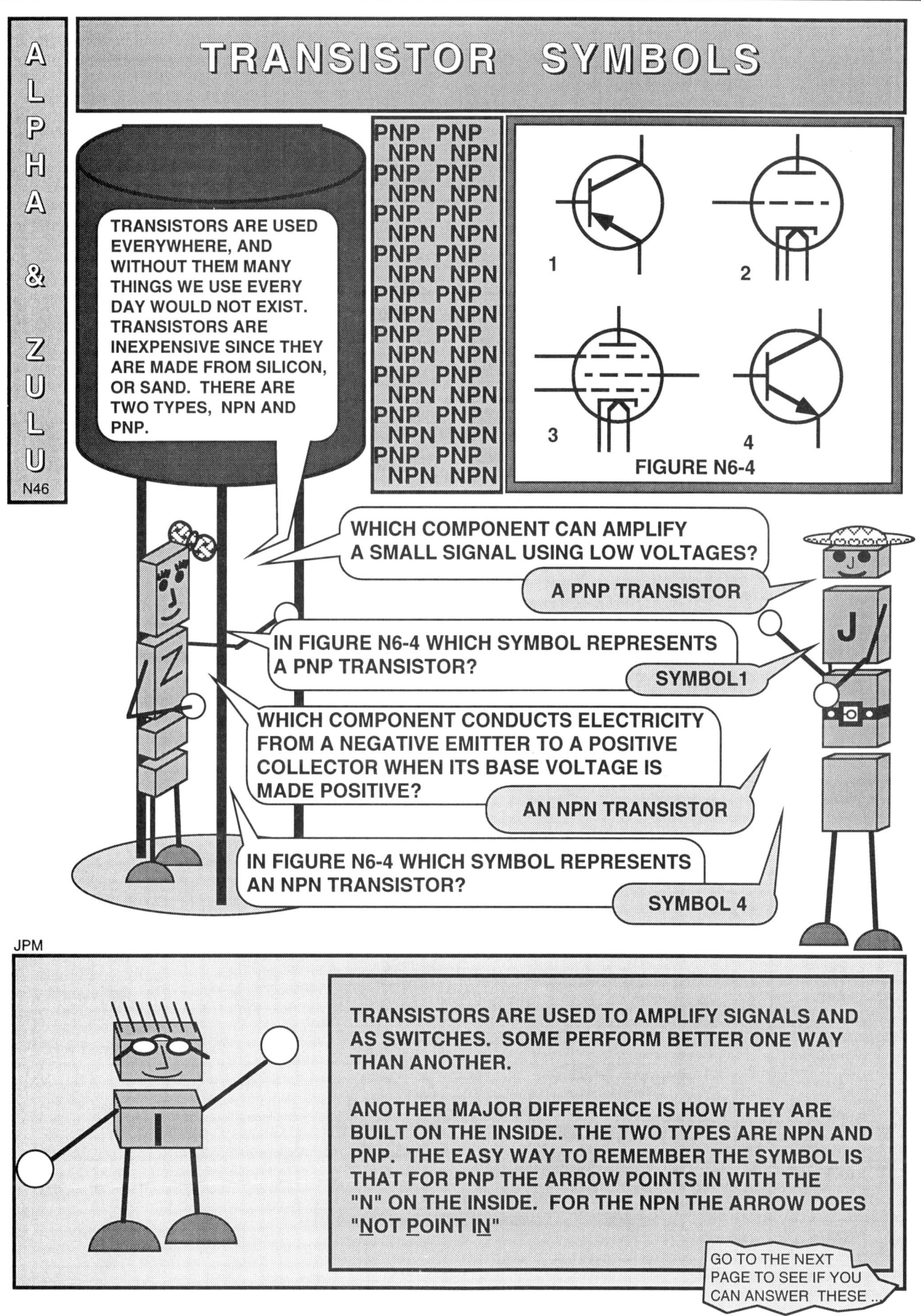

Riding the Airwaves with Alpha & Zulu

SEE IF YOU CAN ANSWER THESE QUESTIONS WITHOUT LOOKING BACK TO THE LAST 'TOON'. WRITE YOUR ANSWERS IN THE ANSWER BOX BELOW. GOOD LUCK.

PRACTICE TEST PAGE N46

N6B01
Which component can amplify a small signal using low voltages?
- A. A PNP transistor
- B. A variable resistor
- C. An electrolytic capacitor
- D. A multiple-cell battery

N6B07
In Figure N6-4 which symbol represents an NPN transistor?
- A. Symbol 1
- B. Symbol 2
- C. Symbol 3
- D. Symbol 4

N6B02
Which component conducts electricity from a negative emitter to a positive collector when its base voltage is made positive?
- A. A variable resistor
- B. An NPN transistor
- C. A triode vacuum tube
- D. A multiple-cell battery

N6B08
In Figure N6-4 which symbol represents a PNP transistor?
- A. Symbol 1
- B. Symbol 2
- C. Symbol 3
- D. Symbol 4

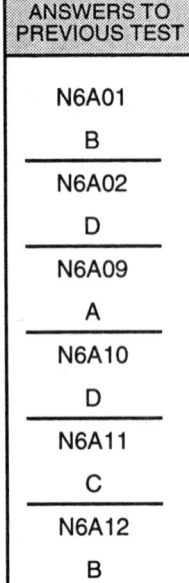

ANSWERS TO PREVIOUS TEST

N6A01	B
N6A02	D
N6A09	A
N6A10	D
N6A11	C
N6A12	B

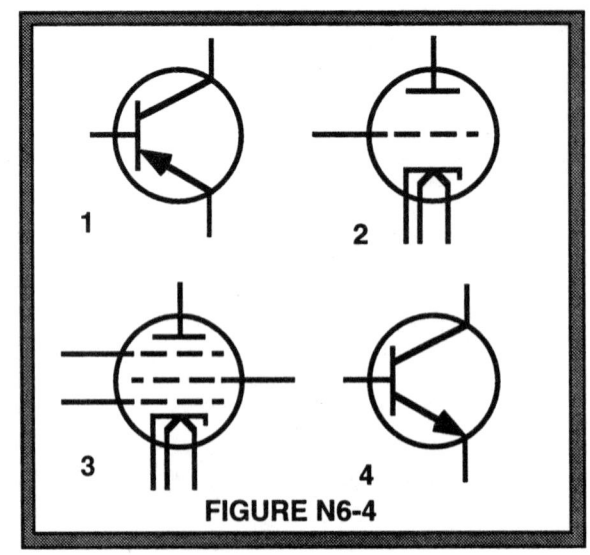

FIGURE N6-4

YOUR ANSWERS TO THIS TEST

N6B01

N6B02

N6B07

N6B08

108   artsci inc

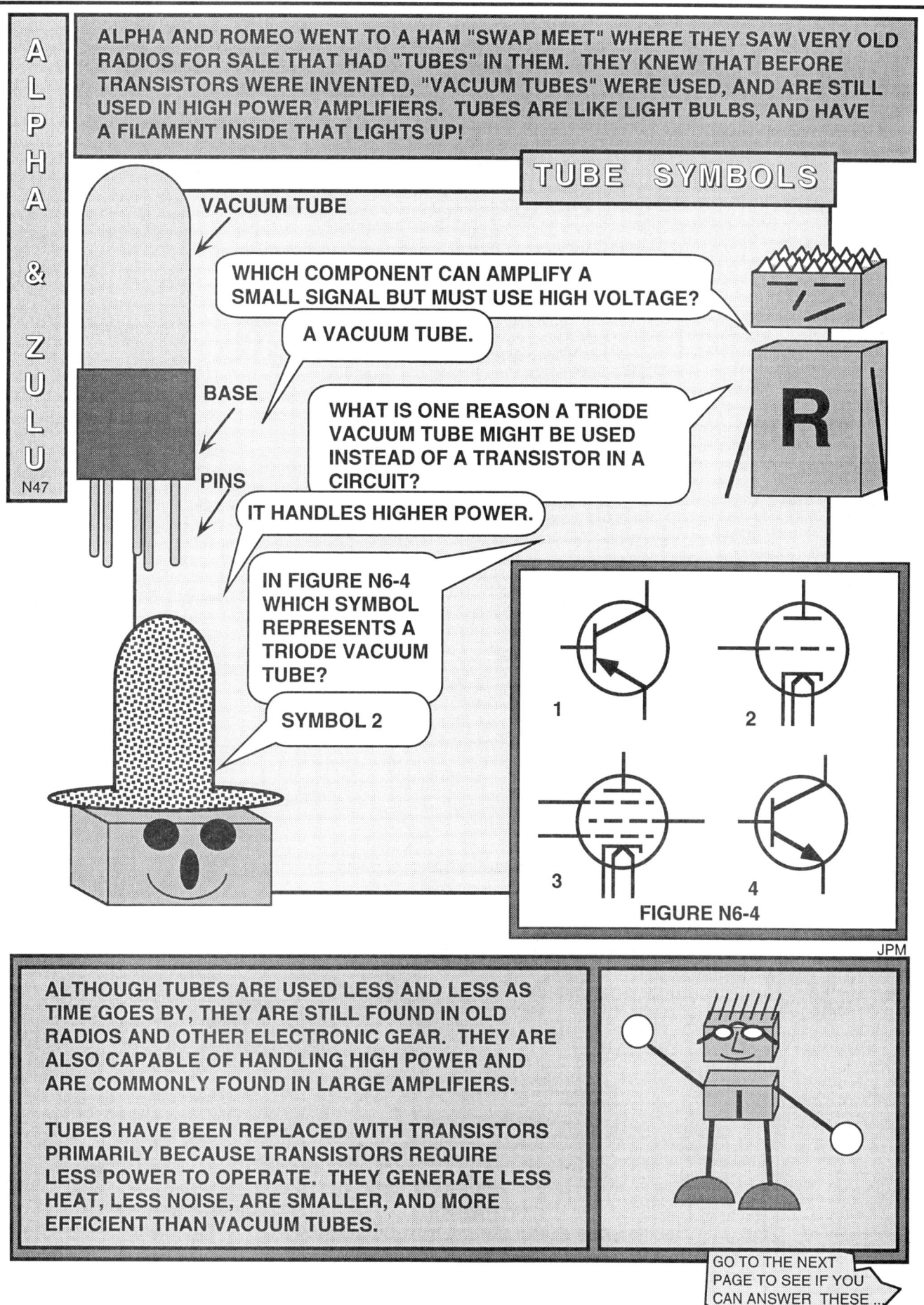

Riding the Airwaves with Alpha & Zulu

SEE IF YOU CAN ANSWER THESE QUESTIONS WITHOUT LOOKING BACK TO THE LAST 'TOON'. WRITE YOUR ANSWERS IN THE ANSWER BOX BELOW. GOOD LUCK.

PRACTICE TEST PAGE N47

N6B09
In Figure N6-4 which symbol represents a triode vacuum tube?
- A. Symbol 1
- B. Symbol 2
- C. Symbol 3
- D. Symbol 4

N6B10
What is one reason a triode vacuum tube might be used instead of a transistor in a circuit?
- A. It handles higher power
- B. It uses lower voltages
- C. It uses less current
- D. It is much smaller

N6B11
Which component can amplify a small signal but must use high voltages?
- A. A transistor
- B. An electrolytic capacitor
- C. A vacuum tube
- D. A multiple-cell battery

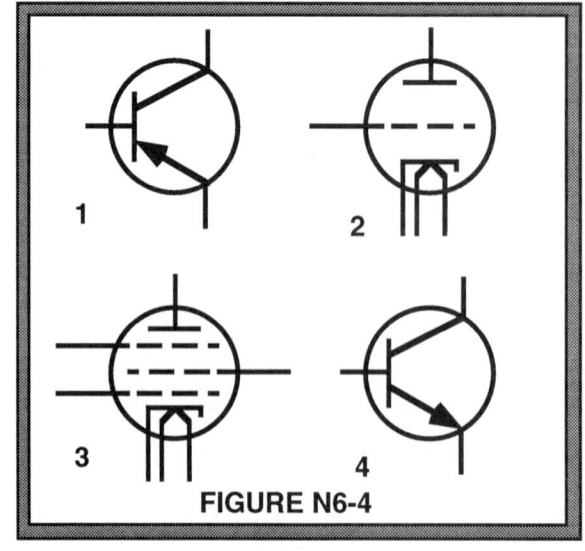

FIGURE N6-4

ANSWERS TO PREVIOUS TEST

N6B01	A
N6B02	B
N6B07	D
N6B08	A

YOUR ANSWERS TO THIS TEST

N6B09	
N6B10	
N6B11	

110 artsci inc

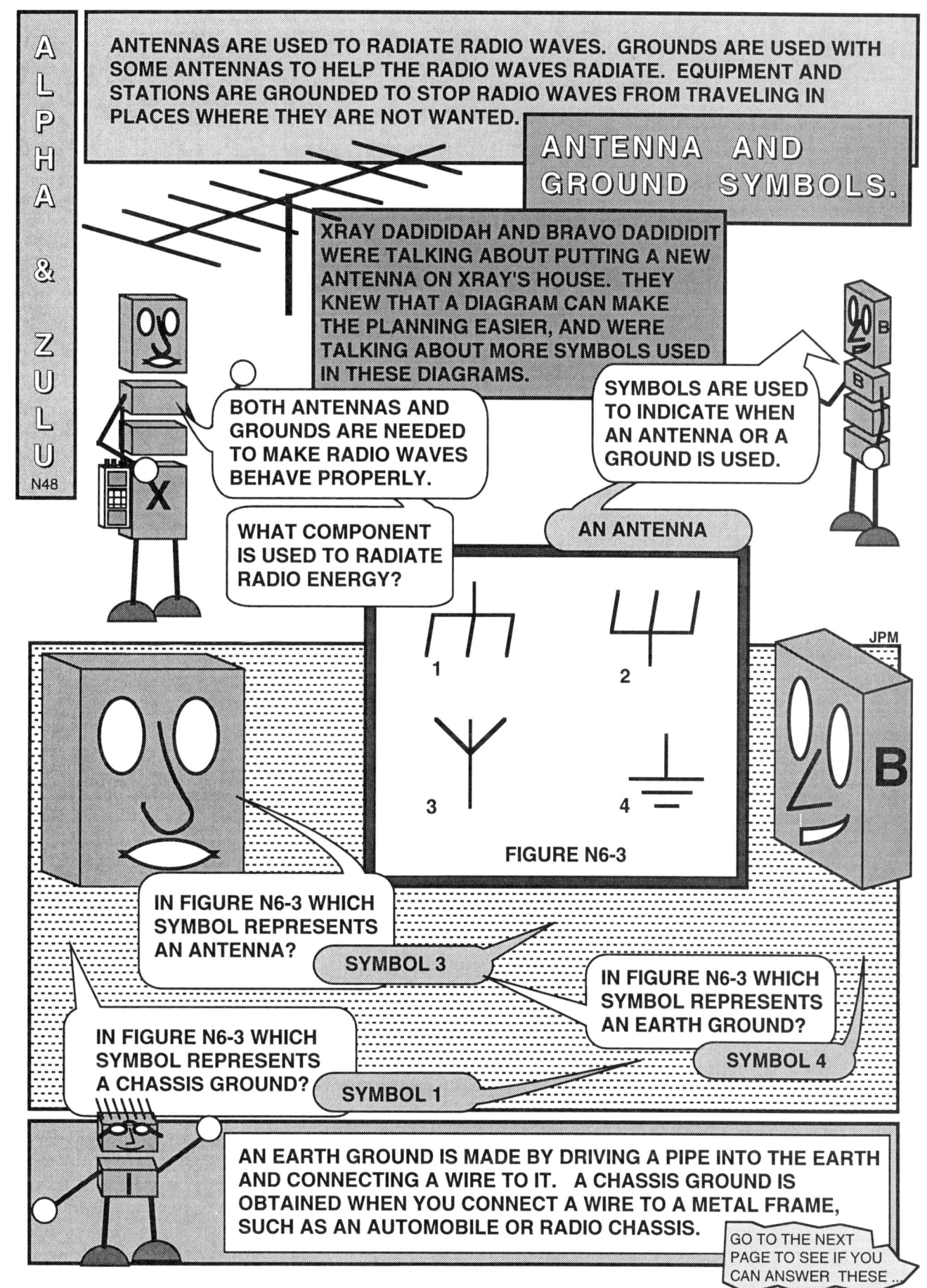

Riding the Airwaves with Alpha & Zulu

SEE IF YOU CAN ANSWER THESE QUESTIONS WITHOUT LOOKING BACK TO THE LAST 'TOON'. WRITE YOUR ANSWERS IN THE ANSWER BOX BELOW. GOOD LUCK.

PRACTICE TEST PAGE N48

N6B03
Which component is used to radiate radio energy?
A. An antenna
B. An earth ground
C. A chassis ground
D. A potentiometer

N6B05
In Figure N6-3 which symbol represents a chassis ground?
A. Symbol 1
B. Symbol 2
C. Symbol 3
D. Symbol 4

N6B04
In Figure N6-3 which symbol represents an earth ground?
A. Symbol 1
B. Symbol 2
C. Symbol 3
D. Symbol 4

N6B06
In Figure N6-3 which symbol represents an antenna?
A. Symbol 1
B. Symbol 2
C. Symbol 3
D. Symbol 4

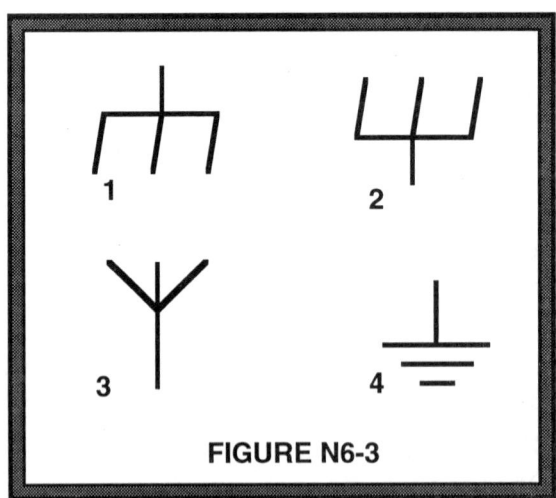

FIGURE N6-3

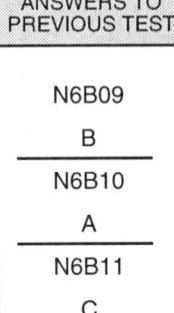

ANSWERS TO PREVIOUS TEST
N6B09
B
N6B10
A
N6B11
C

YOUR ANSWERS TO THIS TEST
N6B03
N6B04
N6B05
N6B06

Riding the Airwaves with Alpha & Zulu

SEE IF YOU CAN ANSWER THESE QUESTIONS WITHOUT LOOKING BACK TO THE LAST 'TOON'. WRITE YOUR ANSWERS IN THE ANSWER BOX BELOW. GOOD LUCK.

PRACTICE TEST PAGE N49

N7A07
If your mobile transceiver works in your car but not in your home, what should you check first?
- A. The power supply
- B. The speaker
- C. The microphone
- D. The SWR meter

N7A12
What device converts household current to 12 VDC?
- A. A catalytic converter
- B. A low-pass filter
- C. A power supply
- D. An RS-232 interface

N7A13
Which of these usually needs a heavy-duty power supply?
- A. An SWR meter
- B. A receiver
- C. A transceiver
- D. An antenna switch

N8A10
What may cause a buzzing or hum in the signal of an HF transmitter?
- A. Using an antenna which is the wrong length
- B. Energy from another transmitter
- C. Bad design of the transmitter's RF power output circuit
- D. A bad filter capacitor in the transmitter's power supply

ANSWERS TO PREVIOUS TEST
N6B03
A
N6B04
D
N6B05
A
N6B06
C

YOUR ANSWERS TO THIS TEST
N7A07
N7A12
N7A13
N8A10

Riding the Airwaves with Alpha & Zulu

SEE IF YOU CAN ANSWER THESE QUESTIONS WITHOUT LOOKING BACK TO THE LAST 'TOON'. WRITE YOUR ANSWERS IN THE ANSWER BOX BELOW. GOOD LUCK.

PRACTICE TEST PAGE N50

N4D01
What is meant by receiver overload?
A. Too much voltage from the power supply
B. Too much current from the power supply
C. Interference caused by strong signals from a nearby transmitter
D. Interference caused by turning the volume up too high

N4D02
What is one way to tell if radio-frequency interference to a receiver is caused by front-end overload?
A. If connecting a low-pass filter to the transmitter greatly cuts down the interference
B. If the interference is about the same no matter what frequency is used for the transmitter
C. If connecting a low-pass filter to the receiver greatly cuts down the interference
D. If grounding the receiver makes the problem worse

N4D03
If your neighbor reports television interference whenever you are transmitting from your amateur station, no matter what frequency band you use, what is probably the cause of the interference?
A. Too little transmitter harmonic suppression
B. Receiver VR tube discharge
C. Receiver overload
D. Incorrect antenna length

N4D05
What type of filter should be connected to a TV receiver as the first step in trying to prevent RF overload from an amateur HF station transmission?
A. Low-pass
B. High-pass
C. Band pass
D. Notch

ANSWERS TO PREVIOUS TEST
N7A07
A
N7A12
C
N7A13
C
N8A10
D

YOUR ANSWERS TO THIS TEST
N4D01
N4D02
N4D03
N4D05

artsci inc

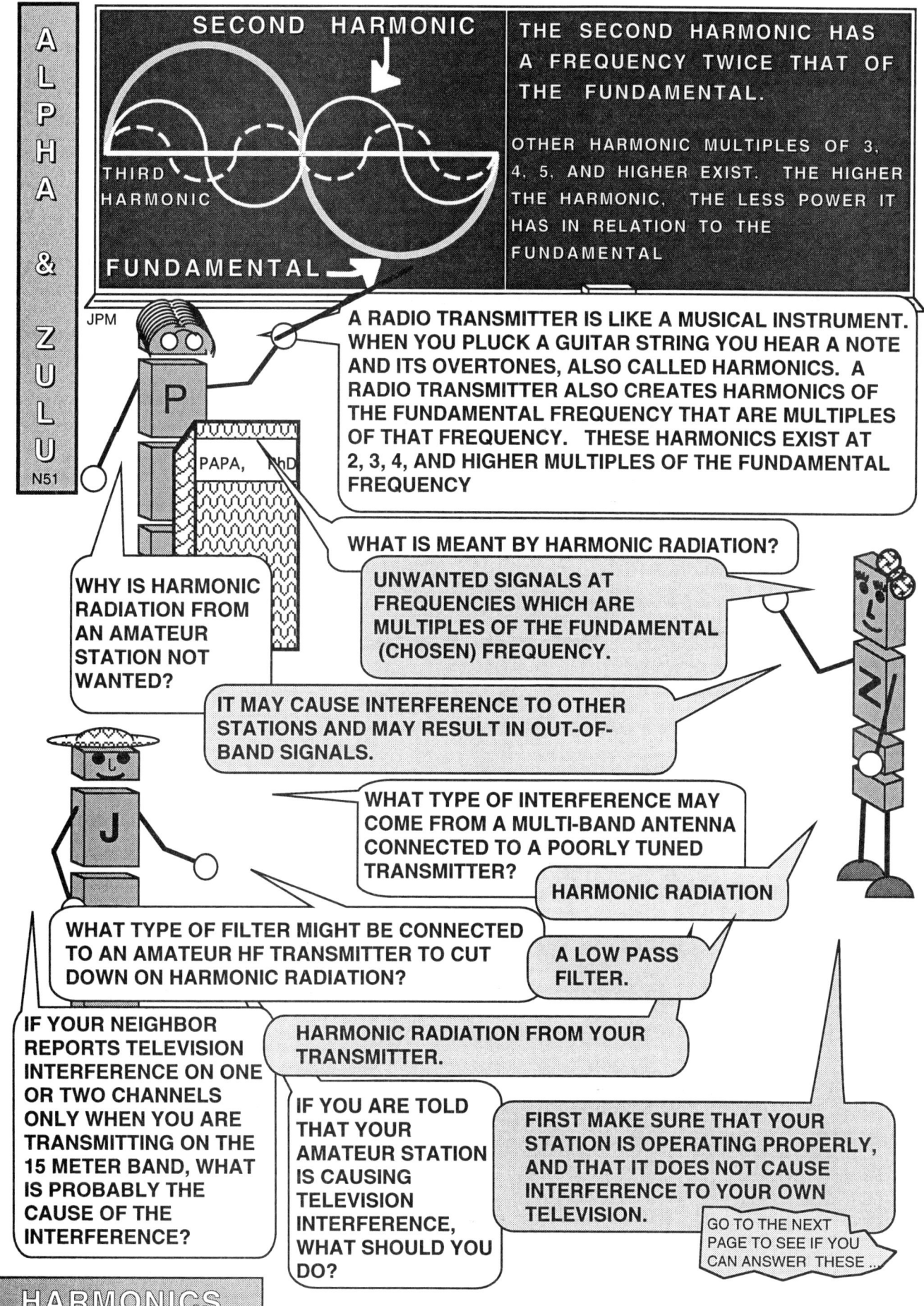

Riding the Airwaves with Alpha & Zulu

SEE IF YOU CAN ANSWER THESE QUESTIONS WITHOUT LOOKING BACK TO THE LAST 'TOON'. WRITE YOUR ANSWERS IN THE ANSWER BOX BELOW. GOOD LUCK.

PRACTICE TEST PAGE N51

N4D04
If your neighbor reports television interference on one or two channels only when you are transmitting on the 15-meter band, what is probably the cause of the interference?
- A. Too much low-pass filtering on the transmitter
- B. De-ionization of the ionosphere near your neighbor's TV antenna
- C. TV receiver front-end overload
- D. Harmonic radiation from your transmitter

N4D06
What type of filter might be connected to an amateur HF transmitter to cut down on harmonic radiation?
- A. A key-click filter
- B. A low-pass filter
- C. A high-pass filter
- D. A CW filter

N4D07
What is meant by harmonic radiation?
- A. Unwanted signals at frequencies which are multiples of the fundamental (chosen) frequency
- B. Unwanted signals that are combined with a 60-Hz hum
- C. Unwanted signals caused by sympathetic vibrations from a nearby transmitter
- D. Signals which cause skip propagation to occur

N4D08
Why is harmonic radiation from an amateur station not wanted?
- A. It may cause interference to other stations and may result in out-of-band signals
- B. It uses large amounts of electric power
- C. It may cause sympathetic vibrations in nearby transmitters
- D. It may cause auroras in the air

N4D09
What type of interference may come from a multi-band antenna connected to a poorly tuned transmitter?
- A. Harmonic radiation
- B. Auroral distortion
- C. Parasitic excitation
- D. Intermodulation

N4D11
If you are told that your amateur station is causing television interference, what should you do?
- A. First make sure that your station is operating properly, and that it does not cause interference to your own television
- B. Immediately turn off your transmitter and contact the nearest FCC office for assistance
- C. Connect a high-pass filter to the transmitter output and a low-pass filter to the antenna-input terminals of the television
- D. Continue operating normally, because you have no reason to worry about the interference

ANSWERS TO PREVIOUS TEST
N4D01
C
N4D02
B
N4D03
C
N4D05
B

YOUR ANSWERS TO THIS TEST
N4D04
N4D07
N4D06
N4D08
N4D09
N4D11

Riding the Airwaves with Alpha & Zulu

SEE IF YOU CAN ANSWER THESE QUESTIONS WITHOUT LOOKING BACK TO THE LAST 'TOON'. WRITE YOUR ANSWERS IN THE ANSWER BOX BELOW. GOOD LUCK.

PRACTICE TEST PAGE N52

N8B01
How does the frequency of a harmonic compare to the desired transmitting frequency?
- A. It is slightly more than the desired frequency
- B. It is slightly less than the desired frequency
- C. It is exactly two, or three, or more times the desired frequency
- D. It is much less than the desired frequency

N8B02
What is the fourth harmonic of a 7160-kHz signal?
- A. 28,640 kHz
- B. 35,800 kHz
- C. 28,160 kHz
- D. 1790 kHz

N8B03
If you are told your station was heard on 21,375 kHz, but at the time you were operating on 7125 kHz, what is one reason this could happen?
- A. Your transmitter's power-supply filter capacitor was bad
- B. You were sending CW too fast
- C. Your transmitter was radiating harmonic signals
- D. Your transmitter's power-supply filter choke was bad

ANSWERS TO PREVIOUS TEST
N4D04
D
N4D07
A
N4D06
B
N4D08
A
N4D09
A
N4D11
A

YOUR ANSWERS TO THIS TEST
N8B01
N8B02
N8B03

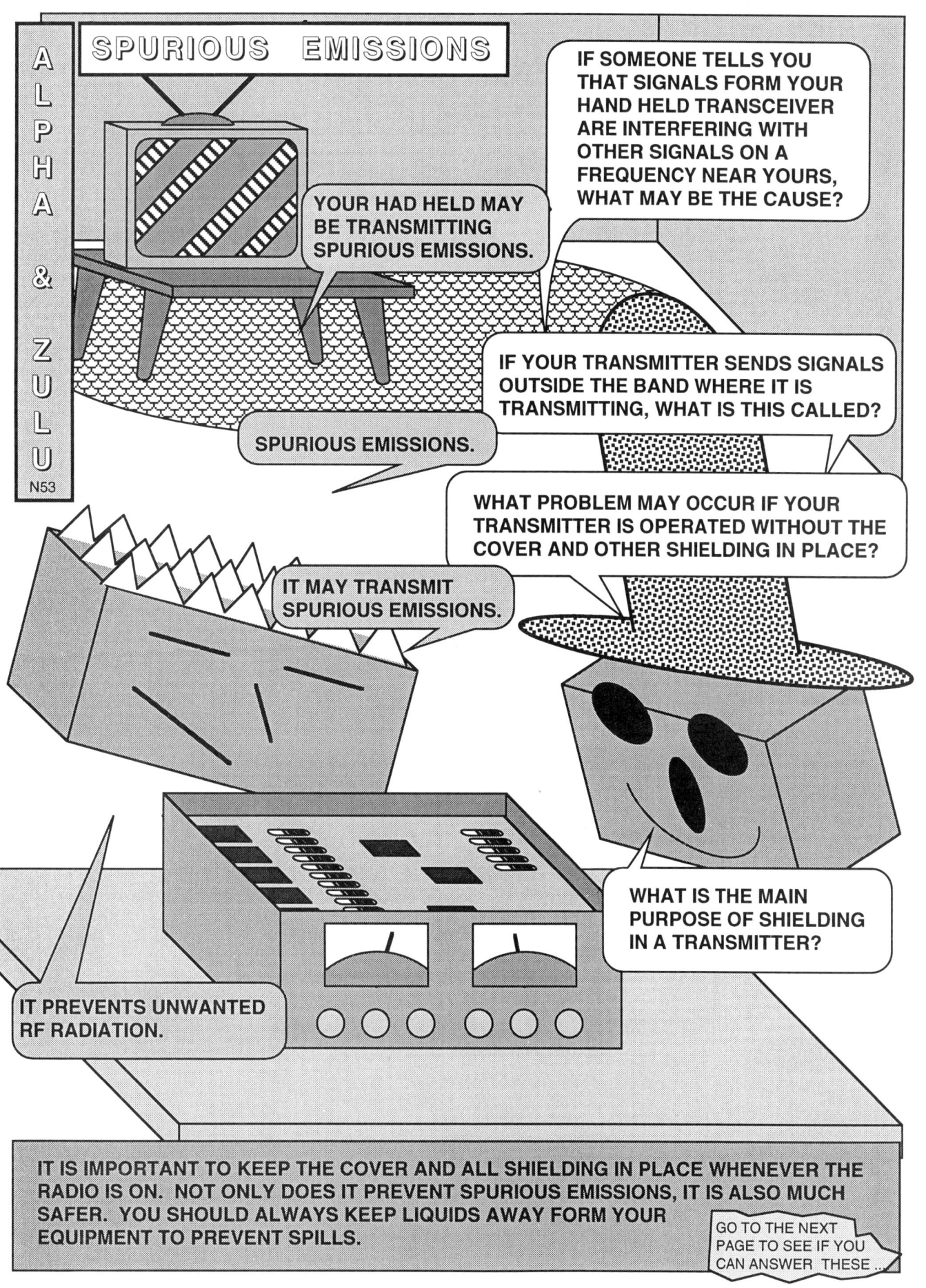

Riding the Airwaves with Alpha & Zulu

SEE IF YOU CAN ANSWER THESE QUESTIONS WITHOUT LOOKING BACK TO THE LAST 'TOON'. WRITE YOUR ANSWERS IN THE ANSWER BOX BELOW. GOOD LUCK.

PRACTICE TEST PAGE N53

N8B04
If someone tells you that signals from your hand held transceiver are interfering with other signals on a frequency near yours, what may be the cause?
- A. You may need a power amplifier for your hand held
- B. Your hand held may have chirp from weak batteries
- C. You may need to turn the volume up on your hand held
- D. Your hand held may be transmitting spurious emissions

N8B05
If your transmitter sends signals outside the band where it is transmitting, what is this called?
- A. Off-frequency emissions
- B. Transmitter chirping
- C. Side tones
- D. Spurious emissions

N8B06
What problem may occur if your transmitter is operated without the cover and other shielding in place?
- A. It may transmit spurious emissions
- B. It may transmit a chirpy signal
- C. It may transmit a weak signal
- D. It may interfere with other stations operating near its frequency

N4D10
What is the main purpose of shielding in a transmitter?
- A. It gives the low-pass filter a solid support
- B. It helps the sound quality of transmitters
- C. It prevents unwanted RF radiation
- D. It helps keep electronic parts warmer and more stable

ANSWERS TO PREVIOUS TEST

N8B01
C

N8B02
A

N8B03
C

YOUR ANSWERS TO THIS TEST

N8B04

N8B05

N8B06

N4D10

artsci inc

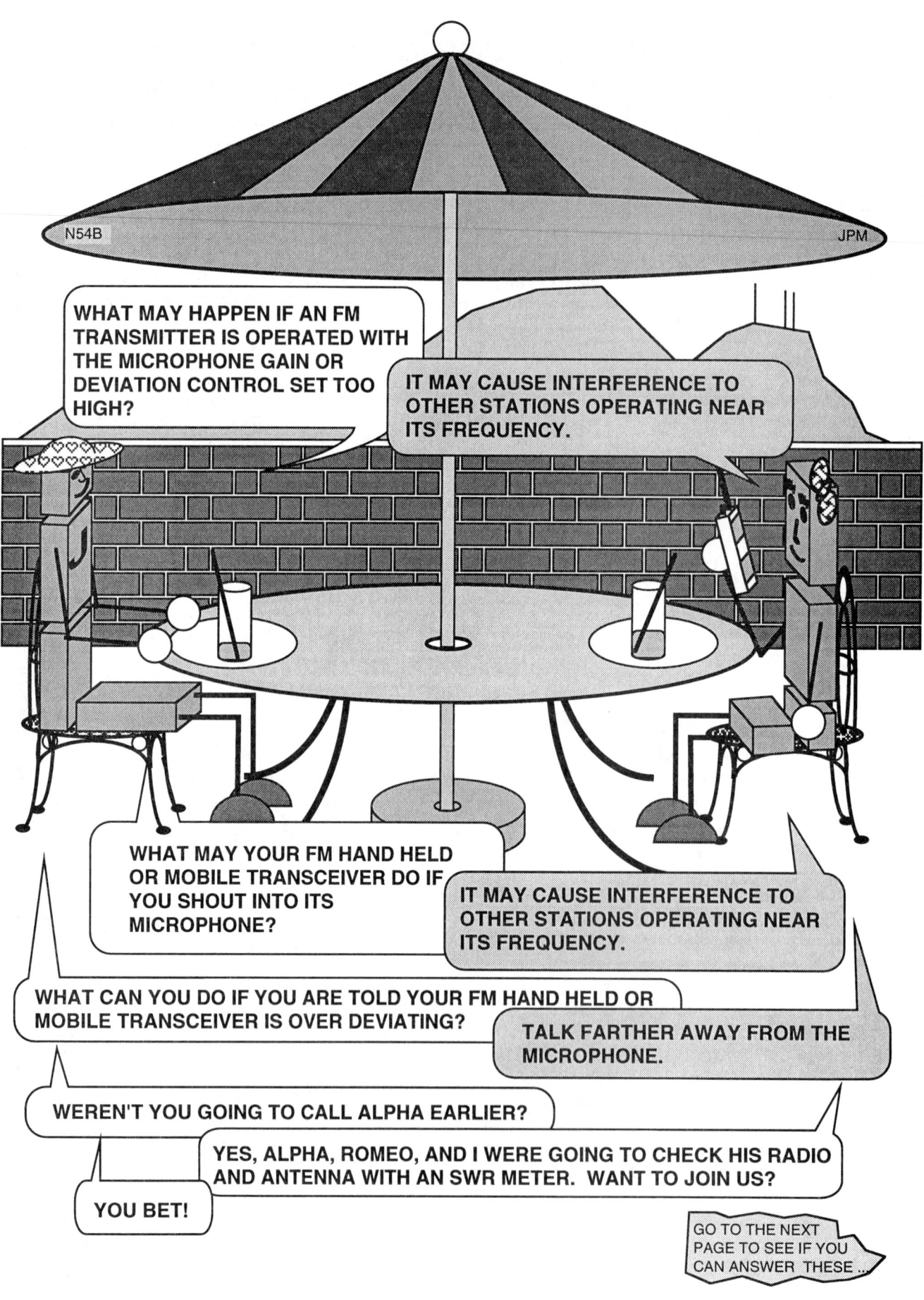

Riding the Airwaves with Alpha & Zulu

SEE IF YOU CAN ANSWER THESE QUESTIONS WITHOUT LOOKING BACK TO THE LAST 'TOON'. WRITE YOUR ANSWERS IN THE ANSWER BOX BELOW. GOOD LUCK.

PRACTICE TEST PAGE N54

N8B07
What may happen if an SSB transmitter is operated with the microphone gain set too high?
 A. It may cause digital interference to computer equipment
 B. It may cause splatter interference to other stations operating near its frequency
 C. It may cause atmospheric interference in the air around the antenna
 D. It may cause interference to other stations operating on a higher frequency band

N8B09
What may happen if an FM transmitter is operated with the microphone gain or deviation control set too high?
 A. It may cause digital interference to computer equipment
 B. It may cause interference to other stations operating near its frequency
 C. It may cause atmospheric interference in the air around the antenna
 D. It may cause interference to other stations operating on a higher frequency band

N8B08
What may happen if an SSB transmitter is operated with too much speech processing?
 A. It may cause digital interference to computer equipment
 B. It may cause splatter interference to other stations operating near its frequency
 C. It may cause atmospheric interference in the air around the antenna
 D. It may cause interference to other stations operating on a higher frequency band

N8B10
What may your FM hand held or mobile transceiver do if you shout into its microphone?
 A. It may cause digital interference to computer equipment
 B. It may cause interference to other stations operating near its frequency
 C. It may cause atmospheric interference in the air around the antenna
 D. It may cause interference to other stations operating on a higher frequency band

N8B11
What can you do if you are told your FM hand held or mobile transceiver is over deviating?
 A. Talk louder into the microphone
 B. Let the transceiver cool off
 C. Change to a higher power level
 D. Talk farther away from the microphone

ANSWERS TO PREVIOUS TEST
N8B04
D
N8B05
D
N8B06
A
N4D10
C

YOUR ANSWERS TO THIS TEST
N8B07
N8B08
N8B09
N8B10
N8B11

artsci inc

RADIO MATCHING

THE PURPOSE OF THIS GAME IS TO MATCH THE RADIO TERMS ON THE LEFT SIDE WITH THEIR DESCRIPTION ON THE RIGHT. TO DO THIS, PLACE THE LETTER OF THE DESCRIPTION ON THE LINE IN FRONT OF THE TERM. THERE ARE MORE DESCRIPTIONS THAN TERMS, SO THERE WILL BE ONE DESCRIPTION LEFT OVER. GOOD LUCK.

___ AMATEUR OPERATOR

___ AMPLIFIER

___ BEAM ANTENNA

___ CHIRP

___ COAXIAL CABLE

___ DIPOLE ANTENNA

___ DIRECT CURRENT

___ FREQUENCY

___ GROUND WAVES

___ HARMONICS

___ INSULATOR

___ MICROPHONE

A. ELECTRON FLOW IN A SINGLE DIRECTION

B. DIRECTIONAL ANTENNA

C. DOES NOT CONDUCT ELECTRICITY

D. COMMON TV ANTENNA

E. RADIO WAVES THAT TRAVEL ALONG THE EARTH'S SURFACE

F. AUTHORIZED TO BE A CONTROL OPERATOR

G. COMMON HALF WAVE ANTENNA

H. DEVICE TO INCREASE POWER

I. A SIGNAL WHICH IS A MULTIPLE OF THE DESIRED FREQUENCY

J. DEVICE TO CONVERT SOUND ENERGY INTO ELECTRICAL ENERGY

K. SHIELDED CONDUCTOR

L. CW TRANSMITTER FREQUENCY SHIFT

M. CYCLES PER SECOND

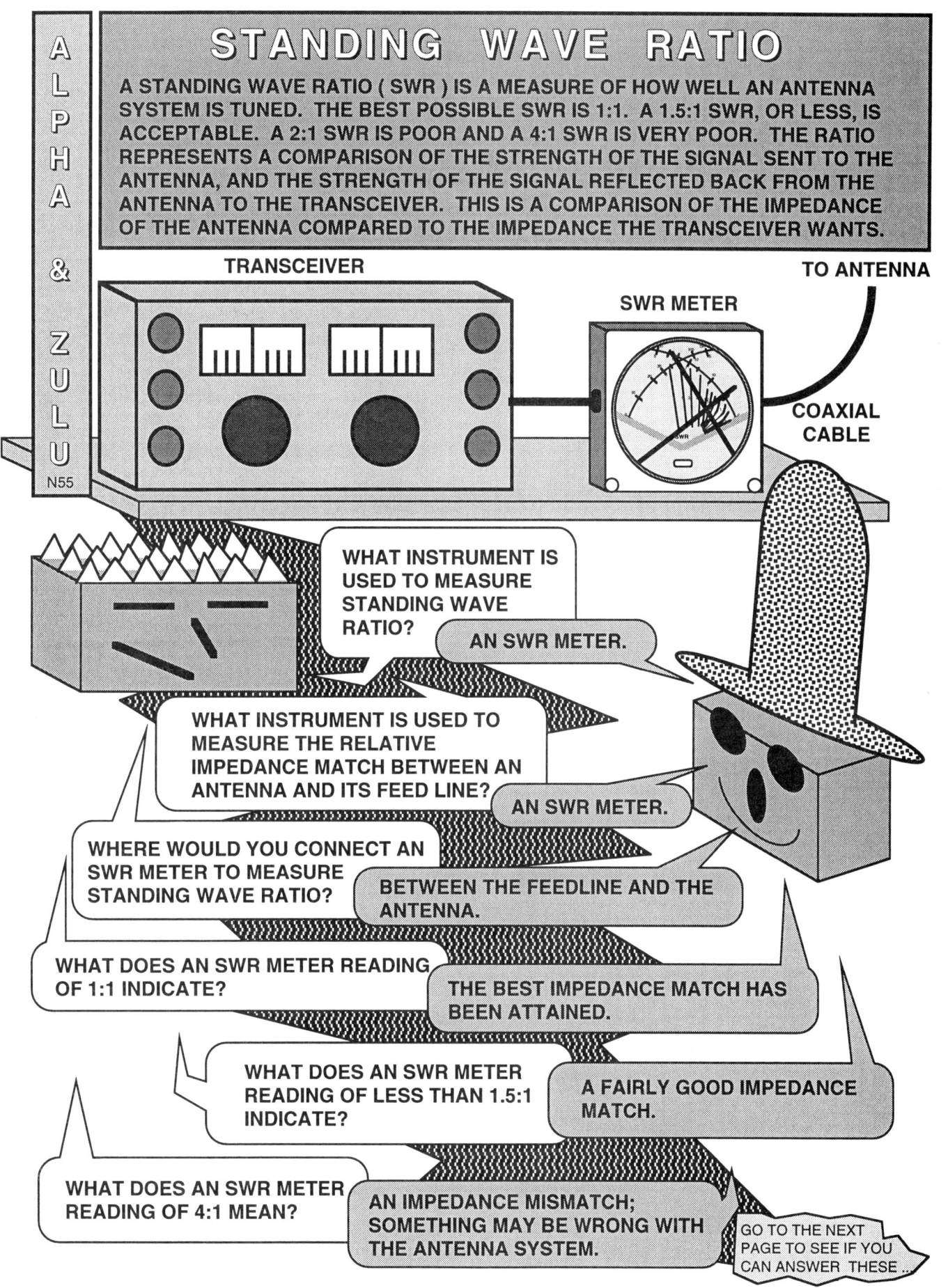

Riding the Airwaves with Alpha & Zulu

SEE IF YOU CAN ANSWER THESE QUESTIONS WITHOUT LOOKING BACK TO THE LAST 'TOON'. WRITE YOUR ANSWERS IN THE ANSWER BOX BELOW. GOOD LUCK.

PRACTICE TEST PAGE N55

N4C01
What instrument is used to measure standing wave ratio?
A. An ohmmeter
B. An ammeter
C. An SWR meter
D. A current bridge

N4C03
Where would you connect an SWR meter to measure standing wave ratio?
A. Between the feed line and the antenna
B. Between the transmitter and the power supply
C. Between the transmitter and the receiver
D. Between the transmitter and the ground

N4C02
What instrument is used to measure the relative impedance match between an antenna and its feed line?
A. An ammeter
B. An ohmmeter
C. A voltmeter
D. An SWR meter

N4C04
What does an SWR reading of 1:1 mean?
A. An antenna for another frequency band is probably connected
B. The best impedance match has been attained
C. No power is going to the antenna
D. The SWR meter is broken

N4C05
What does an SWR reading of less than 1.5:1 mean?
A. An impedance match which is too low
B. An impedance mismatch; something may be wrong with the antenna system
C. A fairly good impedance match
D. An antenna gain of 1.5

N4C06
What does an SWR reading of 4:1 mean?
A. An impedance match which is too low
B. An impedance match which is good, but not the best
C. An antenna gain of 4
D. An impedance mismatch; something may be wrong with the antenna system

ANSWERS TO PREVIOUS TEST

N8B07 — B
N8B08 — B
N8B09 — B
N8B10 — B
N8B11 — D

YOUR ANSWERS TO THIS TEST

N4C01 ___
N4C02 ___
N4C03 ___
N4C04 ___
N4C05 ___
N4C06 ___

Riding the Airwaves with Alpha & Zulu

SEE IF YOU CAN ANSWER THESE QUESTIONS WITHOUT LOOKING BACK TO THE LAST 'TOON'. WRITE YOUR ANSWERS IN THE ANSWER BOX BELOW. GOOD LUCK.

PRACTICE TEST PAGE N56

N4C07
What kind of SWR reading may mean poor electrical contact between parts of an antenna system?
- A. A jumpy reading
- B. A very low reading
- C. No reading at all
- D. A negative reading

N4C08
What does a very high SWR reading mean?
- A. The antenna is the wrong length, or there may be an open or shorted connection somewhere in the feed line
- B. The signals coming from the antenna are unusually strong, which means very good radio conditions
- C. The transmitter is putting out more power than normal, showing that it is about to go bad
- D. There is a large amount of solar radiation, which means very poor radio conditions

N4C09
If an SWR reading at the low frequency end of an amateur band is 2.5:1, and is 5:1 at the high frequency end of the same band, what does this tell you about your 1/2-wavelength dipole antenna?
- A. The antenna is broadbanded
- B. The antenna is too long for operation on the band
- C. The antenna is too short for operation on the band
- D. The antenna is just right for operation on the band

N4C10
If an SWR reading at the low frequency end of an amateur band is 5:1, and 2.5:1 at the high frequency end of the same band, what does this tell you about your 1/2-wavelength dipole antenna?
- A. The antenna is broadbanded
- B. The antenna is too long for operation on the band
- C. The antenna is too short for operation on the band
- D. The antenna is just right for operation on the band

N4C11
If you use a 3-30 MHz RF-power meter at UHF frequencies, how accurate will its readings be?
- A. They may not be accurate at all
- B. They will be accurate enough to get by
- C. They will be accurate but the readings must be divided by two
- D. They will be accurate but the readings must be multiplied by two

ANSWERS TO PREVIOUS TEST

N4C01	C
N4C02	D
N4C03	A
N4C04	B
N4C05	C
N4C06	D

YOUR ANSWERS TO THIS TEST

N4C07	
N4C08	
N4C09	
N4C10	
N4C11	

artsci inc

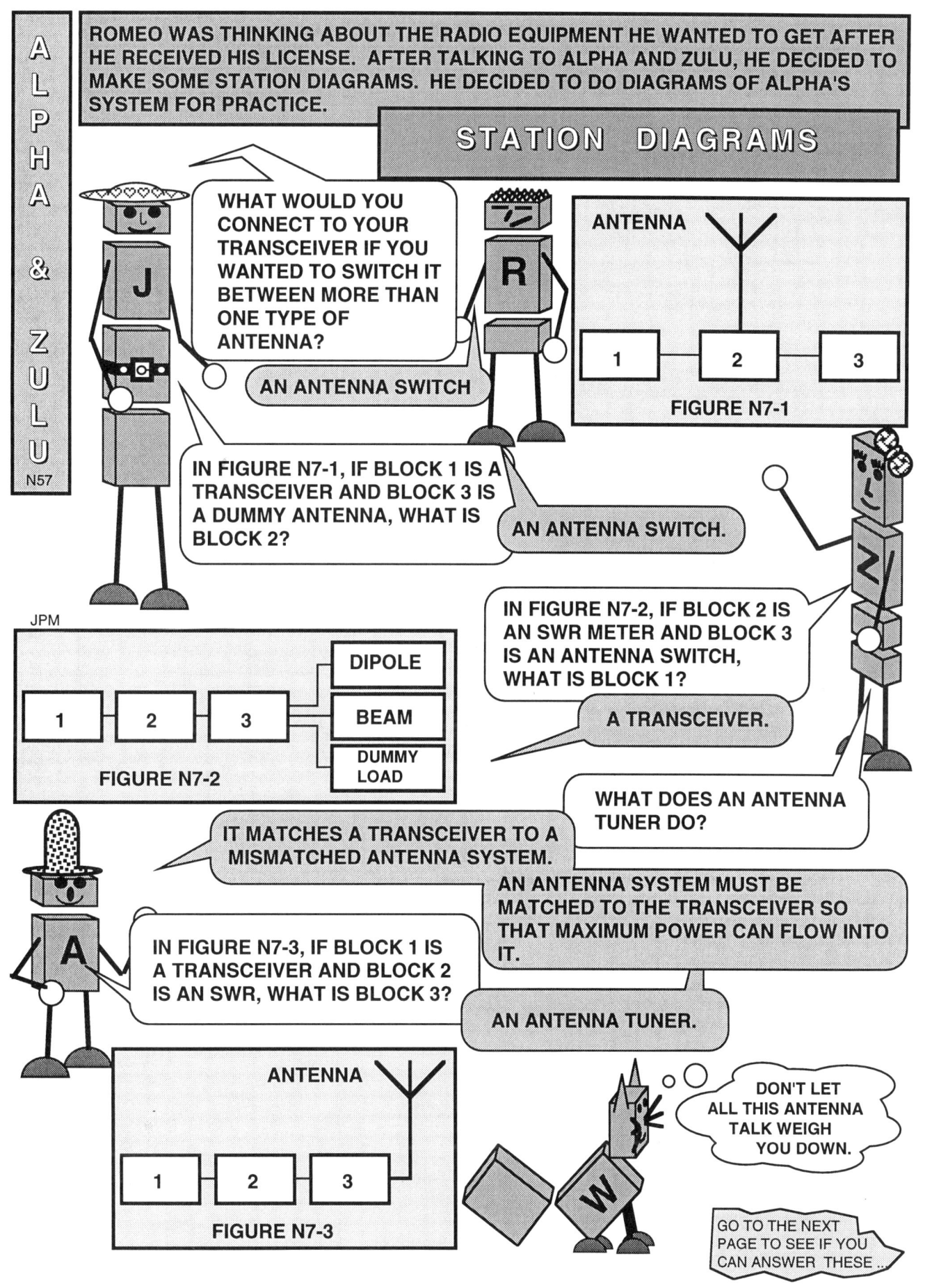

Riding the Airwaves with Alpha & Zulu

SEE IF YOU CAN ANSWER THESE QUESTIONS WITHOUT LOOKING BACK TO THE LAST 'TOON'. WRITE YOUR ANSWERS IN THE ANSWER BOX BELOW. GOOD LUCK.

PRACTICE TEST PAGE N57

N7A01
What would you connect to your transceiver if you wanted to switch it between more than one type of antenna?
  A. A terminal-node switch
  B. An antenna switch
  C. A telegraph key switch
  D. A high-pass filter

N7A08
What does an antenna tuner do?
  A. It matches a transceiver to a mismatched antenna system
  B. It helps a receiver automatically tune in stations that are far away
  C. It switches an antenna system to a transceiver when sending, and to a receiver when listening
  D. It switches a transceiver between different kinds of antennas connected to one feed line

N7A09
In Figure N7-1, if block 1 is a transceiver and block 3 is a dummy antenna what is block 2?
  A. A terminal-node switch
  B. An antenna switch
  C. A telegraph key switch
  D. A high-pass filter

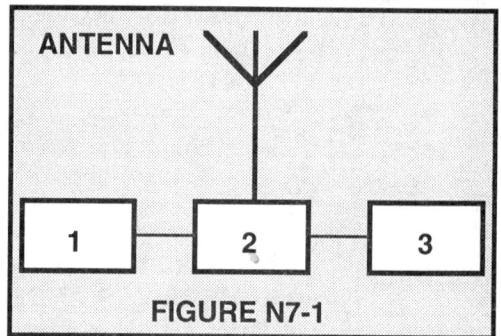
FIGURE N7-1

N7A10
In Figure N7-2, if block 2 is an SWR meter and block 3 is an antenna switch, what is block 1?
  A. A transceiver
  B. A high-pass filter
  C. An antenna tuner
  D. A modem

N7A11
In Figure N7-3, if block 1 is a transceiver and block 2 is an SWR meter, what is block 3?
  A. An antenna switch
  B. An antenna tuner
  C. A key-click filter
  D. A terminal-node controller

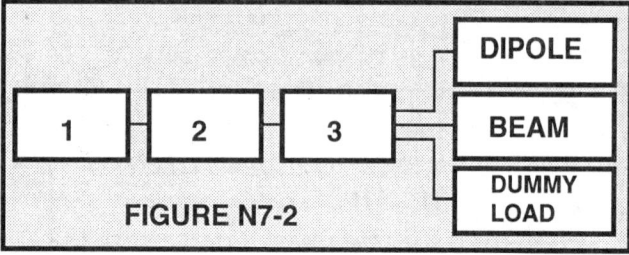
FIGURE N7-2

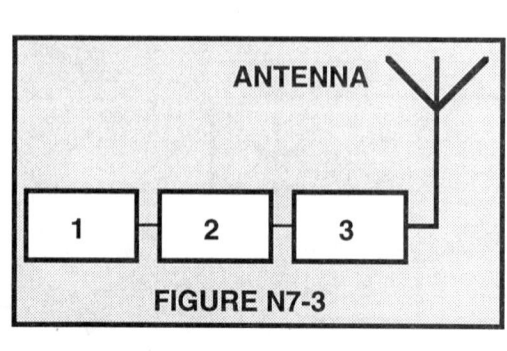
FIGURE N7-3

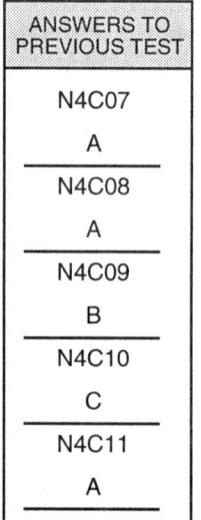

ANSWERS TO PREVIOUS TEST
N4C07 A
N4C08 A
N4C09 B
N4C10 C
N4C11 A

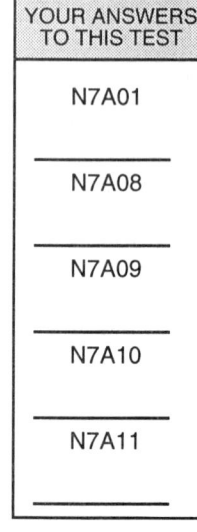

YOUR ANSWERS TO THIS TEST
N7A01
N7A08
N7A09
N7A10
N7A11

artsci inc

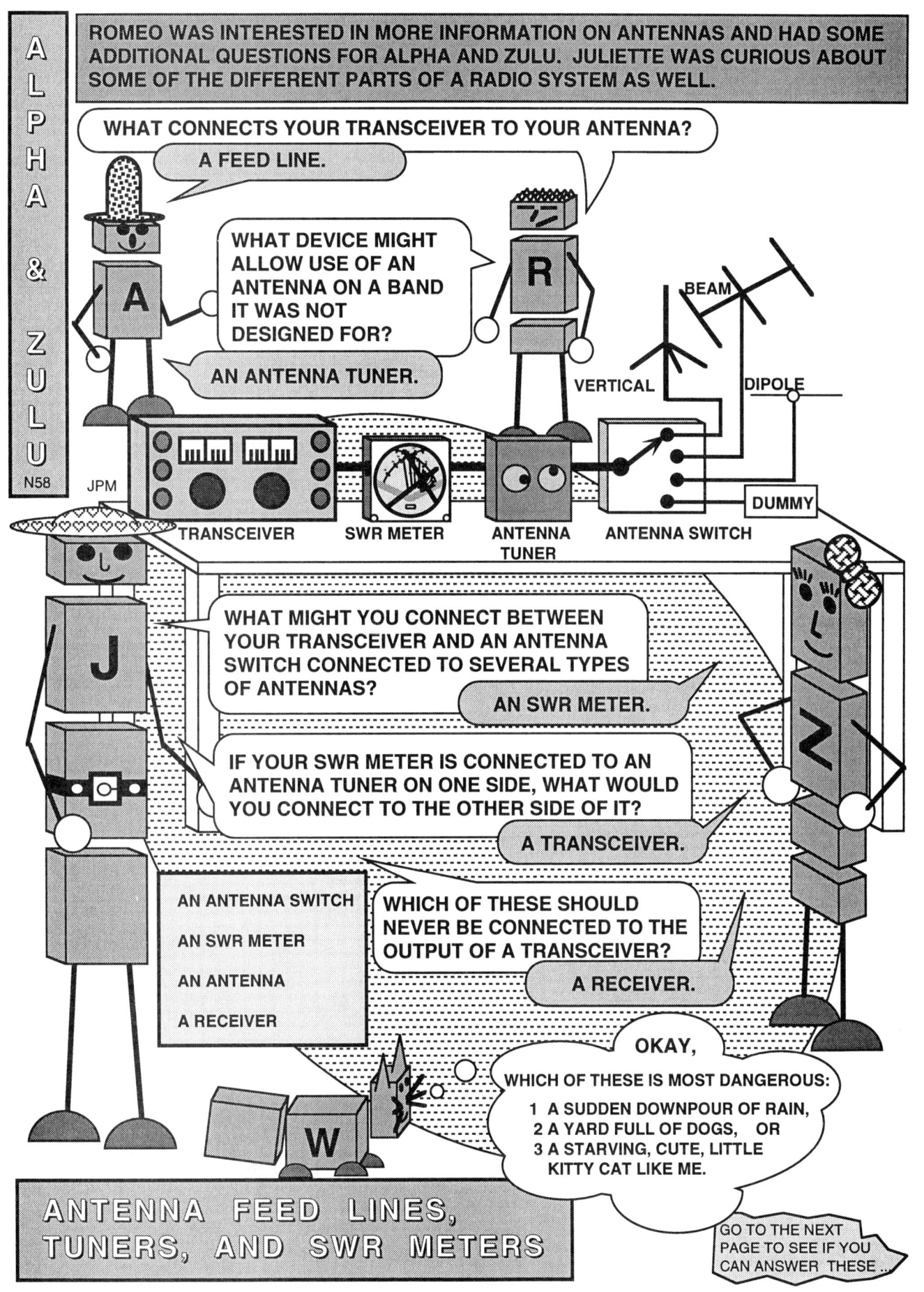

Riding the Airwaves with Alpha & Zulu

SEE IF YOU CAN ANSWER THESE QUESTIONS WITHOUT LOOKING BACK TO THE LAST 'TOON'. WRITE YOUR ANSWERS IN THE ANSWER BOX BELOW. GOOD LUCK.

PRACTICE TEST PAGE N58

N7A03
What connects your transceiver to your antenna?
A. A dummy load
B. A ground wire
C. The power cord
D. A feed line

N7A02
What device might allow use of an antenna on a band it was not designed for?
A. An SWR meter
B. A low-pass filter
C. An antenna tuner
D. A high-pass filter

N7A04
What might you connect between your transceiver and an antenna switch connected to several types of antennas?
A. A high-pass filter
B. An SWR meter
C. A key click filter
D. A mixer

N7A05
If your SWR meter is connected to an antenna tuner on one side, what would you connect to the other side of it?
A. A power supply
B. An antenna
C. An antenna switch
D. A transceiver

N7A06
Which of these should never be connected to the output of a transceiver?
A. An antenna switch
B. An SWR meter
C. An antenna
D. A receiver

ANSWERS TO PREVIOUS TEST
N7A01
B
N7A08
A
N7A09
B
N7A10
A
N7A11
B

YOUR ANSWERS TO THIS TEST
N7A03
N7A02
N7A04
N7A05
N7A06

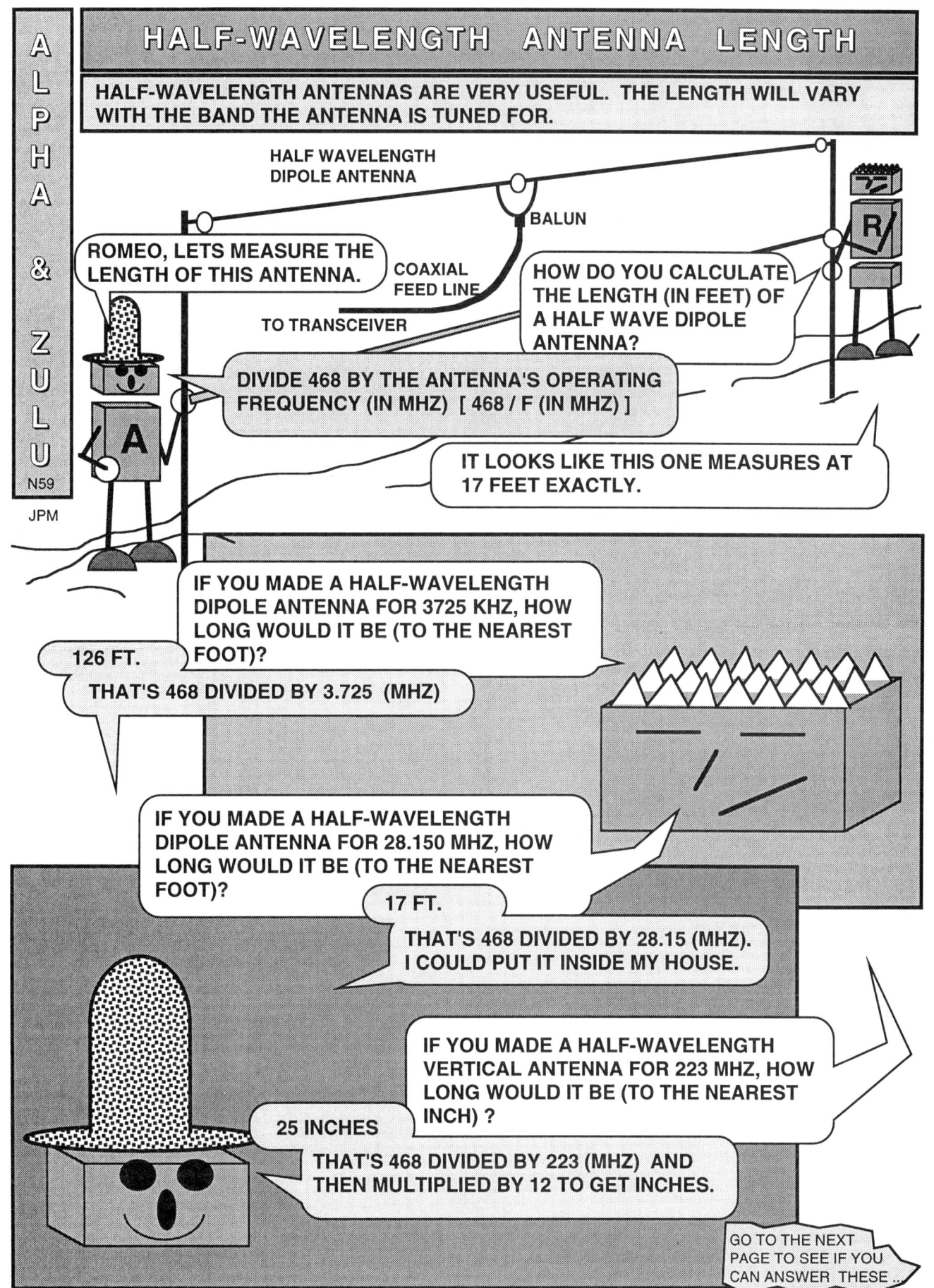

SEE IF YOU CAN ANSWER THESE QUESTIONS WITHOUT LOOKING BACK TO THE LAST 'TOON'. WRITE YOUR ANSWERS IN THE ANSWER BOX BELOW. GOOD LUCK.

PRACTICE TEST PAGE N59

N9A01
How do you calculate the length (in feet) of a half-wavelength dipole antenna?
 A. Divide 150 by the antenna's operating frequency (in MHz) [150/f(in MHz)]
 B. Divide 234 by the antenna's operating frequency (in MHz) [234/f (in MHz)]
 C. Divide 300 by the antenna's operating frequency (in MHz) [300/f (in MHz)]
 D. Divide 468 by the antenna's operating frequency (in MHz) [468/f (in MHz)]

N9A04
If you made a half-wavelength dipole antenna for 28.150 MHz, how long would it be (to the nearest foot)?
 A. 22 ft
 B. 11 ft
 C. 17 ft
 D. 34 ft

N9A03
If you made a half-wavelength dipole antenna for 3725 kHz, how long would it be (to the nearest foot)?
 A. 126 ft
 B. 81 ft
 C. 63 ft
 D. 40 ft

N9A07
If you made a half-wavelength vertical antenna for 223 MHz, how long would it be (to the nearest inch)?
 A. 112 inches
 B. 50 inches
 C. 25 inches
 D. 12 inches

ANSWERS TO PREVIOUS TEST
N7A03
D
N7A02
C
N7A04
B
N7A05
D
N7A06
D

YOUR ANSWERS TO THIS TEST
N9A01
N9A03
N9A04
N9A07

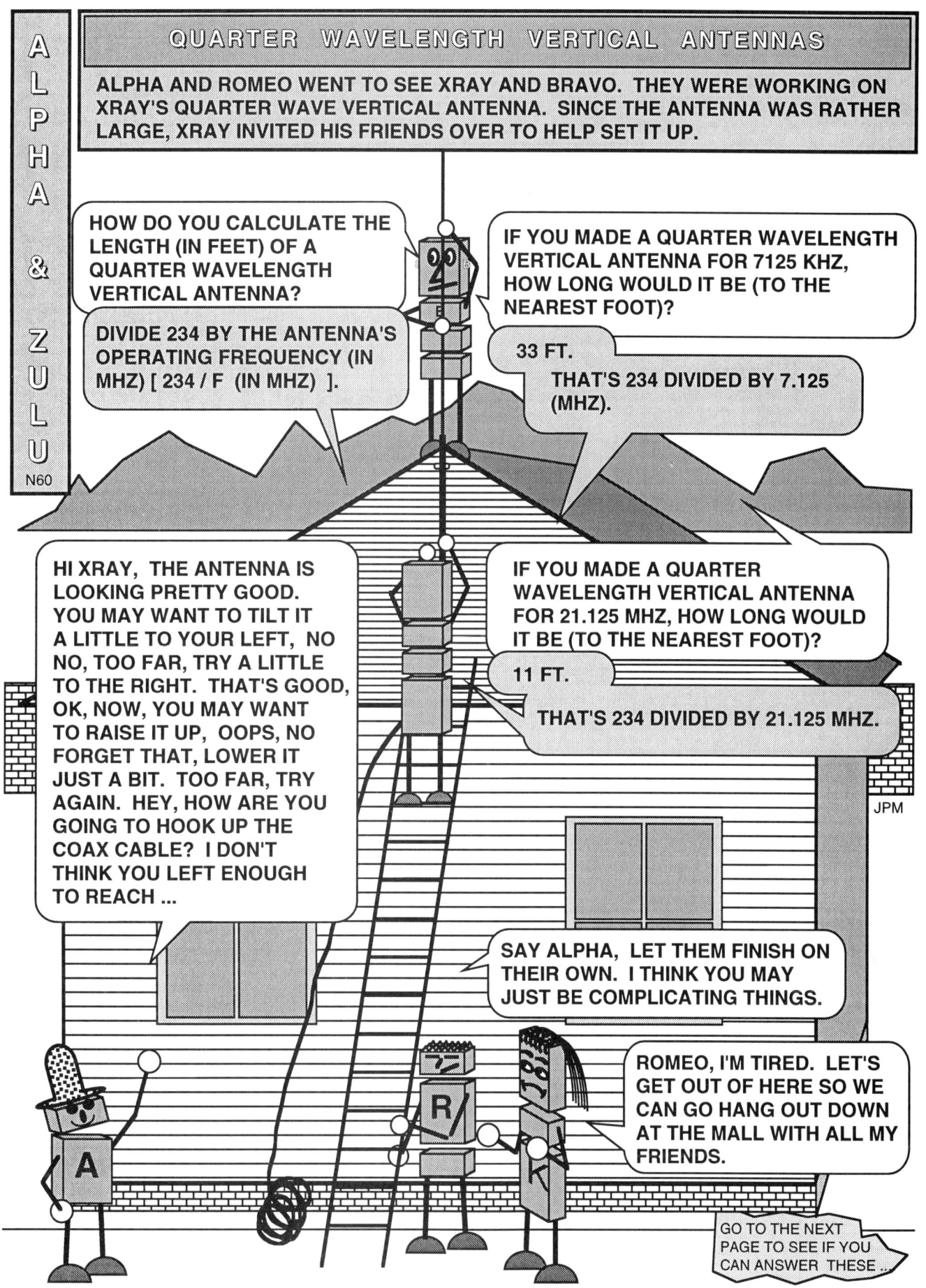

Riding the Airwaves with Alpha & Zulu

SEE IF YOU CAN ANSWER THESE QUESTIONS WITHOUT LOOKING BACK TO THE LAST 'TOON'. WRITE YOUR ANSWERS IN THE ANSWER BOX BELOW. GOOD LUCK.

PRACTICE TEST PAGE N60

N9A02
How do you calculate the length (in feet) of a quarter-wavelength vertical antenna?
- A. Divide 150 by the antenna's operating frequency (in MHz) [150/f (in MHz)]
- B. Divide 234 by the antenna's operating frequency (in MHz) [234/f (in MHz)]
- C. Divide 300 by the antenna's operating frequency (in MHz) [300/f (in MHz)]
- D. Divide 468 by the antenna's operating frequency (in MHz) [468/f (in MHz)]

N9A05
If you made a quarter-wavelength vertical antenna for 7125 kHz, how long would it be (to the nearest foot)?
- A. 11 ft
- B. 16 ft
- C. 21 ft
- D. 33 ft

N9A06
If you made a quarter-wavelength vertical antenna for 21.125 MHz, how long would it be (to the nearest foot)?
- A. 7 ft
- B. 11 ft
- C. 14 ft
- D. 22 ft

ANSWERS TO PREVIOUS TEST

N9A01 — D
N9A03 — A
N9A04 — C
N9A07 — C

YOUR ANSWERS TO THIS TEST

N9A02 ___
N9A05 ___
N9A06 ___

artsci inc

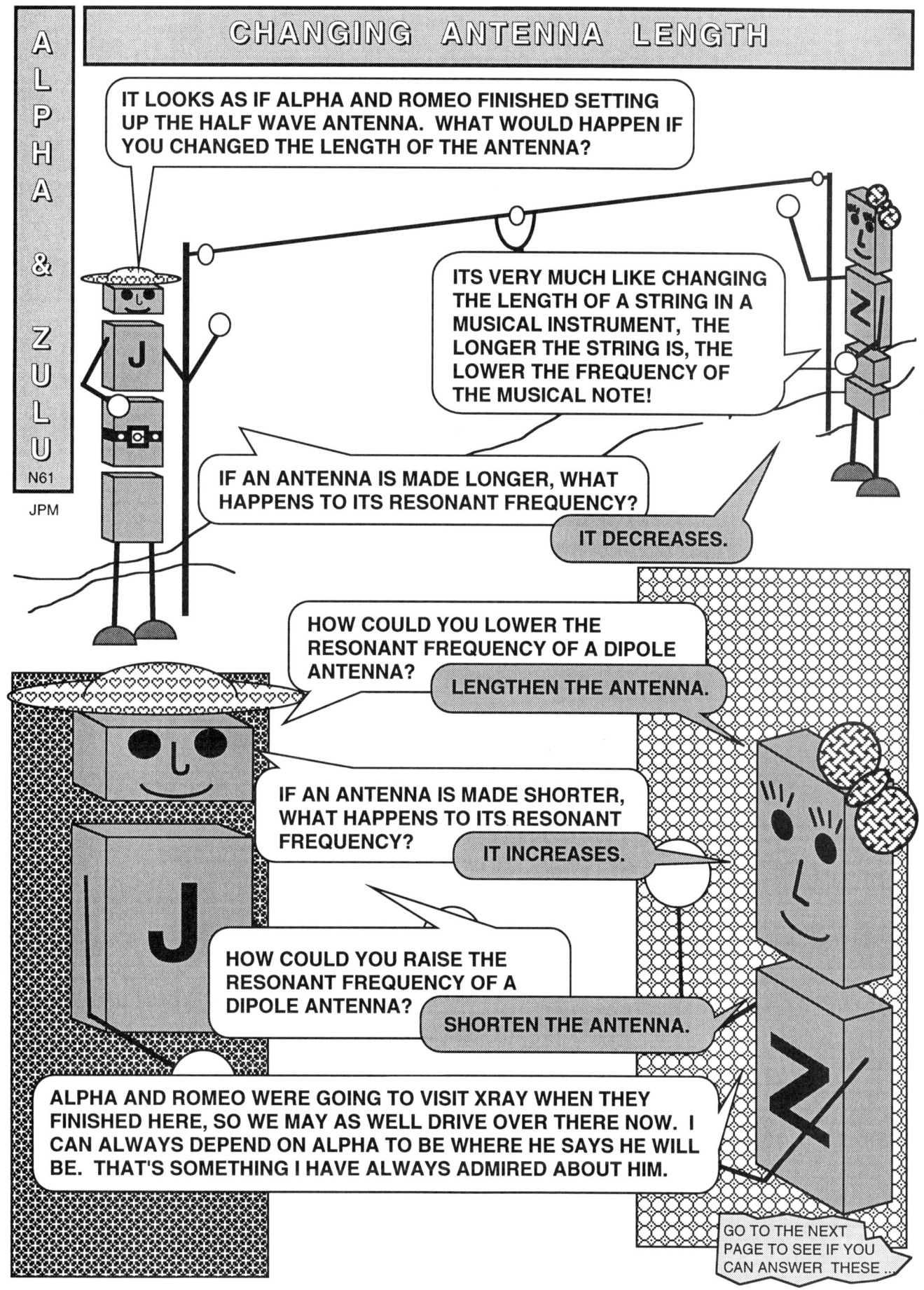

Riding the Airwaves with Alpha & Zulu

SEE IF YOU CAN ANSWER THESE QUESTIONS WITHOUT LOOKING BACK TO THE LAST 'TOON'. WRITE YOUR ANSWERS IN THE ANSWER BOX BELOW. GOOD LUCK.

PRACTICE TEST PAGE N61

N9A08
If an antenna is made longer, what happens to its resonant frequency?
- A. It decreases
- B. It increases
- C. It stays the same
- D. It disappears

N9A09
If an antenna is made shorter, what happens to its resonant frequency?
- A. It decreases
- B. It increases
- C. It stays the same
- D. It disappears

N9A10
How could you lower the resonant frequency of a dipole antenna?
- A. Lengthen the antenna
- B. Shorten the antenna
- C. Use less feed line
- D. Use a smaller size feed line

N9A11
How could you raise the resonant frequency of a dipole antenna?
- A. Lengthen the antenna
- B. Shorten the antenna
- C. Use more feed line
- D. Use a larger size feed line

ANSWERS TO PREVIOUS TEST

N9A02	B
N9A05	D
N9A06	B

YOUR ANSWERS TO THIS TEST

N9A08	
N9A09	
N9A10	
N9A11	

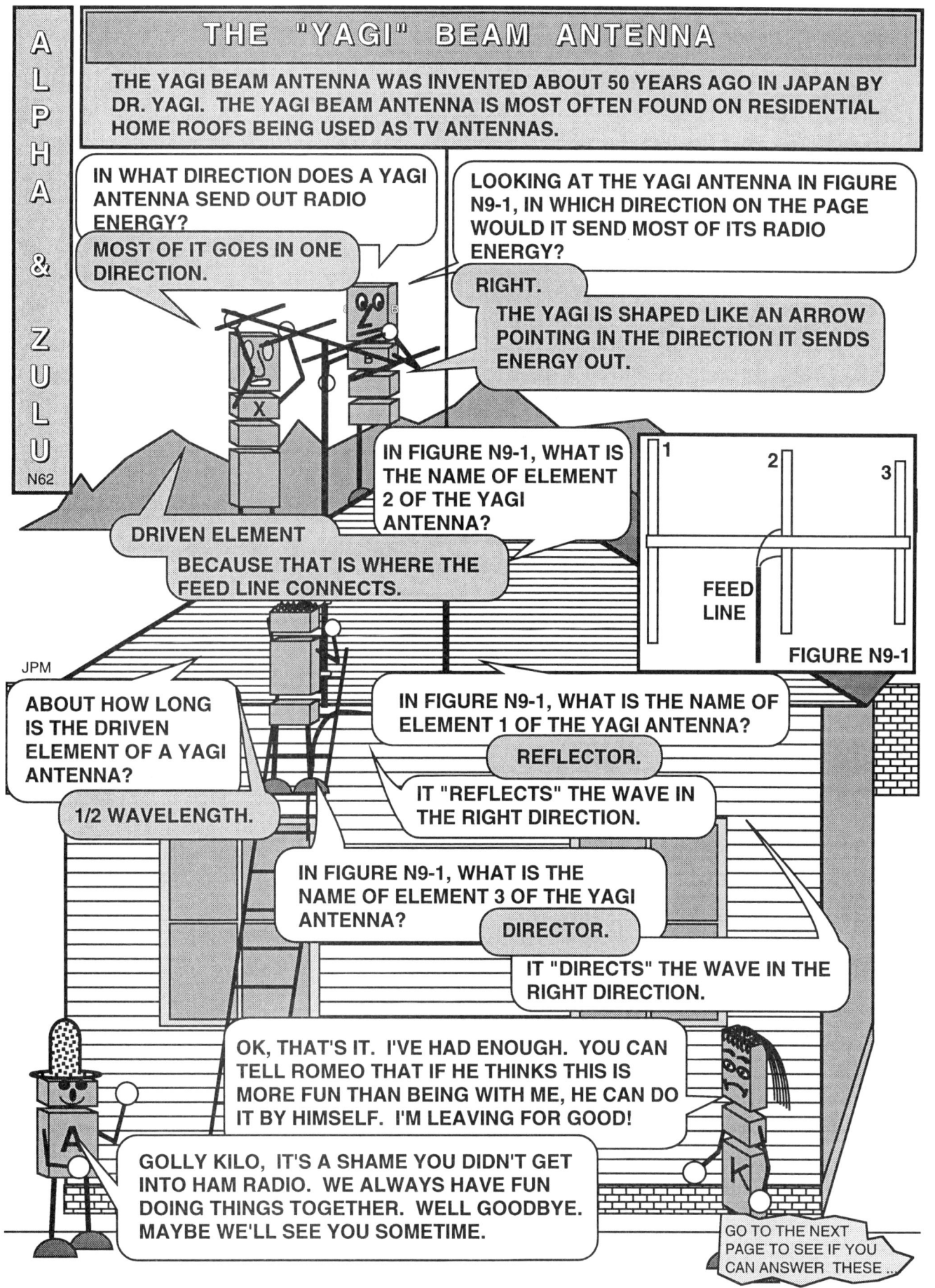

Riding the Airwaves with Alpha & Zulu

SEE IF YOU CAN ANSWER THESE QUESTIONS WITHOUT LOOKING BACK TO THE LAST 'TOON'. WRITE YOUR ANSWERS IN THE ANSWER BOX BELOW. GOOD LUCK.

PRACTICE TEST PAGE N62

N9B01
In what direction does a Yagi antenna send out radio energy?
A. It goes out equally in all directions
B. Most of it goes in one direction
C. Most of it goes equally in two opposite directions
D. Most of it is aimed high into the air

N9B02
About how long is the driven element of a Yagi antenna?
A. 1/4 wavelength
B. 1/3 wavelength
C. 1/2 wavelength
D. 1 wavelength

N9B03
In Figure N9-1, what is the name of element 2 of the Yagi antenna?
A. Director
B. Reflector
C. Boom
D. Driven element

N9B04
In Figure N9-1, what is the name of element 3 of the Yagi antenna?
A. Director
B. Reflector
C. Boom
D. Driven element

N9B05
In Figure N9-1, what is the name of element 1 of the Yagi antenna?
A. Director
B. Reflector
C. Boom
D. Driven element

N9B06
Looking at the Yagi antenna in Figure N9-1, in which direction on the page would it send most of its radio energy?
A. Left
B. Right
C. Top
D. Bottom

ANSWERS TO PREVIOUS TEST
N9A08
A
N9A09
B
N9A10
A
N9A11
B

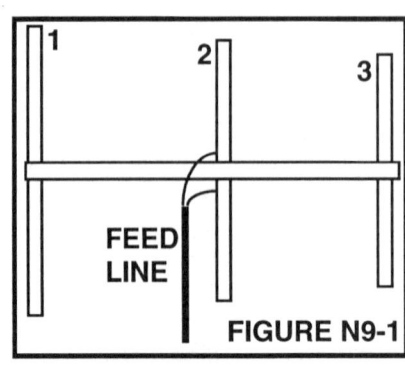

FIGURE N9-1

YOUR ANSWERS TO THIS TEST
N9B01
N9B02
N9B03
N9B04
N9B05
N9B06

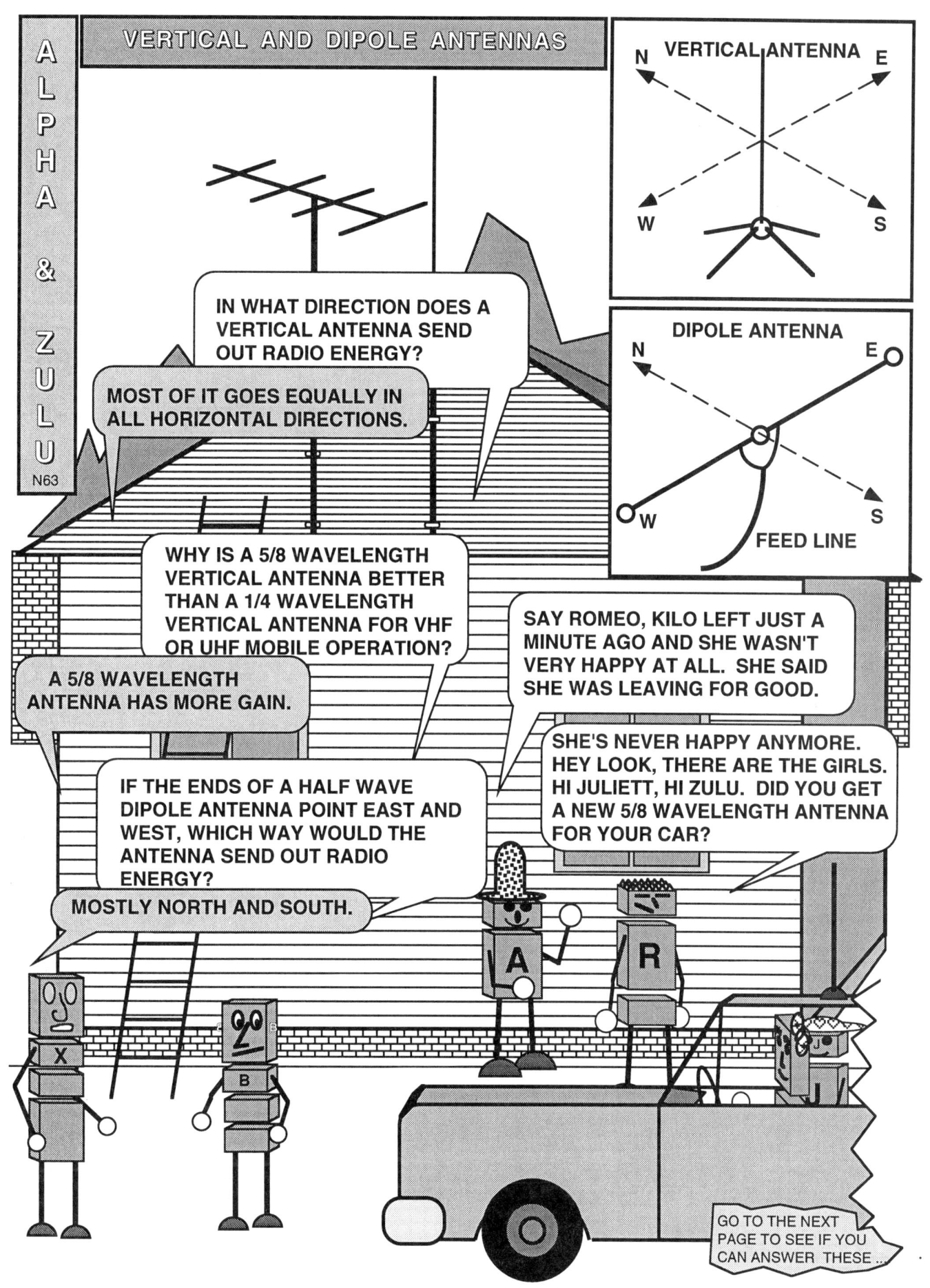

SEE IF YOU CAN ANSWER THESE QUESTIONS WITHOUT LOOKING BACK TO THE LAST 'TOON'. WRITE YOUR ANSWERS IN THE ANSWER BOX BELOW. GOOD LUCK.

PRACTICE TEST PAGE N63

N9B07
Why is a 5/8-wavelength vertical antenna better than a 1/4-wavelength vertical antenna for VHF or UHF mobile operations?
- A. A 5/8-wavelength antenna can handle more power
- B. A 5/8-wavelength antenna has more gain
- C. A 5/8-wavelength antenna has less corona loss
- D. A 5/8-wavelength antenna is easier to install on a car

N9B08
In what direction does a vertical antenna send out radio energy?
- A. Most of it goes in two opposite directions
- B. Most of it goes high into the air
- C. Most of it goes equally in all horizontal directions
- D. Most of it goes in one direction

N9B09
If the ends of a half-wave dipole antenna point east and west, which way would the antenna send out radio energy?
- A. Equally in all directions
- B. Mostly up and down
- C. Mostly north and south
- D. Mostly east and west

ANSWERS TO PREVIOUS TEST

N9B01	B
N9B02	C
N9B03	D
N9B04	A
N9B05	B
N9B06	B

YOUR ANSWERS TO THIS TEST

N9B07	
N9B08	
N9B09	

artsci inc

Riding the Airwaves with Alpha & Zulu

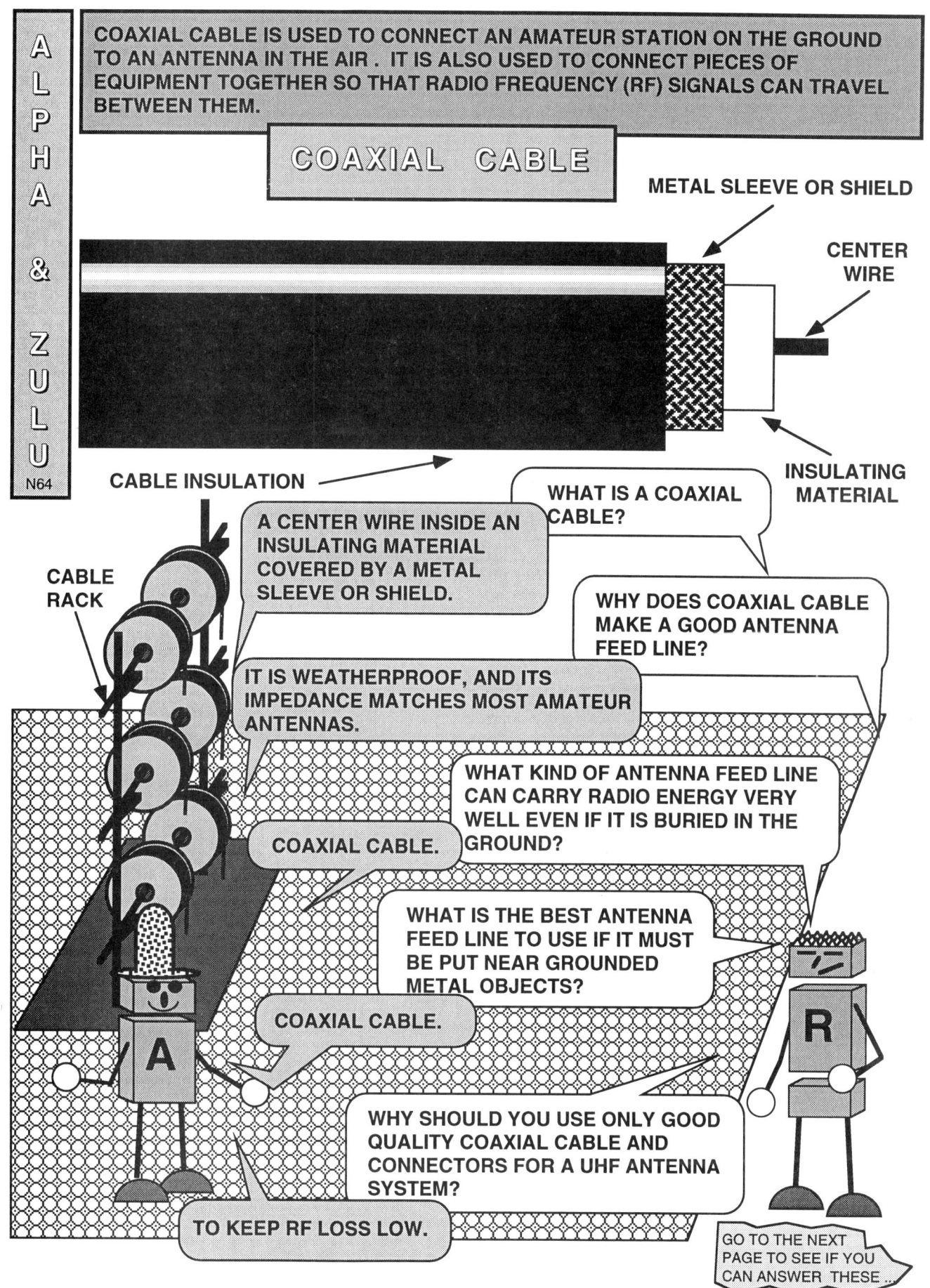

Riding the Airwaves with Alpha & Zulu

SEE IF YOU CAN ANSWER THESE QUESTIONS WITHOUT LOOKING BACK TO THE LAST 'TOON'. WRITE YOUR ANSWERS IN THE ANSWER BOX BELOW. GOOD LUCK.

PRACTICE TEST PAGE N64

N9C01
What is a coaxial cable?
A. Two wires side-by-side in a plastic ribbon
B. Two wires side-by-side held apart by insulating rods
C. Two wires twisted around each other in a spiral
D. A center wire inside an insulating material covered by a metal sleeve or shield

N9C02
Why does coaxial cable make a good antenna feed line?
A. You can make it at home, and its impedance matches most amateur antennas
B. It is weatherproof, and its impedance matches most amateur antennas
C. It is weatherproof, and its impedance is higher than that of most amateur antennas
D. It can be used near metal objects, and its impedance is higher than that of most amateur antennas

N9C03
Which kind of antenna feed line can carry radio energy very well even if it is buried in the ground?
A. Twin lead
B. Coaxial cable
C. Parallel conductor
D. Twisted pair

N9C04
What is the best antenna feed line to use if it must be put near grounded metal objects?
A. Coaxial cable
B. Twin lead
C. Twisted pair
D. Ladder-line

N4B04
Why should you use only good quality coaxial cable and connectors for a UHF antenna system?
A. To keep RF loss low
B. To keep television interference high
C. To keep the power going to your antenna system from getting too high
D. To keep the standing wave ratio of your antenna system high

ANSWERS TO PREVIOUS TEST
N9B07
B
N9B08
C
N9B09
C

YOUR ANSWERS TO THIS TEST
N9C01
N9C02
N9C03
N9C04
N4B04

146 artsci inc

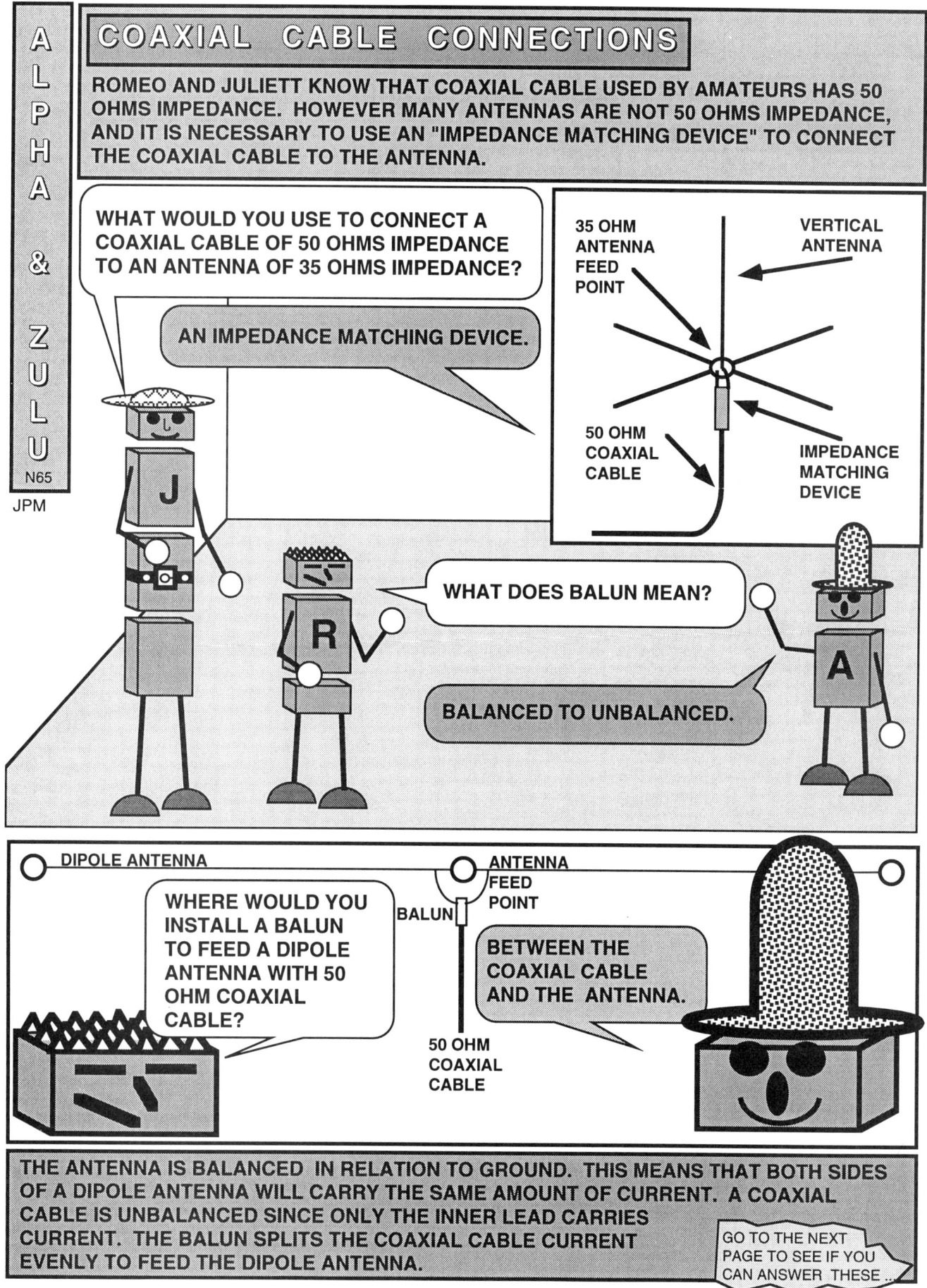

Riding the Airwaves with Alpha & Zulu

SEE IF YOU CAN ANSWER THESE QUESTIONS WITHOUT LOOKING BACK TO THE LAST 'TOON'. WRITE YOUR ANSWERS IN THE ANSWER BOX BELOW. GOOD LUCK.

PRACTICE TEST PAGE N65

N9C09
What would you use to connect a coaxial cable of 50-ohms impedance to an antenna of 35-ohms impedance?
 A. A terminating resistor
 B. An SWR meter
 C. An impedance matching device
 D. A low-pass filter

N9C10
What does balun mean?
 A. Balanced antenna network
 B. Balanced unloader
 C. Balanced unmodulator
 D. Balanced to unbalanced

N9C11
Where would you install a balun to feed a dipole antenna with 50-ohm coaxial cable?
 A. Between the coaxial cable and the antenna
 B. Between the transmitter and the coaxial cable
 C. Between the antenna and the ground
 D. Between the coaxial cable and the ground

ANSWERS TO PREVIOUS TEST
N9C01
D
N9C02
B
N9C03
B
N9C04
A
N4B04
A

YOUR ANSWERS TO THIS TEST
N9C09
N9C10
N9C11

artsci inc

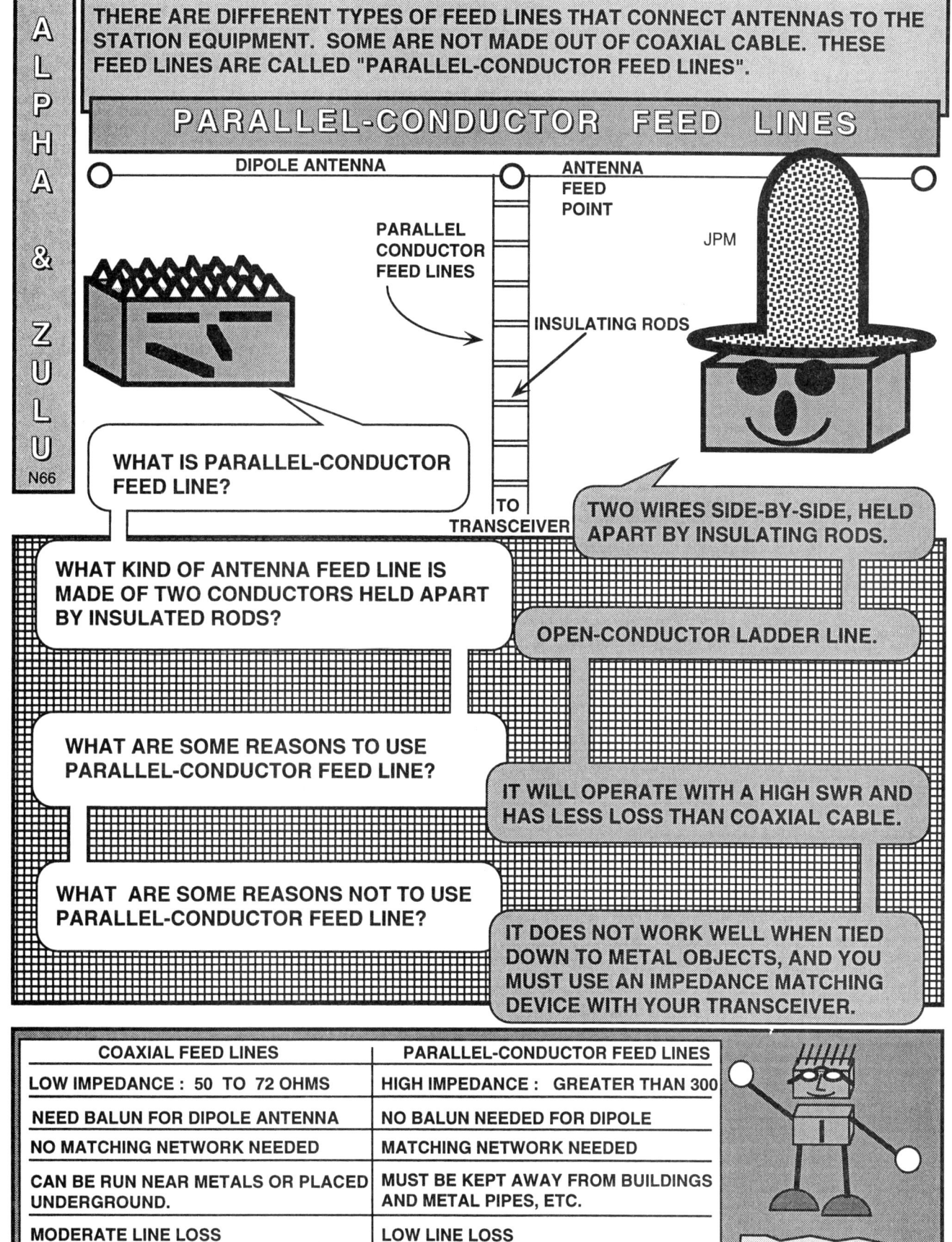

Riding the Airwaves with Alpha & Zulu

SEE IF YOU CAN ANSWER THESE QUESTIONS WITHOUT LOOKING BACK TO THE LAST 'TOON'. WRITE YOUR ANSWERS IN THE ANSWER BOX BELOW. GOOD LUCK.

PRACTICE TEST PAGE N66

N9C05
What is parallel-conductor feed line?
A. Two wires twisted around each other in a spiral
B. Two wires side-by-side held apart by insulating rods
C. A center wire inside an insulating material which is covered by a metal sleeve or shield
D. A metal pipe which is as wide or slightly wider than a wavelength of the signal it carries

N9C07
What are some reasons not to use parallel-conductor feed line?
A. It does not work well when tied down to metal objects, and you must use an impedance matching device with your transceiver
B. It is difficult to make at home, and it does not work very well with a high SWR
C. It does not work well when tied down to metal objects, and it cannot operate under high power
D. You must use an impedance matching device with your transceiver, and it does not work very well with a high SWR

N9C06
What are some reasons to use parallel-conductor feed line?
A. It has low impedance, and will operate with a high SWR
B. It will operate with a high SWR, and it works well when tied down to metal objects
C. It has a low impedance, and has less loss than coaxial cable
D. It will operate with a high SWR, and has less loss than coaxial cable

N9C08
What kind of antenna feed line is made of two conductors held apart by insulated rods?
A. Coaxial cable
B. Open-conductor ladder line
C. Twin lead in a plastic ribbon
D. Twisted pair

ANSWERS TO PREVIOUS TEST	
N9C09	C
N9C10	D
N9C11	A

YOUR ANSWERS TO THIS TEST
N9C05
N9C06
N9C07
N9C08

PRACTICE TEST PAGE N67

GOOD LUCK.

N4B01
What should you do for safety when operating at 1270 MHz?
A. Make sure that an RF leakage filter is installed at the antenna feed point
B. Keep antenna away from your eyes when RF is applied
C. Make sure the standing wave ratio is low before you conduct a test
D. Never use a shielded horizontally polarized antenna

N4B02
What should you do for safety if you put up a UHF transmitting antenna?
A. Make sure the antenna will be in a place where no one can get near it when you are transmitting
B. Make sure that RF field screens are in place
C. Make sure the antenna is near the ground to keep its RF energy pointing in the correct direction
D. Make sure you connect an RF leakage filter at the antenna feed point

N4B05
Why should you make sure the antenna of a hand held transceiver is not close to your head when transmitting?
A. To help the antenna radiate energy equally in all directions
B. To reduce your exposure to the radio-frequency energy
C. To use your body to reflect the signal in one direction
D. To keep static charges from building up

N4B06
Microwave oven radiation is similar to what type of amateur station RF radiation?
A. Signals in the 3.5 MHz range
B. Signals in the 21 MHz range
C. Signals in the 50 MHz range
D. Signals in the 1270 MHz range

N4B10
For safety, how high should you place a horizontal wire antenna?
A. High enough so that no one can touch any part of it from the ground
B. As close to the ground as possible
C. Just high enough so you can easily reach it for adjustments or repairs
D. Above high-voltage electrical lines

N9B11
Why should your outside antennas be high enough so that no one can touch them while you are transmitting?
A. Touching the antenna might cause television interference
B. Touching the antenna might cause RF burns
C. Touching the antenna might radiate harmonics
D. Touching the antenna might reflect the signal back to the transmitter and cause damage

N9B10
How should you hold the antenna of a hand held transceiver while you are transmitting?
A. Away from your head and away from others
B. Pointed towards the station you are contacting
C. Pointed away from the station you are contacting
D. Pointed down to bounce the signal off the ground

ANSWERS TO PREVIOUS TEST

N9C05	B
N9C06	D
N9C07	A
N9C08	B

YOUR ANSWERS TO THIS TEST

N4B01
N4B02
N4B05
N4B06
N4B10
N9B11
N9B10

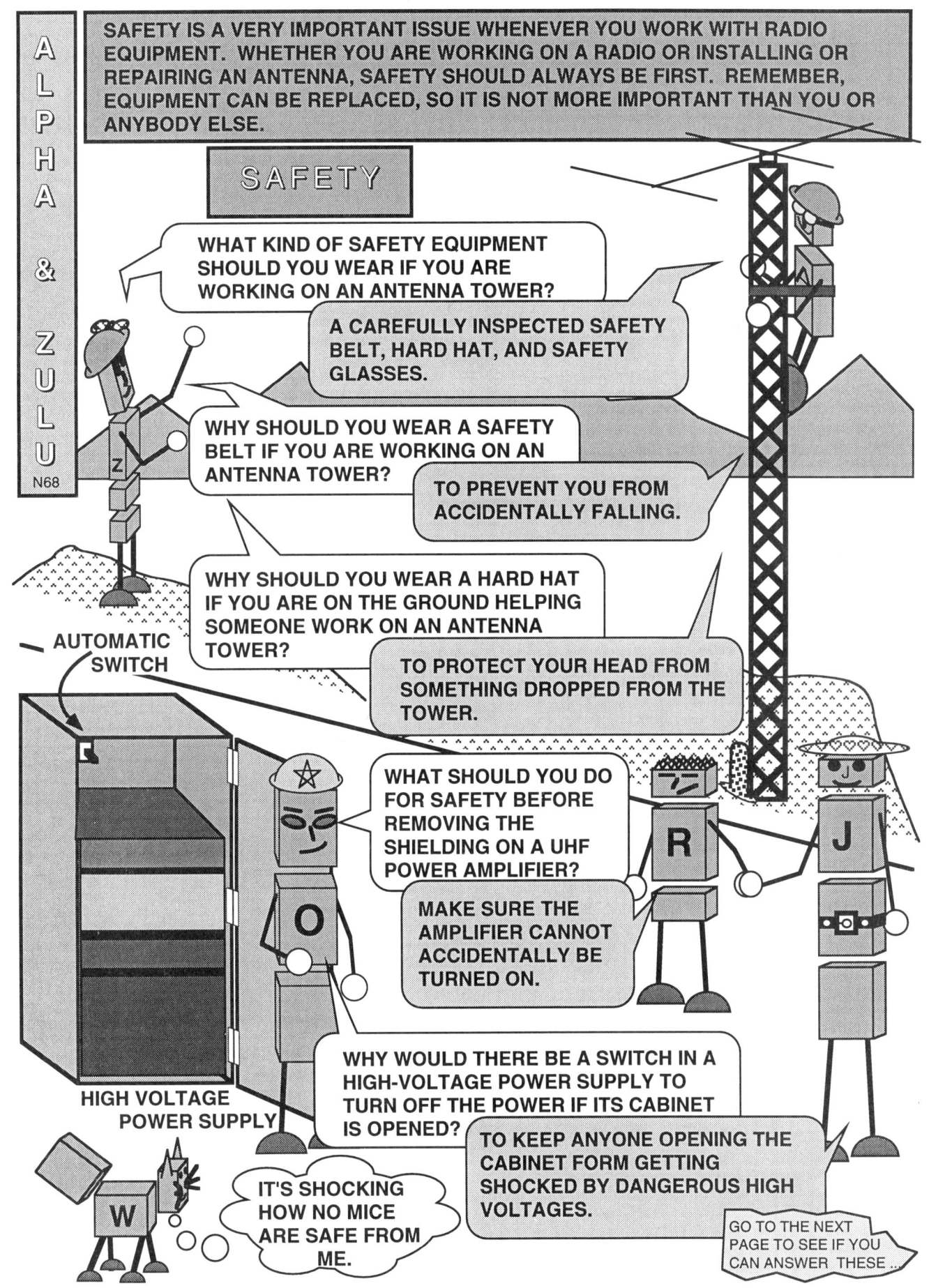

Riding the Airwaves with Alpha & Zulu

SEE IF YOU CAN ANSWER THESE QUESTIONS WITHOUT LOOKING BACK TO THE LAST 'TOON'. WRITE YOUR ANSWERS IN THE ANSWER BOX BELOW. GOOD LUCK.

PRACTICE TEST PAGE N68

N4B03
What should you do for safety before removing the shielding on a UHF power amplifier?
- A. Make sure all RF screens are in place at the antenna feed line
- B. Make sure the antenna feed line is properly grounded
- C. Make sure the amplifier cannot accidentally be turned on
- D. Make sure that RF leakage filters are connected

N4B07
Why would there be a switch in a high-voltage power supply to turn off the power if its cabinet is opened?
- A. To keep dangerous RF radiation from leaking out through an open cabinet
- B. To keep dangerous RF radiation from coming in through an open cabinet
- C. To turn the power supply off when it is not being used
- D. To keep anyone opening the cabinet from getting shocked by dangerous high voltages

N4B08
What kind of safety equipment should you wear if you are working on an antenna tower?
- A. A grounding chain
- B. A reflective vest of approved color
- C. A flashing red, yellow or white light
- D. A carefully inspected safety belt, hard hat and safety glasses

N4B09
Why should you wear a safety belt if you are working on an antenna tower?
- A. To safely hold your tools so they don't fall and injure someone on the ground
- B. To keep the tower from becoming unbalanced while you are working
- C. To safely bring any tools you might use up and down the tower
- D. To prevent you from accidentally falling

N4B11
Why should you wear a hard hat if you are on the ground helping someone work on an antenna tower?
- A. So you won't be hurt if the tower should accidentally fall
- B. To keep RF energy away from your head during antenna testing
- C. To protect your head from something dropped from the tower
- D. So someone passing by will know that work is being done on the tower and will stay away

ANSWERS TO PREVIOUS TEST
N4B01 B
N4B02 A
N4B05 B
N4B06 D
N4B10 A
N9B11 B
N9B10 A

YOUR ANSWERS TO THIS TEST
N4B03
N4B07
N4B08
N4B09
N4B11

154   artsci inc

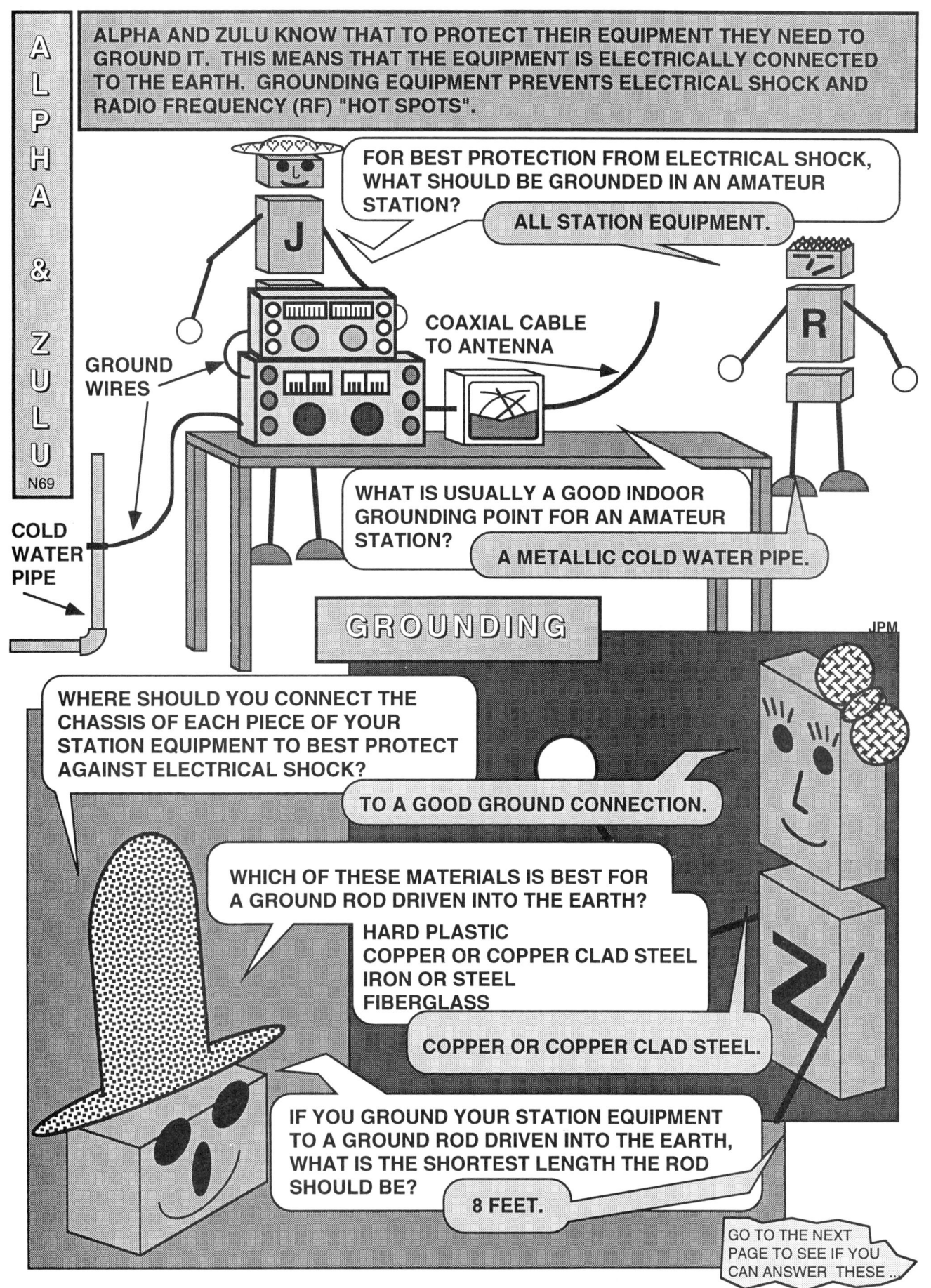

Riding the Airwaves with Alpha & Zulu

SEE IF YOU CAN ANSWER THESE QUESTIONS WITHOUT LOOKING BACK TO THE LAST 'TOON'. WRITE YOUR ANSWERS IN THE ANSWER BOX BELOW. GOOD LUCK.

PRACTICE TEST PAGE N69

N4A07
For best protection from electrical shock, what should be grounded in an amateur station?
- A. The power supply primary
- B. All station equipment
- C. The antenna feed line
- D. The AC power mains

N4A08
What is usually a good indoor grounding point for an amateur station?
- A. A metallic cold water pipe
- B. A plastic cold water pipe
- C. A window screen
- D. A metallic natural gas pipe

N4A09
Where should you connect the chassis of each piece of your station equipment to best protect against electrical shock?
- A. To insulated shock mounts
- B. To the antenna
- C. To a good ground connection
- D. To a circuit breaker

N4A10
Which of these materials is best for a ground rod driven into the earth?
- A. Hard plastic
- B. Copper or copper-clad steel
- C. Iron or steel
- D. Fiberglass

N4A11
If you ground your station equipment to a ground rod driven into the earth, what is the shortest length the rod should be?
- A. 4 feet
- B. 6 feet
- C. 8 feet
- D. 10 feet

ANSWERS TO PREVIOUS TEST
N4B03
C
N4B07
D
N4B08
D
N4B09
D
N4B11
C

YOUR ANSWERS TO THIS TEST
N4A07
N4A08
N4A09
N4A10
N4A11

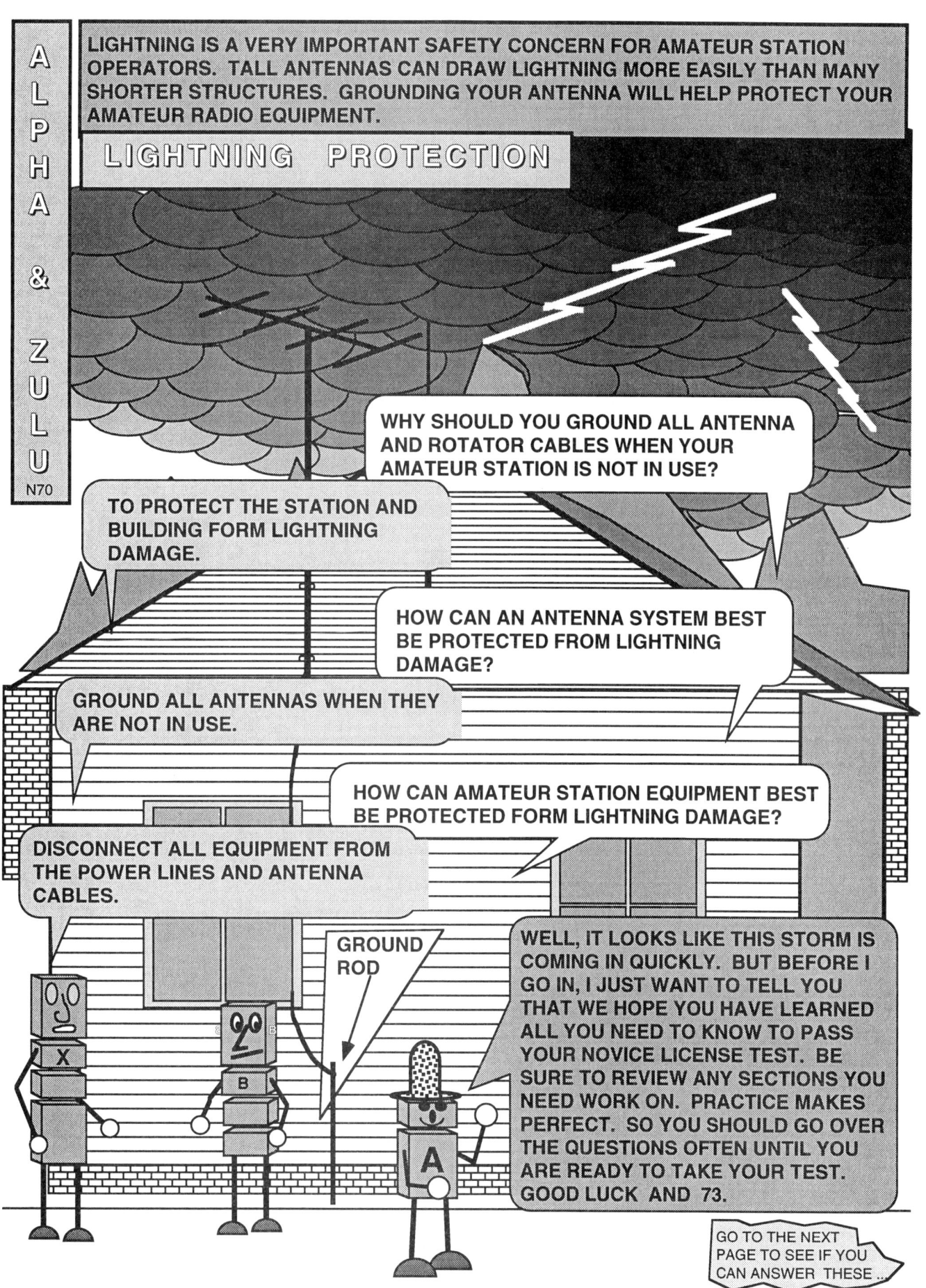

Riding the Airwaves with Alpha & Zulu

WELL, HERE'S THE LAST TEST IN THE NOVICE SECTION. SEE IF YOU CAN ANSWER THESE QUESTIONS WITHOUT LOOKING BACK TO THE LAST 'TOON'. WRITE YOUR ANSWERS IN THE ANSWER BOX BELOW. GOOD LUCK.

PRACTICE TEST PAGE N70

N4A04
Why should you ground all antenna and rotator cables when your amateur station is not in use?
- A. To lock the antenna system in one position
- B. To avoid radio frequency interference
- C. To save electricity
- D. To protect the station and building from lightning damage

N4A05
How can an antenna system best be protected from lightning damage?
- A. Install a balun at the antenna feed point
- B. Install an RF choke in the antenna feed line
- C. Ground all antennas when they are not in use
- D. Install a fuse in the antenna feed line

N4A06
How can amateur station equipment best be protected from lightning damage?
- A. Use heavy insulation on the wiring
- B. Never turn off the equipment
- C. Disconnect the ground system from all radios
- D. Disconnect all equipment from the power lines and antenna cables

ANSWERS TO PREVIOUS TEST
N4A07
B
N4A08
A
N4A09
C
N4A10
B
N4A11
C

YOUR ANSWERS TO THIS TEST
N4A04
N4A05
N4A06

artsci inc

Riding the Airwaves with Alpha & Zulu

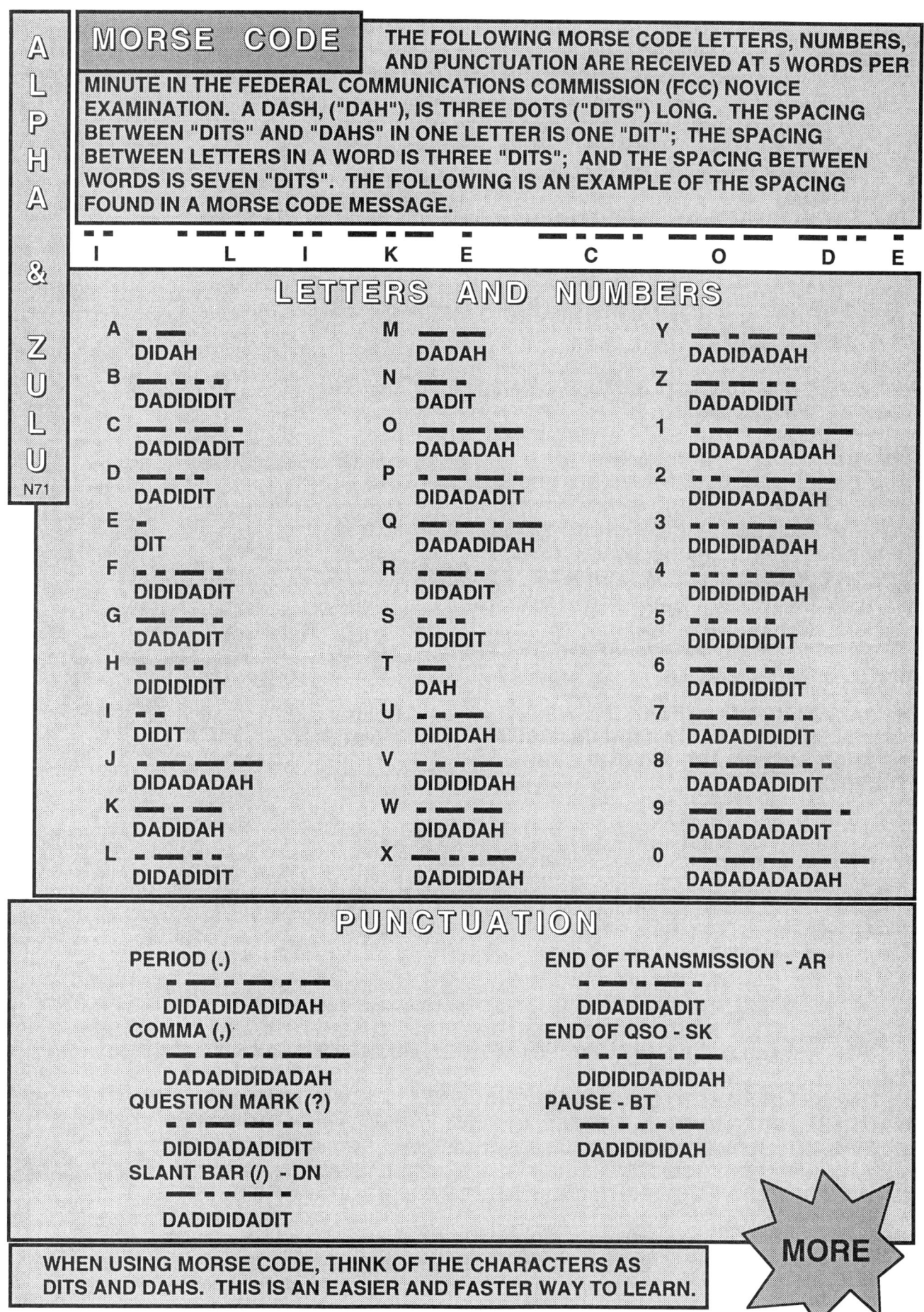

MORSE CODE SHOULD NOT BE LEARNED IN THE NORMAL LETTER SEQUENCE OF THE ALPHABET! THERE ARE SEVERAL EXCELLENT AUDIO CASSETTE TAPES AVAILABLE FOR LEARNING THE MORSE CODE, AND ONE IN PARTICULAR IS RECOMMENDED. IT IS CALLED: "SPACECODE CASSETTE", AND IS AVAILABLE FROM

    MEDIA MENTORS INC.
    P.O. BOX 131646,
    STATEN ISLAND, NY, 10313-0006
    (718) 983-1416

CONTACT MEDIA MENTORS INC. FOR THE CURRENT PRICE OF THE TAPE.

IT IS VERY IMPORTANT THAT THE STUDENT OBTAIN A CODE PRACTICE OSCILLATOR (CPO). THE NOVICE EXAMINATION DOES NOT REQUIRE THE ABILITY TO SEND THE MORSE CODE, BUT IT IS MUCH EASIER TO LEARN TO RECEIVE IF THE STUDENT ALSO HAS THE ABILITY TO SEND!

AN INEXPENSIVE CPO CAN BE BUILT USING THE FOLLOWING LIST OF PARTS. IT IS A GOOD BEGINNERS PROJECT:

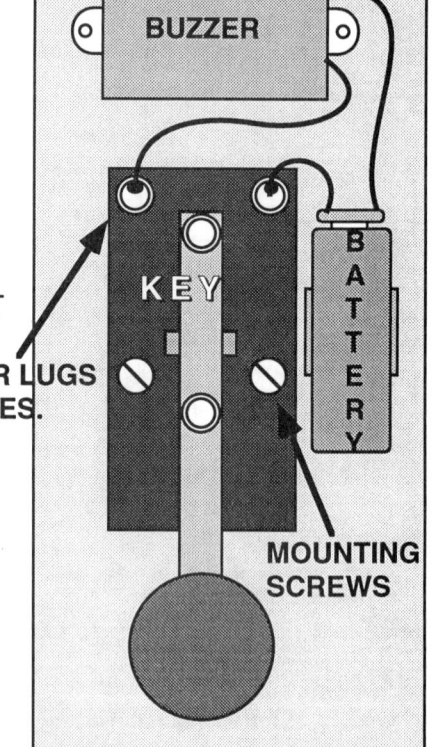

KEY	AMECO #K-1	$8.00
BUZZER	RADIO SHACK #273-055	$2.50
BATTERY CLIP	RADIO SHACK #270-325	$0.25
BATTERY HOLDER ...	RADIO SHACK #270-326	$0.50
9 VOLT BATTERY	RADIO SHACK #23-553	$1.90
SOLDER LUGS	LOCAL HARDWARE STORE	$0.25
MOUNTING SCREWS..	LOCAL HARDWARE STORE	$0.10
4"x6"x3/4" PARTICLE BOARD		
	......LOCAL HARDWARE STORE	$0.25
	TOTAL	$13.75

THE PARTS ARE MOUNTED ON THE HEAVY PARTICLE BOARD. THIS WILL HELP KEEP THE CPO FROM SLIDING AROUND WHILE THE STUDENT IS PRACTICING.

WHEN SENDING THE CODE, ONLY THE ELBOW SHOULD BE ON THE TABLE, WITH THE WRIST ARCHED ABOVE THE TABLE. THE INDEX AND CENTER FINGERS SHOULD BE ON THE TOP FRONT, AND THE THUMB ON THE SIDE OF THE KEY KNOB. THE OTHER TWO FINGERS SHOULD BE FOLDED OUT OF THE WAY.

NEVER TAP THE KEY! IT IS IMPOSSIBLE TO SEND MORSE CODE RAPIDLY OR CORRECTLY BY TAPPING. THE FINGERS SHOULD LIGHTLY GRASP THE KEY KNOB AT ALL TIMES, AND THE WRIST SHOULD MOVE IN A SMOOTH RHYTHM, MUCH LIKE PLAYING A GUITAR.

JUST AS IT IS EASIEST TO LEARN A LANGUAGE BY LISTENING TO IT IN USE, MORSE CODE SHOULD BE LISTENED TO ON THE AIR. AT FIRST YOU WILL PICK OUT ONLY A FEW LETTERS, BUT SLOWLY WORDS AND SENTENCES WILL BECOME RECOGNIZABLE! INEXPENSIVE "HAM RECEIVER" RADIO KITS ARE AVAILABLE. THE 40 METER NOVICE BAND IS ACTIVE BOTH DAY AND NIGHT, AND WILL PROVIDE GOOD MORSE CODE PRACTICE.

THERE ARE MANY PUBLICATIONS AVAILABLE ON THE MORSE CODE, AS WELL AS DETAILED THEORY, AND A GOOD SOURCE IS THE AMERICAN RADIO RELAY LEAGUE (ARRL), 225 MAIN ST., NEWINGTON, CT, 06111. WRITE TO ARRL FOR A LIST OF THEIR PUBLICATIONS, PARTICULARLY FOR BEGINNERS.

MORSE CODE PRACTICE

DaDiDaDit DaDaDiDah   DaDiDaDit DaDaDiDah   DaDiDaDit DaDaDiDah   DaDiDit Dit
DiDah DiDah DaDiDiDiDit DiDah DiDah DiDah   DiDah DiDah DaDiDiDiDit DiDah DiDah DiDah
                                                  DiDah DiDah DaDiDiDiDit DiDah DiDah DiDah

___ ___ ___ ___ ___ ___ ___ ___ ___ ___ ___ ___

DiDah DiDah DaDiDiDiDit DiDah DiDah DiDah
DiDah DiDah DaDiDiDiDit DiDah DiDah DiDah   DaDiDit Dit
DiDah DiDaDiDit DiDiDiDiDah DiDaDiDah DiDaDiDit
DiDah DiDaDiDit DiDiDiDiDah DiDaDiDah DiDaDiDit
DiDah DiDaDiDit DiDiDiDiDah DiDaDiDit DiDaDiDit

___ ___   ___ ___   ___ ___
___ ___   ___ ___   ___ ___
___ ___   ___ ___

DaDit DiDah DaDah Dit DiDiDiDit Dit DiDaDit Dit DiDit DiDiDit DiDah DiDaDit DiDaDaDit DiDiDiDit DiDah DiDah DiDaDiDit DiDaDaDit DiDiDiDit DiDah	___ ___ ___ ___ ___ ___ ___ ___ ___ ___
___ ___ ___ ___ ___ ___ ___ ___ ___ ___ ___ ___	DiDiDiDit DiDit DiDah DiDaDit DiDaDaDit DiDiDiDit DiDah DaDit DiDah DaDah Dit DiDiDiDit Dit DiDaDit Dit DiDit DiDiDit DiDaDit DiDit DaDah DiDah DiDaDiDit DiDit DaDah DiDah
DiDiDah DiDaDit DiDaDit DiDiDit Dah DiDiDiDit DiDiDiDiDit DaDaDaDaDit DiDiDiDit DiDiDiDiDit DaDaDaDaDit	___ ___ ___ ___ ___ ___ ___ ___
___ ___ ___ ___ ___ ___ ___ ___ ___ ___	Dah DaDit DaDiDiDah DiDiDah DiDaDit DiDaDit DiDiDit Dah DiDiDit DiDiDiDit DaDaDaDaDit DiDiDiDit DiDiDiDit DaDaDaDaDit
DaDah DaDaDah Dah DiDiDiDit Dit DiDaDit DiDit DiDiDit DaDiDaDit DiDah DiDaDit DiDaDiDit DiDit DaDit DaDaDit DiDiDit DiDah DiDiDiDah Dit Dah DaDaDah DaDaDit DaDaDah DaDiDiDit DiDiDaDah DiDah DiDaDit DiDiDiDit DiDaDit DiDaDiDit DaDiDit Dit DiDah DiDah DaDiDiDiDit DiDah DiDah DiDah	___ ___ ___ ___ ___ ___ ___ ___ ___ ___

THE UNITED STATES IS DIVIDED INTO 10 CALL SIGN ZONES. YOUR CALL SIGN WILL HAVE THE NUMBER OF THE ZONE YOU LIVE IN.

HOW WELL CAN YOU READ A MAP?

SEE IF YOU CAN HELP ME PUT THE STATES INTO THEIR PROPER ZONES

ZONE 0

ZONE 1

ZONE 2

ZONE 3

DISTRICT OF COLUMBIA

ZONE 4

PUERTO RICO

VIRGIN ISLANDS

ZONE 5

ZONE 6

ZONE 7

ZONE 8

ZONE 9

ALPHA AND ZULU ALWAYS ENJOY MAKING NEW FRIENDS. IT IS EASY FOR THEM BECAUSE THEY DO NOT TRY TO HIDE WHEN THEY MEET NEW PEOPLE. IN FACT, THEY WOULD LIKE TO GET TO KNOW YOU BETTER, BUT THEY CAN'T SEEM TO COME OUT OF THE PAGE COMPLETELY. IF YOU CONNECT THE DOTS, YOU CAN HELP BRING THEM OUT. BE CAREFUL THOUGH, SOMETIMES YOU DON'T WANT TO CONNECT ALL THE DOTS, SO LOOK AHEAD, AND TRY TO FIGURE OUT WHETHER YOU WANT TO DRAW THE NEXT LINE. HAVE FUN.

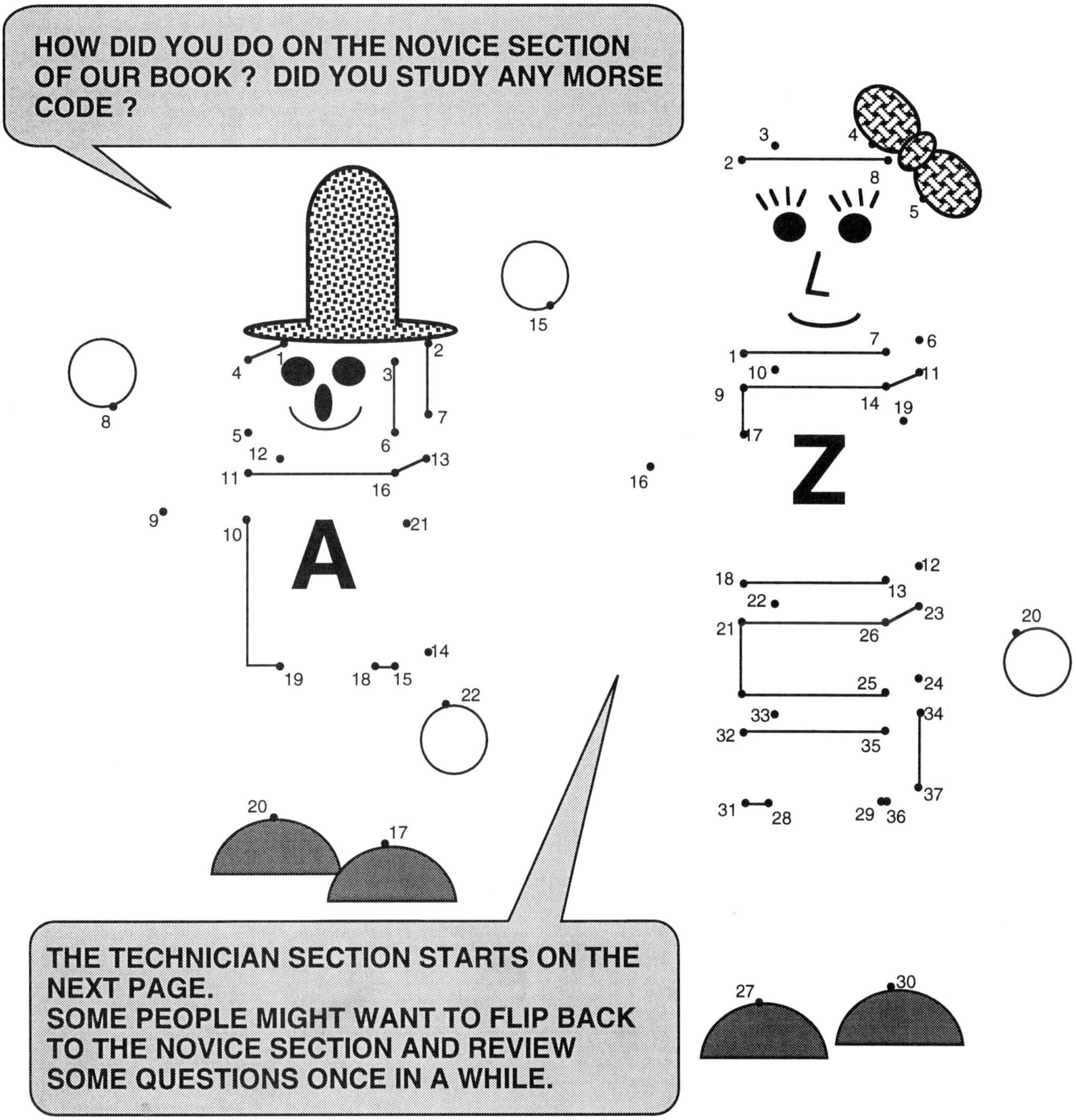

HOW DID YOU DO ON THE NOVICE SECTION OF OUR BOOK ? DID YOU STUDY ANY MORSE CODE ?

THE TECHNICIAN SECTION STARTS ON THE NEXT PAGE.
SOME PEOPLE MIGHT WANT TO FLIP BACK TO THE NOVICE SECTION AND REVIEW SOME QUESTIONS ONCE IN A WHILE.

SOMETIMES WE LEARN MORE WHEN WE MAKE A MISTAKE THAN WHEN WE DO SOMETHING RIGHT !

DID YOU HAVE TO USE AN ERASER ON THIS PAGE? IF YOU DID, YOU PROBABLY DID NOT READ THE INSTRUCTIONS FIRST! ALWAYS READ THE INSTRUCTIONS FIRST, NO MATTER WHAT YOU ARE WORKING WITH. THIS TIME YOUR MISTAKES CAN BE CORRECTED BY AN ERASER, NEXT TIME YOU MAY DAMAGE A EXPENSIVE PIECE OF EQUIPMENT!!!

Notes

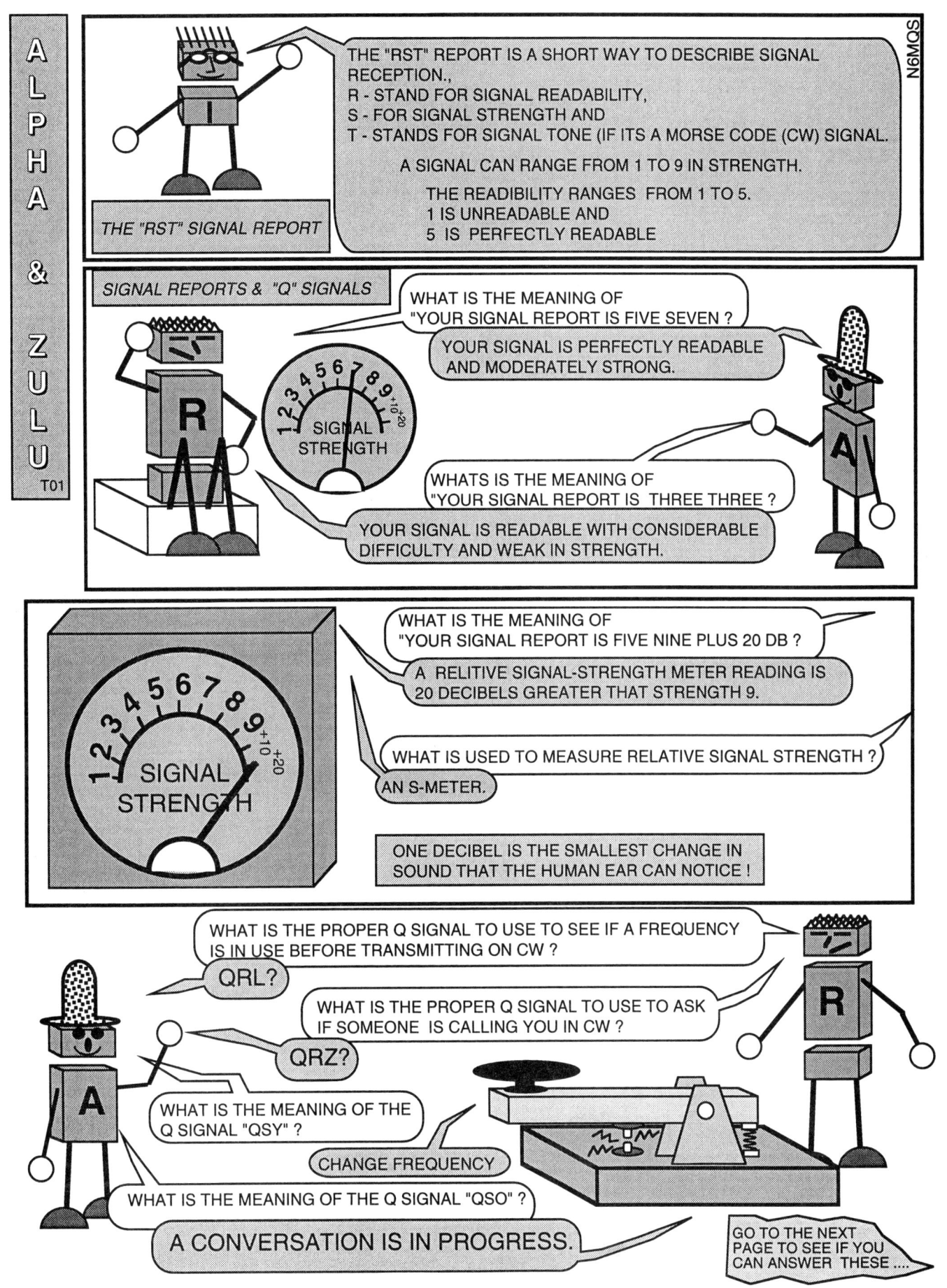

Riding the Airwaves with Alpha & Zulu

SEE IF YOU CAN ANSWER THESE TECHNICIAN QUESTIONS WITHOUT LOOKING BACK TO THE LAST 'TOON'. WRITE YOUR ANSWERS IN THE ANSWER BOX BELOW. GOOD LUCK. DA-DI-DAH

PRACTICE TEST PAGE T01

T2B05
What is the proper Q signal to use to see if a frequency is in use before transmitting on CW?
A. QRV?
B. QRU?
C. QRL?
D. QRZ?

T2B06
What is one meaning of the Q signal "QSY"?
A. Change frequency
B. Send more slowly
C. Send faster
D. Use more power

T2B07
What is one meaning of the Q signal "QSO"?
A. A contact is confirmed
B. A conversation is in progress
C. A contact is ending
D. A conversation is desired

T2B08
What is the proper Q signal to use to ask if someone is calling you on CW?
A. QSL?
B. QRZ?
C. QRL?
D. QRT?

T2B09
What is the meaning of: "Your signal report is five seven..."?
A. Your signal is perfectly readable and moderately strong
B. Your signal is perfectly readable, but weak
C. Your signal is readable with considerable difficulty
D. Your signal is perfectly readable with near pure tone

T2B10
What is the meaning of: "Your signal report is three three..."?
A. The contact is serial number thirty-three
B. The station is located at latitude 33 degrees
C. Your signal is readable with considerable difficulty and weak in strength
D. Your signal is unreadable, very weak in strength

T2B11
What is the meaning of: "Your signal report is five nine plus 20 dB..."?
A. Your signal strength has increased by a factor of 100
B. Repeat your transmission on a frequency 20 kHz higher
C. The bandwidth of your signal is 20 decibels above linearity
D. A relative signal-strength meter reading is 20 decibels greater than strength 9

T4D08
What is used to measure relative signal strength in a receiver?
A. An S meter
B. An RST meter
C. A signal deviation meter
D. An SSB meter

ANSWERS TO PREVIOUS NOVICE TEST

N4A04 — D
N4A05 — C
N4A06 — D

YOUR ANSWERS TO THIS TEST

T2B05
T2B06
T2B07
T2B08
T2B09
T2B10
T2B11
T4D08

Riding the Airwaves with Alpha & Zulu

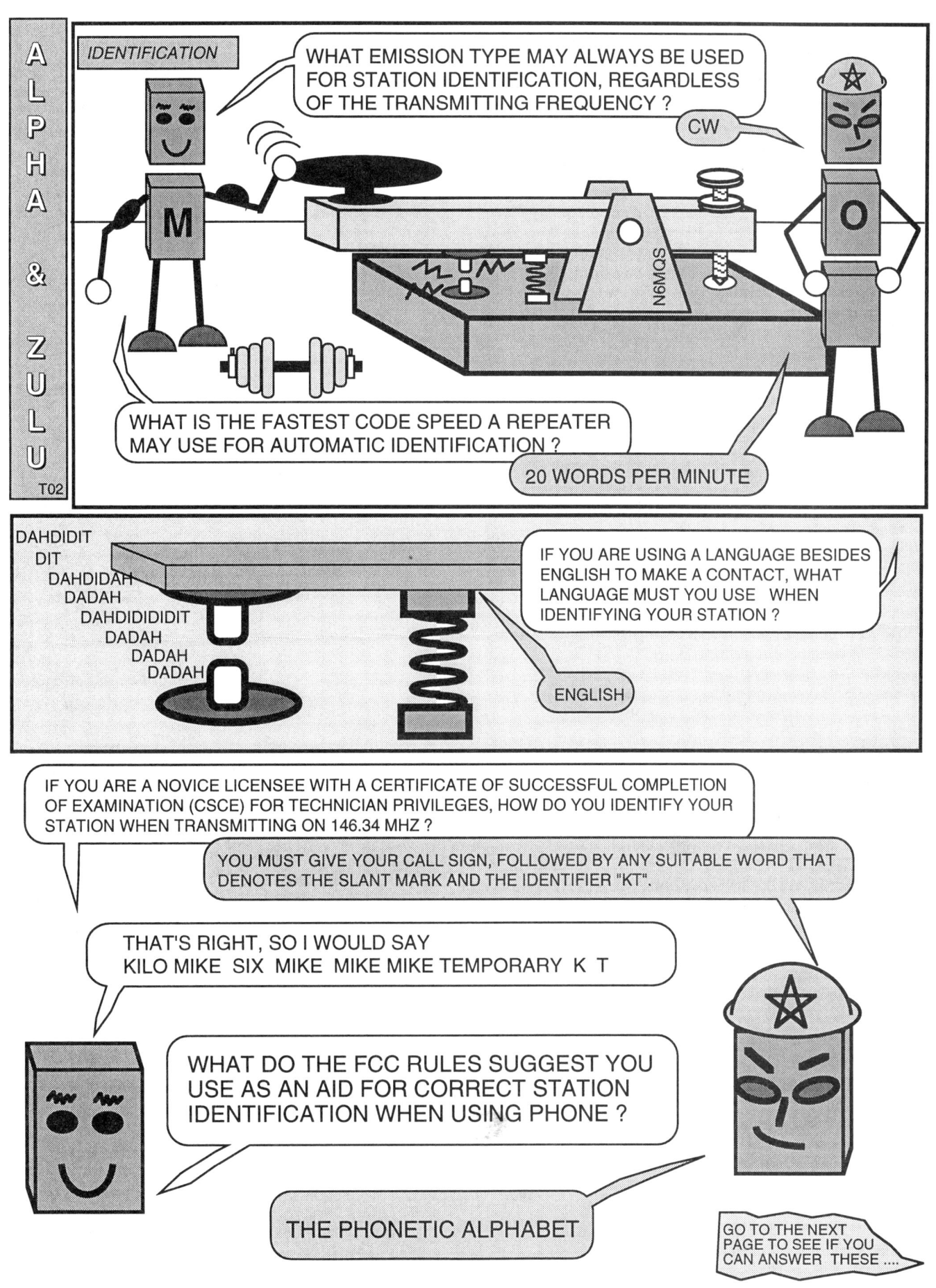

Riding the Airwaves with Alpha & Zulu

SEE IF YOU CAN ANSWER THESE TECHNICIAN QUESTIONS WITHOUT LOOKING BACK TO THE LAST 'TOON'. WRITE YOUR ANSWERS IN THE ANSWER BOX BELOW. GOOD LUCK. DA-DI-DAH

PRACTICE TEST PAGE T02

T1B07 [97.305a]
What emission type may always be used for station identification, regardless of the transmitting frequency?
A. CW
B. RTTY
C. MCW
D. Phone

T1C01 [97.119e1]
If you are a Novice licensee with a Certificate of Successful Completion of Examination (CSCE) for Technician privileges, how do you identify your station when transmitting on 146.34 MHz?
A. You must give your call sign, followed by any suitable word that denotes the slant mark and the identifier "KT"
B. You may not operate on 146.34 until your new license arrives
C. No special form of identification is needed
D. You must give your call sign and the location of the VE examination where you obtained the CSCE

T1D03 [97.119b2]
If you are using a language besides English to make a contact, what language must you use when identifying your station?
A. The language being used for the contact
B. The language being used for the contact, providing the US has a third-party communications agreement with that country
C. English
D. Any language of a country which is a member of the International Telecommunication Union

T1D02 [97.119b1]
What is the fastest code speed a repeater may use for automatic identification?
A. 13 words per minute
B. 20 words per minute
C. 25 words per minute
D. There is no limitation

T1D04 [97.119b2]
What do the FCC rules suggest you use as an aid for correct station identification when using phone?
A. A speech compressor
B. Q signals
C. A phonetic alphabet
D. Unique words of your choice

ANSWERS TO PREVIOUS TEST

T2B05	C
T2B06	A
T2B07	B
T2B08	B
T2B09	A
T2B10	C
T2B11	D
T4D08	A

YOUR ANSWERS TO THIS TEST

T1B07 _____
T1C01 _____
T1D03 _____
T1D02 _____
T1D04 _____

168   artsci inc

Riding the Airwaves with Alpha & Zulu

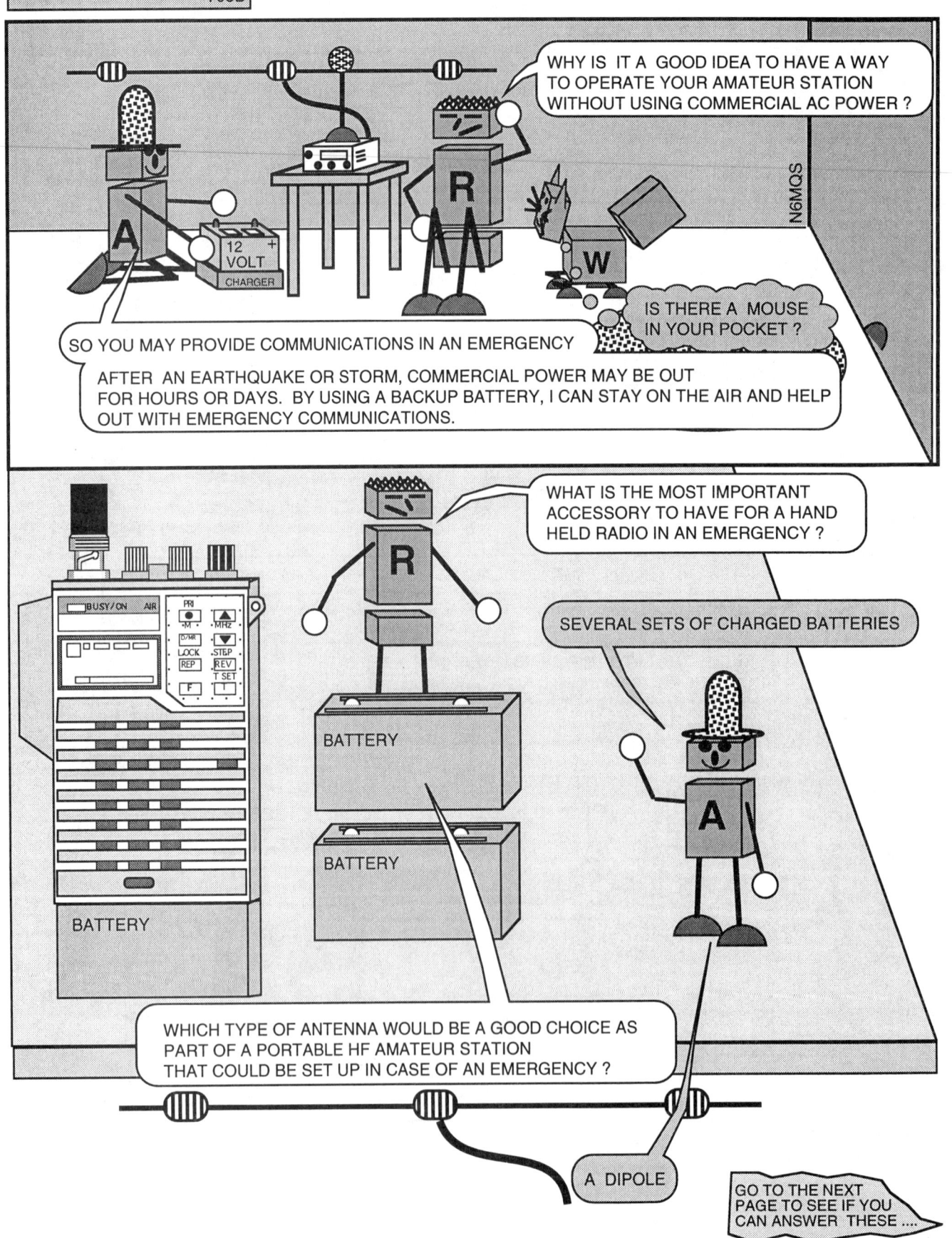

Riding the Airwaves with Alpha & Zulu

HERE'S A TWO PAGE TEST
GOOD LUCK. DA-DI-DAH

PRACTICE TEST PAGE T03A

T1E10 [97.401a]
If a disaster disrupts normal communication systems in an area where the amateur service is regulated by the FCC, what kinds of transmissions may stations make?
A. Those which are necessary to meet essential communication needs and facilitate relief actions
B. Those which allow a commercial business to continue to operate in the affected area
C. Those for which material compensation has been paid to the amateur operator for delivery into the affected area
D. Those which are to be used for program production or news gathering for broadcasting purposes

T1E11 [97.401c]
What information is included in an FCC declaration of a temporary state of communication emergency?
A. A list of organizations authorized to use radio communications in the affected area
B. A list of amateur frequency bands to be used in the affected area
C. Any special conditions and special rules to be observed during the emergency
D. An operating schedule for authorized amateur emergency stations

T2C01
What is the proper distress call to use when operating phone?
A. Say "MAYDAY" several times
B. Say "HELP" several times
C. Say "EMERGENCY" several times
D. Say "SOS" several times

T2C02
What is the proper distress call to use when operating CW?
A. MAYDAY
B. QRRR
C. QRZ
D. SOS

T2C03
What is the proper way to interrupt a repeater conversation to signal a distress call?
A. Say "BREAK" twice, then your call sign
B. Say "HELP" as many times as it takes to get someone to answer
C. Say "SOS," then your call sign
D. Say "EMERGENCY" three times

T2C08
What type of messages concerning a person's well-being are sent into or out of a disaster area?
A. Routine traffic
B. Tactical traffic
C. Formal message traffic
D. Health and Welfare traffic

ANSWERS TO PREVIOUS TEST

T1B07	A
T1C01	A
T1D03	C
T1D02	B
T1D04	C

artsci inc

PRACTICE TEST PAGE T03B

CONTINUES FROM THE LAST PAGE

T2C09
What are messages called which are sent into or out of a disaster area concerning the immediate safety of human life?
A. Tactical traffic
B. Emergency traffic
C. Formal message traffic
D. Health and Welfare traffic

T2C10
Why is it a good idea to have a way to operate your amateur station without using commercial AC power lines?
A. So you may use your station while mobile
B. So you may provide communications in an emergency
C. So you may operate in contests where AC power is not allowed
D. So you will comply with the FCC rules

T2C11
What is the most important accessory to have for a hand held radio in an emergency?
A. An extra antenna
B. A portable amplifier
C. Several sets of charged batteries
D. A microphone headset for hands-free operation

T2C12
Which type of antenna would be a good choice as part of a portable HF amateur station that could be set up in case of an emergency?
A. A three-element quad
B. A three-element Yagi
C. A dipole
D. A parabolic dish

YOUR ANSWERS TO THIS TEST

T1E10

T1E11

T2C01

T2C02

T2C03

T2C08

T2C09

T2C10

T2C11

T2C12

Riding the Airwaves with Alpha & Zulu

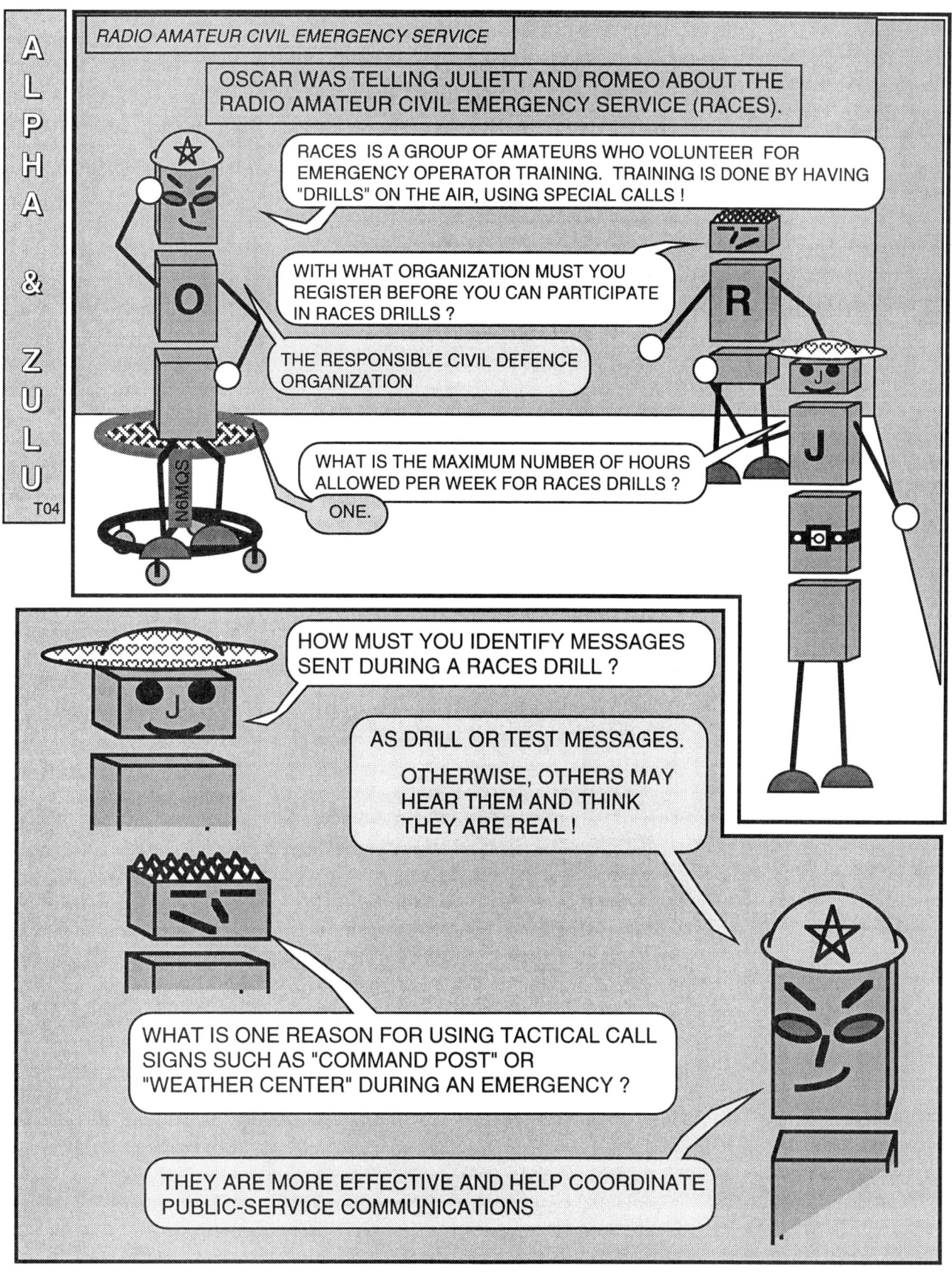

SEE IF YOU CAN ANSWER THESE TECHNICIAN QUESTIONS WITHOUT LOOKING BACK TO THE LAST 'TOON'. WRITE YOUR ANSWERS IN THE ANSWER BOX BELOW. GOOD LUCK. DA-DI-DAH

PRACTICE TEST PAGE T04

T2C04
With what organization must you register before you can participate in RACES drills?
- A. A local Amateur Radio club
- B. A local racing organization
- C. The responsible civil defense organization
- D. The Federal Communications Commission

T2C05
What is the maximum number of hours allowed per week for RACES drills?
- A. One
- B. Six, but not more than one hour per day
- C. Eight
- D. As many hours as you want

T2C06
How must you identify messages sent during a RACES drill?
- A. As emergency messages
- B. As amateur traffic
- C. As official government messages
- D. As drill or test messages

T2C07
What is one reason for using tactical call signs such as "command post" or "weather center" during an emergency?
- A. They keep the general public informed about what is going on
- B. They are more efficient and help coordinate public-service communications
- C. They are required by the FCC
- D. They increase goodwill between amateurs

ANSWERS TO PREVIOUS TEST

T1E10	A
T1E11	C
T2C01	A
T2C02	D
T2C03	A
T2C08	D
T2C09	B
T2C10	B
T2C11	C
T2C12	C

YOUR ANSWERS TO THIS TEST

T2C04	
T2C05	
T2C06	
T2C07	

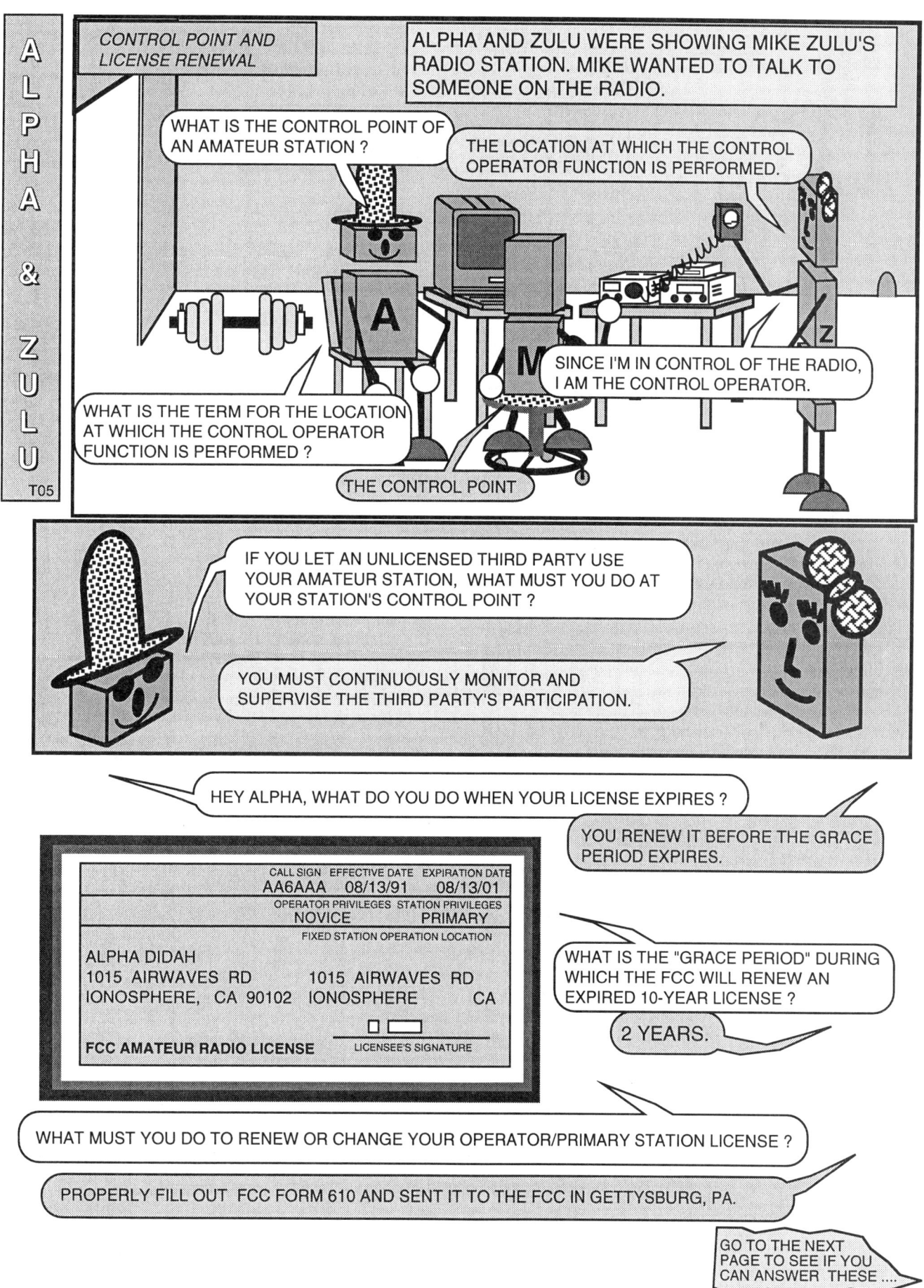

SEE IF YOU CAN ANSWER THESE TECHNICIAN QUESTIONS WITHOUT LOOKING BACK TO THE LAST 'TOON'. WRITE YOUR ANSWERS IN THE ANSWER BOX BELOW. GOOD LUCK. DA-DI-DAH

PRACTICE TEST PAGE T05

T1A01 [97.3a12]
What is the control point of an amateur station?
A. The on/off switch of the transmitter
B. The input/output port of a packet controller
C. The variable frequency oscillator of a transmitter
D. The location at which the control operator function is performed

T1A02 [97.3a12]
What is the term for the location at which the control operator function is performed?
A. The operating desk
B. The control point
C. The station location
D. The manual control location

T1A03 [97.19a/b]
What must you do to renew or change your operator/primary station license?
A. Properly fill out FCC Form 610 and send it to the FCC in Gettysburg, PA
B. Properly fill out FCC Form 610 and send it to the nearest FCC field office
C. Properly fill out FCC form 610 and send it to the FCC in Washington, DC
D. An amateur license never needs changing or renewing

T1A04 [97.19c]
What is the "grace period" during which the FCC will renew an expired 10-year license?
A. 2 years
B. 5 years
C. 10 years
D. There is no grace period

T1E09 [97.115b1]
If you let an unlicensed third party use your amateur station, what must you do at your station's control point?
A. You must continuously monitor and supervise the third party's participation
B. You must monitor and supervise the communication only if contacts are made in countries which have no third-party communications agreement with the US
C. You must monitor and supervise the communication only if contacts are made on frequencies below 30 MHz
D. You must key the transmitter and make the station identification

ANSWERS TO PREVIOUS TEST
T2C04
C
T2C05
A
T2C06
D
T2C07
B

YOUR ANSWERS TO THIS TEST
T1A01
T1A02
T1A03
T1A04
T1E09

PRACTICE TEST PAGE T06

SEE IF YOU CAN ANSWER THESE TECHNICIAN QUESTIONS WITHOUT LOOKING BACK TO THE LAST 'TOON'. WRITE YOUR ANSWERS IN THE ANSWER BOX BELOW. GOOD LUCK. DA-DI-DAH

T1E01 [97.3a10]
What is meant by the term broadcasting?
  A. Transmissions intended for reception by the general public, either direct or relayed
  B. Retransmission by automatic means of programs or signals from non-amateur stations
  C. One-way radio communications, regardless of purpose or content
  D. One-way or two-way radio communications between two or more stations

T1E02 [97.3a10]
Which of the following one-way communications may not be transmitted in the amateur service?
  A. Telecommands to model craft
  B. Broadcasts intended for the general public
  C. Brief transmissions to make adjustments to the station
  D. Morse code practice

T1E03 [97.113b]
What kind of payment is allowed for third-party messages sent by an amateur station?
  A. Any amount agreed upon in advance
  B. Donation of equipment repairs
  C. Donation of amateur equipment
  D. No payment of any kind is allowed

T1E07 [97.113e]
If you wanted to use your amateur station to retransmit communications between a space shuttle and its associated Earth stations, what agency must first give its approval?
  A. The FCC in Washington, DC
  B. The office of your local FCC Engineer In Charge (EIC)
  C. The National Aeronautics and Space Administration
  D. The Department of Defense

T1E08 [97.115a2]
When are third-party messages allowed to be sent to a foreign country?
  A. When sent by agreement of both control operators
  B. When the third party speaks to a relative
  C. They are not allowed under any circumstances
  D. When the US has a third-party agreement with the foreign country or the third party is qualified to be a control operator

---

ANSWERS TO PREVIOUS TEST

T1A01	D
T1A02	B
T1A03	A
T1A04	A
T1E09	A

YOUR ANSWERS TO THIS TEST

T1E01 ____
T1E02 ____
T1E03 ____
T1E07 ____
T1E08 ____

Riding the Airwaves with Alpha & Zulu

SEE IF YOU CAN ANSWER THESE TECHNICIAN QUESTIONS WITHOUT LOOKING BACK TO THE LAST 'TOON'. WRITE YOUR ANSWERS IN THE ANSWER BOX BELOW. GOOD LUCK. DA-DI-DAH

PRACTICE TEST PAGE T07

T2A15
Why should local amateur communications use VHF and UHF frequencies instead of HF frequencies?
- A. To minimize interference on HF bands capable of long distance communication
- B. Because greater output power is permitted on VHF and UHF
- C. Because HF transmissions are not propagated locally
- D. Because signals are louder on VHF and UHF frequencies

T1E05 [97.113d]
When may you send indecent words from your amateur station?
- A. Only when they do not cause interference to other communications
- B. Only when they are not retransmitted through a repeater
- C. Any time, but there is an unwritten rule among amateurs that they should not be used on the air
- D. Never; indecent words are prohibited in amateur transmissions

T2A17
How can on-the-air interference be minimized during a lengthy transmitter testing or loading up procedure?
- A. Choose an unoccupied frequency
- B. Use a dummy load
- C. Use a non-resonant antenna
- D. Use a resonant antenna that requires no loading-up procedure

T1E04 [97.113d]
When may you send obscene words from your amateur station?
- A. Only when they do not cause interference to other communications
- B. Never; obscene words are prohibited in amateur transmissions
- C. Only when they are not retransmitted through a repeater
- D. Any time, but there is an unwritten rule among amateurs that they should not be used on the air

T1E06 [97.113d]
When may you send profane words from your amateur station?
- A. Only when they do not cause interference to other communications
- B. Only when they are not retransmitted through a repeater
- C. Never; profane words are prohibited in amateur transmissions
- D. Any time, but there is an unwritten rule among amateurs that they should not be used on the air

ANSWERS TO PREVIOUS TEST

T1E01	A
T1E02	B
T1E03	D
T1E07	C
T1E08	D

YOUR ANSWERS TO THIS TEST

T2A15	
T2A17	
T1E05	
T1E04	
T1E06	

artsci inc

Riding the Airwaves with Alpha & Zulu

SEE IF YOU CAN ANSWER THESE TECHNICIAN QUESTIONS WITHOUT LOOKING BACK TO THE LAST 'TOON'. WRITE YOUR ANSWERS IN THE ANSWER BOX BELOW. GOOD LUCK. DA-DI-DAH

PRACTICE TEST PAGE T08

T4D01
What device should be connected to a transmitter's output when you are making transmitter adjustments?
 A. A multimeter
 B. A reflectometer
 C. A receiver
 D. A dummy antenna

T4D02
What is a dummy antenna?
 A. An nondirectional transmitting antenna
 B. A nonradiating load for a transmitter
 C. An antenna used as a reference for gain measurements
 D. A flexible antenna usually used on hand held transceivers

T4D03
What is the main component of a dummy antenna?
 A. A wire-wound resistor
 B. An iron-core coil
 C. A noninductive resistor
 D. An air-core coil

T4D04
What device is used in place of an antenna during transmitter tests so that no signal is radiated?
 A. An antenna matcher
 B. A dummy antenna
 C. A low-pass filter
 D. A decoupling resistor

T4D05
Why would you use a dummy antenna?
 A. For off-the-air transmitter testing
 B. To reduce output power
 C. To give comparative signal reports
 D. To allow antenna tuning without causing interference

T4D06
What minimum rating should a dummy antenna have for use with a 100-watt single-sideband phone transmitter?
 A. 100 watts continuous
 B. 141 watts continuous
 C. 175 watts continuous
 D. 200 watts continuous

T4D07
Why might a dummy antenna get warm when in use?
 A. Because it stores electric current
 B. Because it stores radio waves
 C. Because it absorbs static electricity
 D. Because it changes RF energy into heat

ANSWERS TO PREVIOUS TEST

T2A15	A
T2A17	B
T1E05	D
T1E04	B
T1E06	C

YOUR ANSWERS TO THIS TEST

T4D01 ____
T4D02 ____
T4D03 ____
T4D04 ____
T4D05 ____
T4D06 ____
T4D07 ____

artsci inc

Riding the Airwaves with Alpha & Zulu

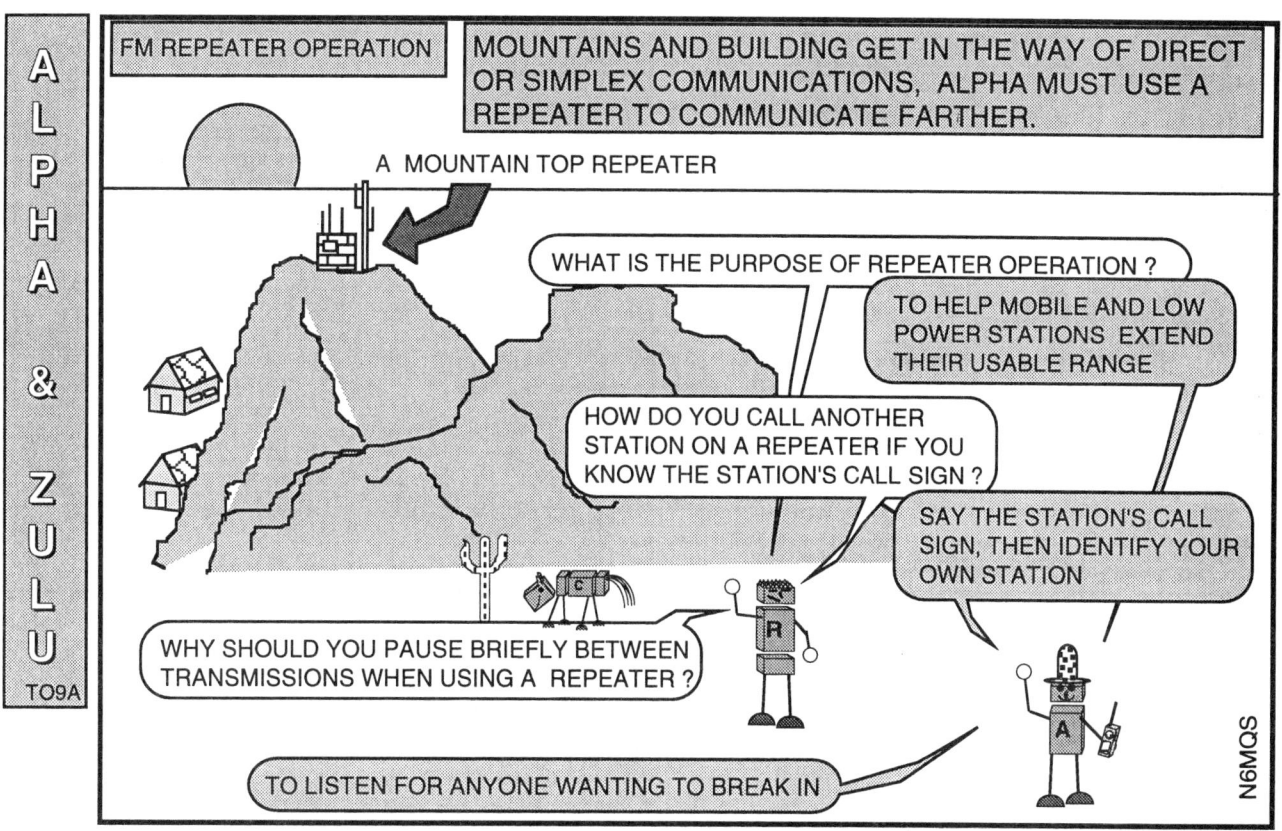

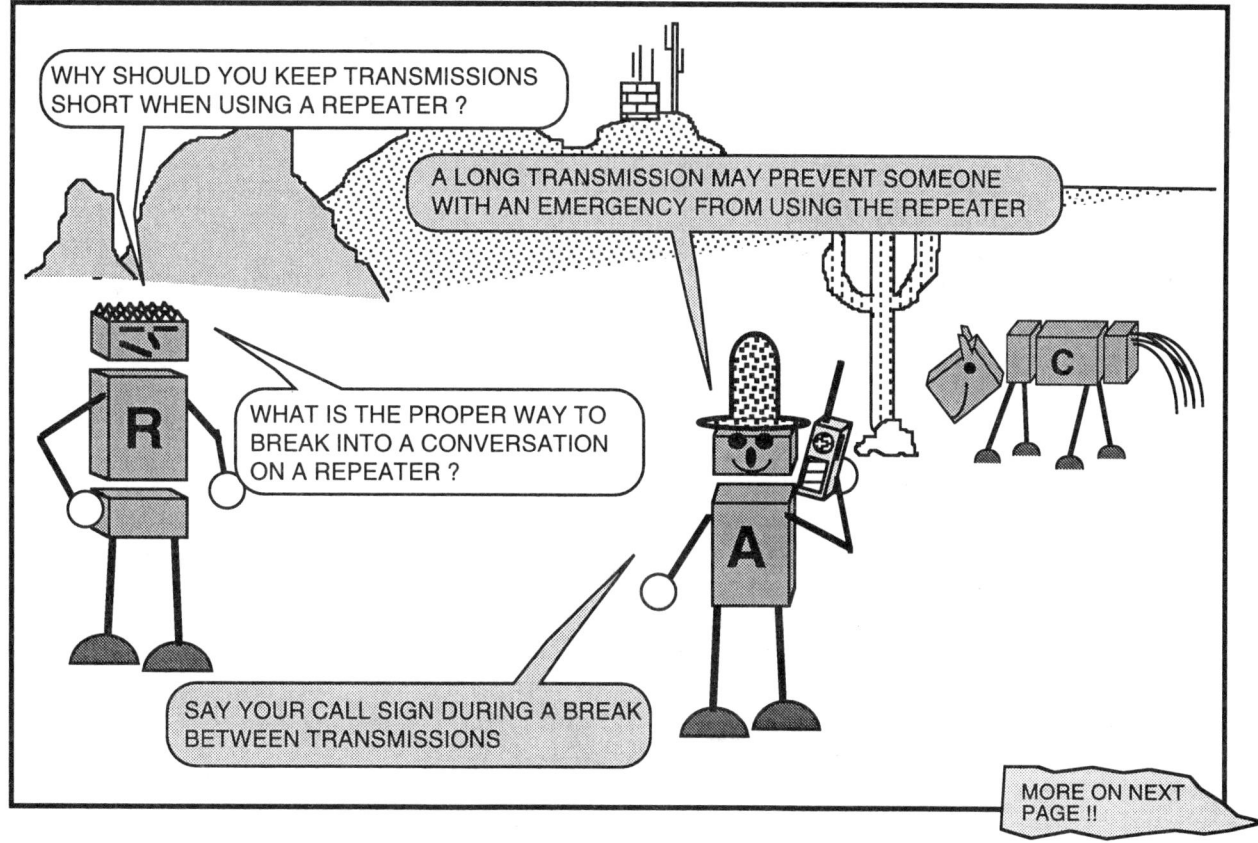

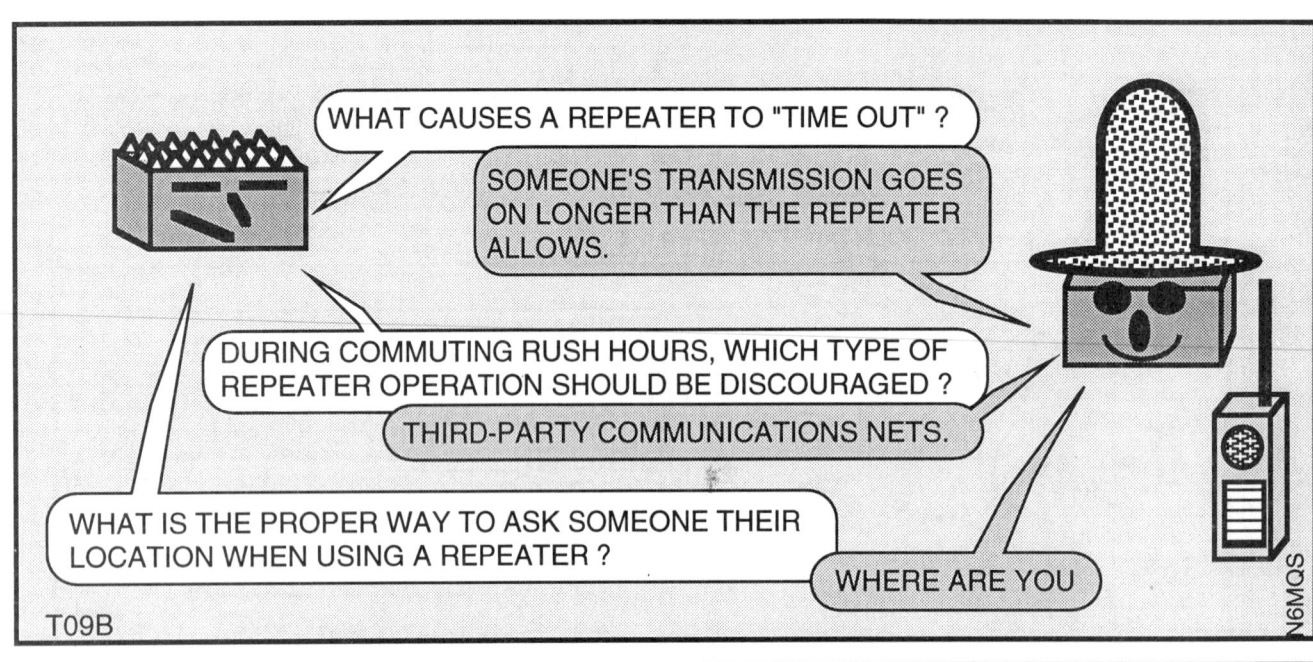

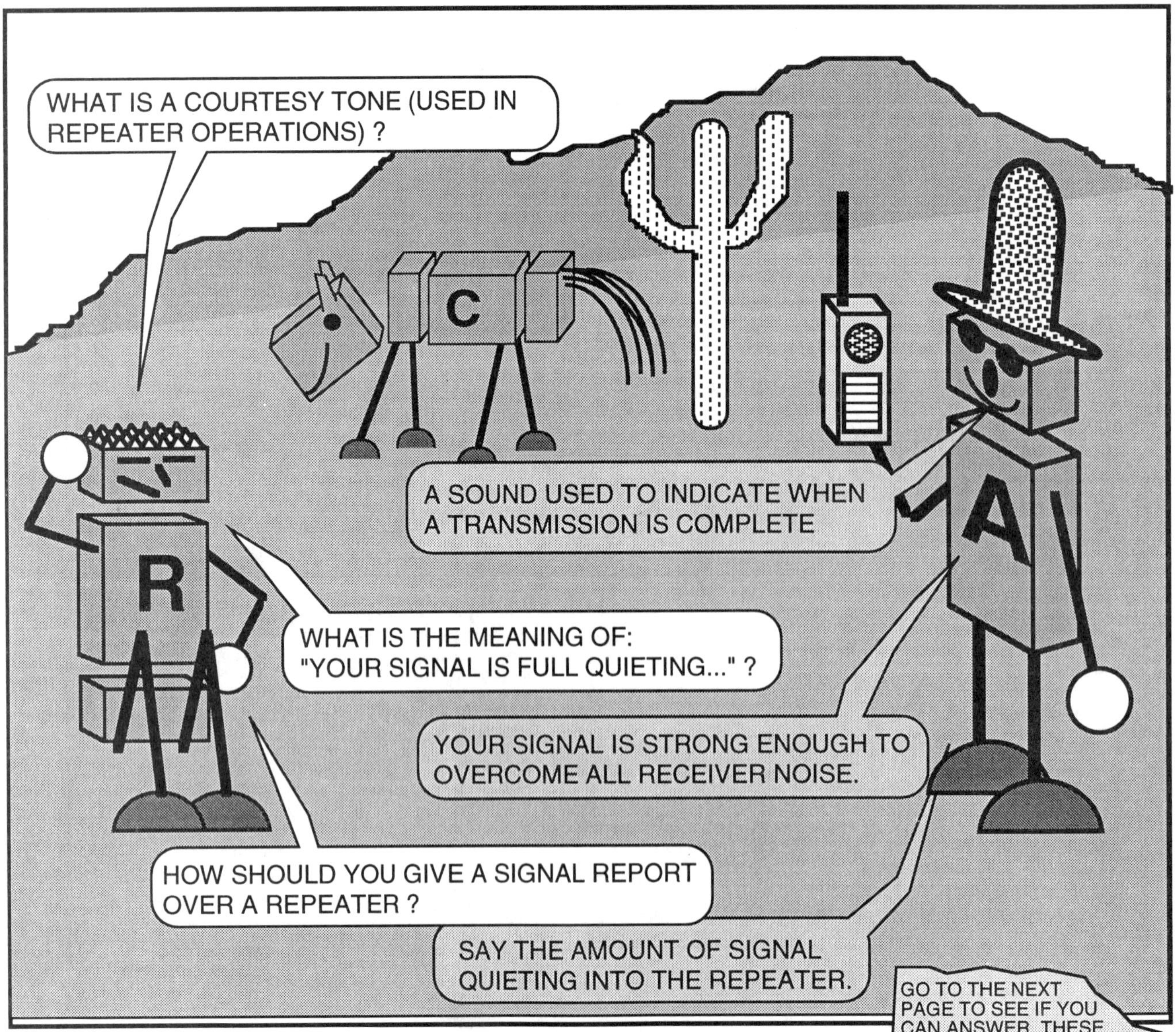

Riding the Airwaves with Alpha & Zulu

HERE'S ANOTHER TWO PAGE TEST
WRITE YOUR ANSWERS IN THE ANSWER BOX ON THE NEXT PAGE.
GOOD LUCK.

PRACTICE TEST PAGE T09A

T2A05
What is the purpose of repeater operation?
- A. To cut your power bill by using someone else's higher power system
- B. To help mobile and low-power stations extend their usable range
- C. To transmit signals for observing propagation and reception
- D. To make calls to stores more than 50 miles away

T2A03
Why should you keep transmissions short when using a repeater?
- A. A long transmission may prevent someone with an emergency from using the repeater
- B. To see if the receiving station operator is still awake
- C. To give any listening non-hams a chance to respond
- D. To keep long distance charges down

T2A01
How do you call another station on a repeater if you know the station's call sign?
- A. Say "break, break 79," then say the station's call sign
- B. Say the station's call sign, then identify your own station
- C. Say "CQ" three times, then say the station's call sign
- D. Wait for the station to call "CQ," then answer it

T2A02
Why should you pause briefly between transmissions when using a repeater?
- A. To check the SWR of the repeater
- B. To reach for pencil and paper for third-party communications
- C. To listen for anyone wanting to break in
- D. To dial up the repeater's autopatch

T2A04
What is the proper way to break into a conversation on a repeater?
- A. Wait for the end of a transmission and start calling the desired party
- B. Shout, "break, break!" to show that you're eager to join the conversation
- C. Turn on an amplifier and override whoever is talking
- D. Say your call sign during a break between transmissions

T2A06
What causes a repeater to "time out"?
- A. The repeater's battery supply runs out
- B. Someone's transmission goes on longer than the repeater allows
- C. The repeater gets too hot and stops transmitting until its circuitry cools off
- D. Something is wrong with the repeater

ANSWERS TO PREVIOUS TEST	
T4D01	D
T4D02	B
T4D03	C
T4D04	B
T4D05	A
T4D06	A
T4D07	D

PRACTICE TEST PAGE T09B

T2A07
During commuting rush hours, which type of repeater operation should be discouraged?
A. Mobile stations
B. Low-power stations
C. Highway traffic information nets
D. Third-party communications nets

T2A18
What is the proper way to ask someone their location when using a repeater?
A. What is your QTH
B. What is your 20
C. Where are you
D. Locations are not normally told by radio

T2A08
What is a courtesy tone (used in repeater operations)?
A. A sound used to identify the repeater
B. A sound used to indicate when a transmission is complete
C. A sound used to indicate that a message is waiting for someone
D. A sound used to activate a receiver in case of severe weather

T2A09
What is the meaning of: "Your signal is full quieting..."?
A. Your signal is strong enough to overcome all receiver noise
B. Your signal has no spurious sounds
C. Your signal is not strong enough to be received
D. Your signal is being received, but no audio is being heard

T2A10
How should you give a signal report over a repeater?
A. Say what your receiver's S-meter reads
B. Always say: "Your signal report is five five..."
C. Say the amount of signal quieting into the repeater
D. Try to imitate the sound quality you are receiving

YOUR ANSWERS TO THIS TEST

T2A05

T2A01

T2A03

T2A02

T2A04

T2A06

T2A07

T2A08

T2A09

T2A10

T2A18

Riding the Airwaves with Alpha & Zulu

ALPHA & ZULU — T10

REPEATERS, ON HIGH MOUNTAIN TOPS, HAVE TO BE 'COORDINATED" TO REDUCE INTERFERENCE. SOME REPEATERS ALLOW ANYONE TO OPERATE ON THEM, OTHERS ARE "CLOSED" ELECTRONICALLY.

FM REPEATER COORDINATION, OPEN & CLOSED REPEATERS

WHAT IS A REPEATER FREQUENCY COORDINATOR?

A PERSON OR GROUP THAT RECOMMENDS FREQUENCIES FOR REPEATER USAGE.

WHAT IS A REPEATER CALLED WHICH IS AVAILABLE FOR ANYONE TO USE?

AN OPEN REPEATER.

HOW MIGHT YOU JOIN A CLOSED REPEATER SYSTEM?

CONTACT THE CONTROL OPERATOR AND ASK TO JOIN

IF A REPEATER IS CAUSING HARMFUL INTERFERENCE TO ANOTHER REPEATER AND A FREQUENCY COORDINATOR HAS NOT RECOMMENDED EITHER STATION, WHO IS PRIMARILY RESPONSIBLE FOR RESOLVING THE INTERFERENCE?

BOTH REPEATER LICENSEES

IF A REPEATER IS CAUSING HARMFUL INTERFERENCE TO ANOTHER REPEATER AND A FREQUENCY COORDINATOR HAS RECOMMENDED THE OPERATION OF ONE STATION ONLY, WHO IS RESPONSIBLE FOR RESOLVING THE INTERFERENCE?

THE LICENSEE OF THE UNRECOMMENDED REPEATER

IF A REPEATER IS CAUSING HARMFUL INTERFERENCE TO ANOTHER AMATEUR REPEATER AND A FREQUENCY COORDINATOR HAS RECOMMENDED THE OPERATION OF BOTH STATIONS, WHO IS RESPONSIBLE FOR RESOLVING THE INTERFERENCE?

BOTH REPEATERS LICENSEES

GO TO THE NEXT PAGE TO SEE IF YOU CAN ANSWER THESE ....

Riding the Airwaves with Alpha & Zulu

SEE IF YOU CAN ANSWER THESE TECHNICIAN QUESTIONS WITHOUT LOOKING BACK TO THE LAST 'TOON'. WRITE YOUR ANSWERS IN THE ANSWER BOX BELOW. GOOD LUCK. DA-DI-DAH

PRACTICE TEST PAGE T10

T2B04
What is a repeater frequency coordinator?
- A. Someone who organizes the assembly of a repeater station
- B. Someone who provides advice on what kind of repeater to buy
- C. The person whose call sign is used for a repeater's identification
- D. A person or group that recommends frequencies for repeater usage

T1D06 [97.205c]
If a repeater is causing harmful interference to another repeater and a frequency coordinator has recommended the operation of one station only, who is responsible for resolving the interference?
- A. The licensee of the unrecommended repeater
- B. Both repeater licensees
- C. The licensee of the recommended repeater
- D. The frequency coordinator

T1D07 [97.205c]
If a repeater is causing harmful interference to another amateur repeater and a frequency coordinator has recommended the operation of both stations, who is responsible for resolving the interference?
- A. The licensee of the repeater which has been recommended for the longest period of time
- B. The licensee of the repeater which has been recommended the most recently
- C. The frequency coordinator
- D. Both repeater licensees

T1D08 [97.205c]
If a repeater is causing harmful interference to another repeater and a frequency coordinator has NOT recommended either station, who is primarily responsible for resolving the interference?
- A. Both repeater licensees
- B. The licensee of the repeater which has been in operation for the longest period of time
- C. The licensee of the repeater which has been in operation for the shortest period of time
- D. The frequency coordinator

T2A11
What is a repeater called which is available for anyone to use?
- A. An open repeater
- B. A closed repeater
- C. An autopatch repeater
- D. A private repeater

T2A16
How might you join a closed repeater system?
- A. Contact the control operator and ask to join
- B. Use the repeater until told not to
- C. Use simplex on the repeater input until told not to
- D. Write the FCC and report the closed condition

ANSWERS TO PREVIOUS TEST	
T2A05	B
T2A01	B
T2A03	A
T2A02	C
T2A04	D
T2A06	B
T2A07	D
T2A08	B
T2A09	A
T2A10	C
T2A18	C

YOUR ANSWERS TO THIS TEST
T2B04
T1D06
T1D07
T1D08
T2A11
T2A16

artsci inc

Riding the Airwaves with Alpha & Zulu

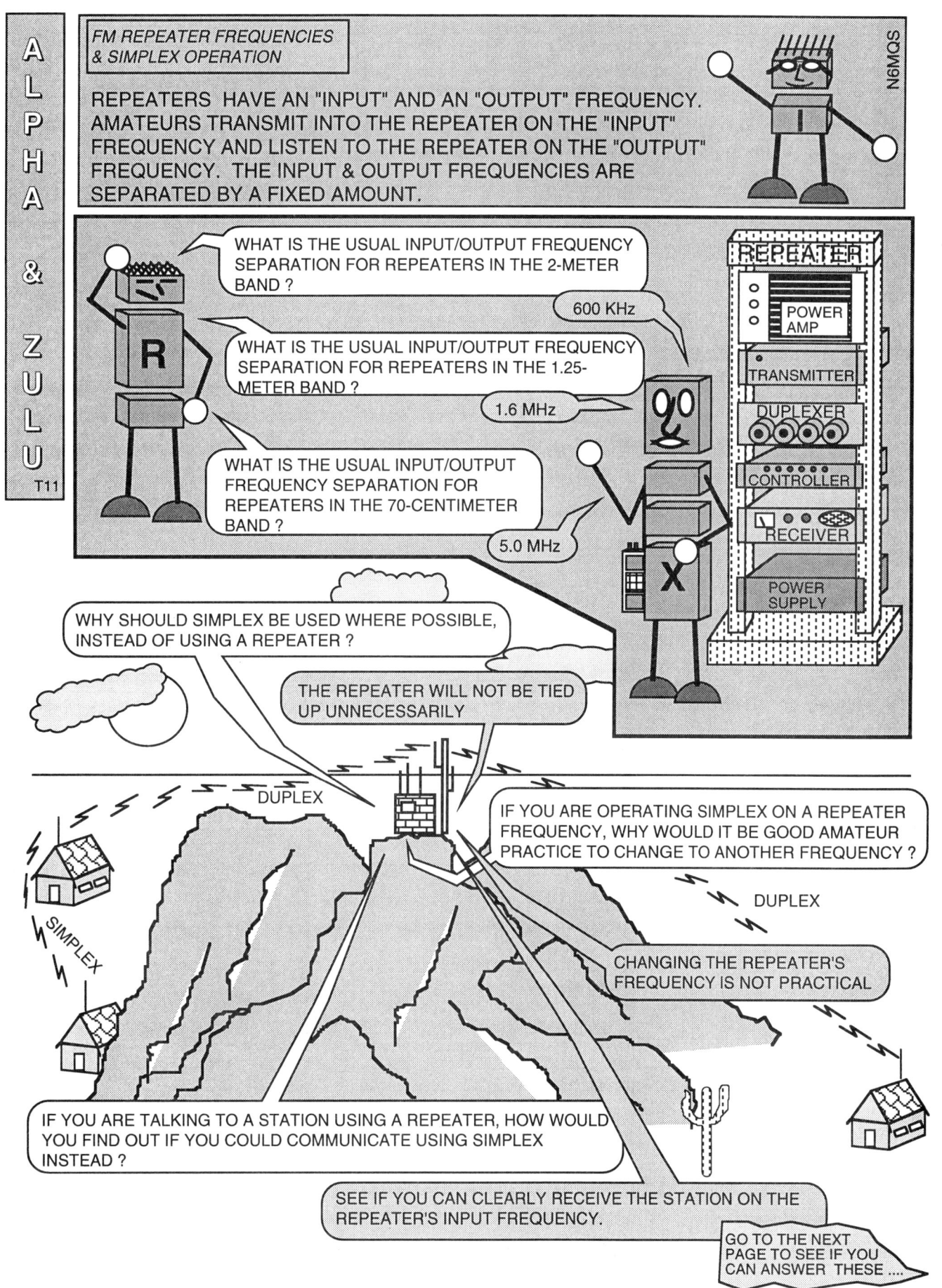

Riding the Airwaves with Alpha & Zulu

SEE IF YOU CAN ANSWER THESE TECHNICIAN QUESTIONS WITHOUT LOOKING BACK TO THE LAST 'TOON'. WRITE YOUR ANSWERS IN THE ANSWER BOX BELOW. GOOD LUCK. DA-DI-DAH

PRACTICE TEST PAGE T11

T2A12
What is the usual input/output frequency separation for repeaters in the 2-meter band?
- A. 600 kHz
- B. 1.0 MHz
- C. 1.6 MHz
- D. 5.0 MHz

T2A13
What is the usual input/output frequency separation for repeaters in the 1.25-meter band?
- A. 600 kHz
- B. 1.0 MHz
- C. 1.6 MHz
- D. 5.0 MHz

T2A14
What is the usual input/output frequency separation for repeaters in the 70-centimeter band?
- A. 600 kHz
- B. 1.0 MHz
- C. 1.6 MHz
- D. 5.0 MHz

T2B01
Why should simplex be used where possible, instead of using a repeater?
- A. Signal range will be increased
- B. Long distance toll charges will be avoided
- C. The repeater will not be tied up unnecessarily
- D. Your antenna's effectiveness will be better tested

T2B02
If you are talking to a station using a repeater, how would you find out if you could communicate using simplex instead?
- A. See if you can clearly receive the station on the repeater's input frequency
- B. See if you can clearly receive the station on a lower frequency band
- C. See if you can clearly receive a more distant repeater
- D. See if a third station can clearly receive both of you

T2B03
If you are operating simplex on a repeater frequency, why would it be good amateur practice to change to another frequency?
- A. The repeater's output power may ruin your station's receiver
- B. There are more repeater operators than simplex operators
- C. Changing the repeater's frequency is not practical
- D. Changing the repeater's frequency requires the authorization of the FCC

ANSWERS TO PREVIOUS TEST

T2B04	D
T1D06	A
T1D07	D
T1D08	A
T2A11	A
T2A16	A

YOUR ANSWERS TO THIS TEST

- T2A12 ____
- T2A13 ____
- T2A14 ____
- T2B01 ____
- T2B02 ____
- T2B03 ____

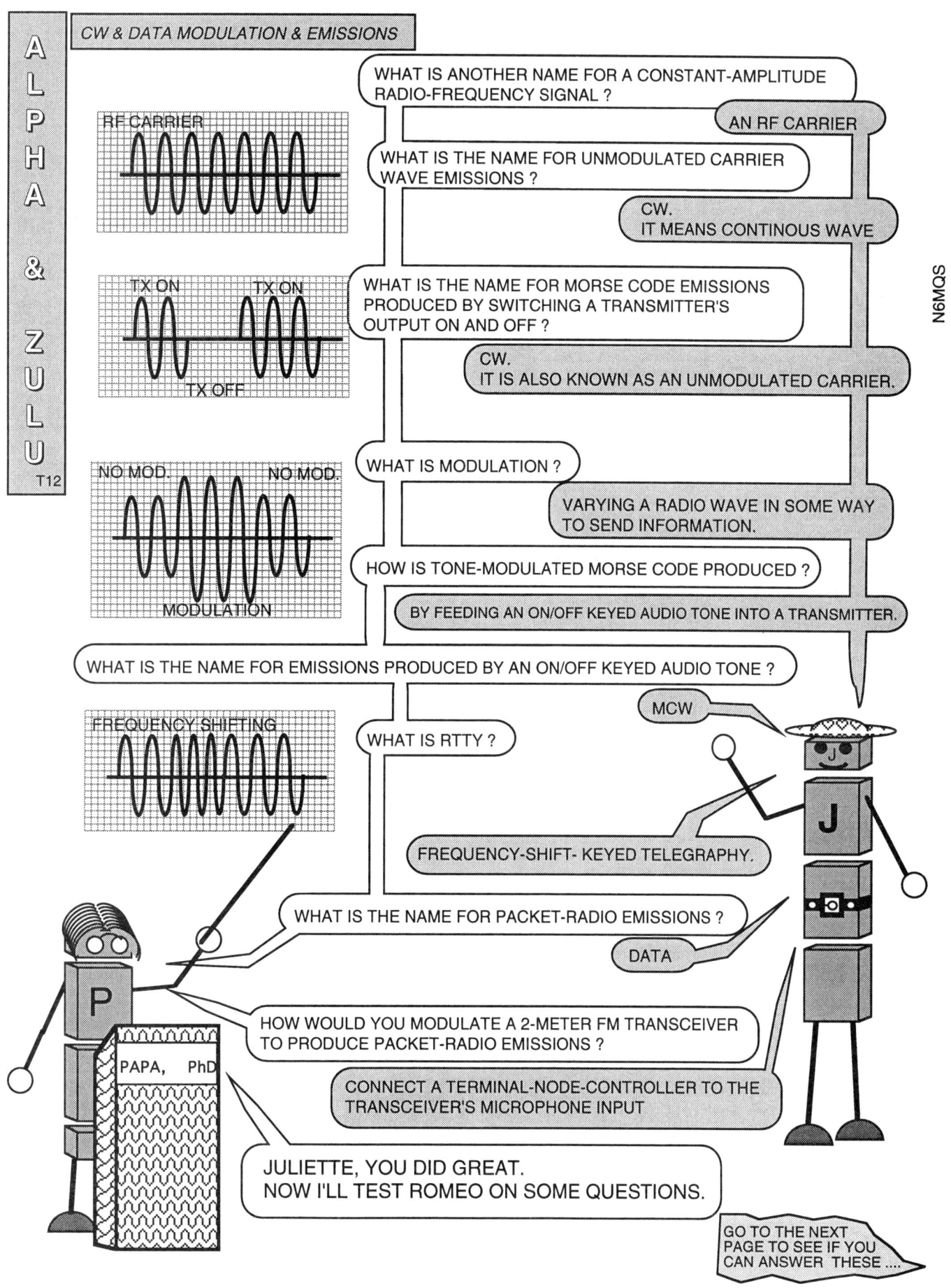

PRACTICE TEST PAGE T12

T8A01
What is the name for unmodulated carrier wave emissions?
- A. Phone
- B. Test
- C. CW
- D. RTTY

T8A02
What is the name for Morse code emissions produced by switching a transmitter's output on and off?
- A. Phone
- B. Test
- C. CW
- D. RTTY

T8A03
What is RTTY?
- A. Amplitude-keyed telegraphy
- B. Frequency-shift-keyed telegraphy
- C. Frequency-modulated telephony
- D. Phase-modulated telephony

T8A04
What is the name for packet-radio emissions?
- A. CW
- B. Data
- C. Phone
- D. RTTY

T8A05
How is tone-modulated Morse code produced?
- A. By feeding a microphone's audio signal into an FM transmitter
- B. By feeding an on/off keyed audio tone into a CW transmitter
- C. By on/off keying of a carrier
- D. By feeding an on/off keyed audio tone into a transmitter

T8A11
What is the name for emissions produced by an on/off keyed audio tone?
- A. RTTY
- B. MCW
- C. CW
- D. Phone

T8B01
What is another name for a constant-amplitude radio-frequency signal?
- A. An RF carrier
- B. An AF carrier
- C. A sideband carrier
- D. A subcarrier

T8B02
What is modulation?
- A. Varying a radio wave in some way to send information
- B. Receiving audio information from a signal
- C. Increasing the power of a transmitter
- D. Suppressing the carrier in a single-sideband transmitter

T8B04
How would you modulate a 2-meter FM transceiver to produce packet-radio emissions?
- A. Connect a terminal-node-controller to interrupt the transceiver's carrier wave
- B. Connect a terminal-node-controller to the transceiver's microphone input
- C. Connect a keyboard to the transceiver's microphone input
- D. Connect a DTMF key pad to the transceiver's microphone input

ANSWERS TO PREVIOUS TEST

T2A12	A
T2A13	C
T2A14	D
T2B01	C
T2B02	A
T2B03	C

YOUR ANSWERS TO THIS TEST

- T8A01
- T8A02
- T8A03
- T8A04
- T8A05
- T8A11
- T8B01
- T8B02
- T8B04

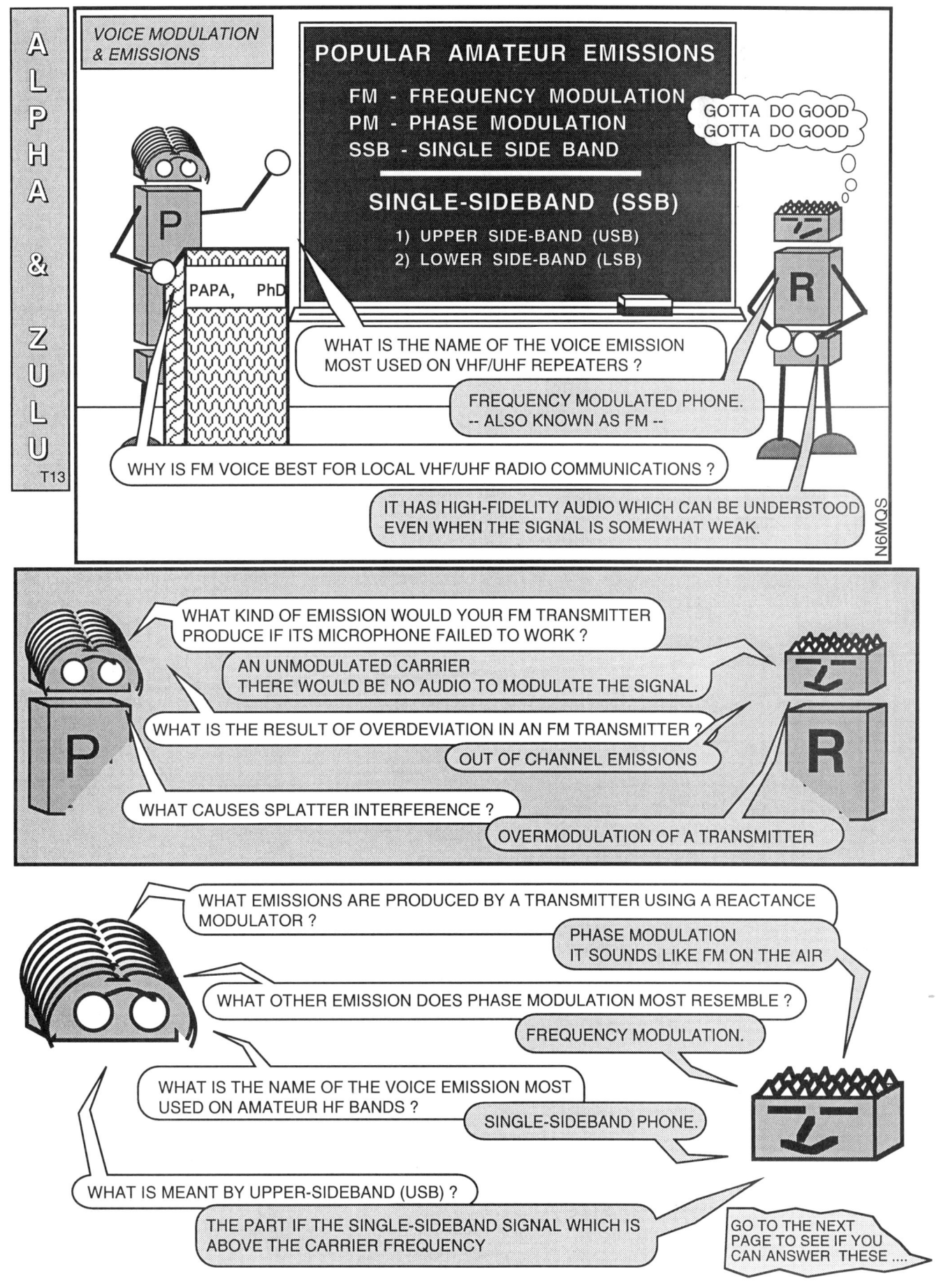

PRACTICE TEST PAGE T13

T8A06
What is the name of the voice emission most used on VHF/UHF repeaters?
- A. Single-sideband phone
- B. Pulse-modulated phone
- C. Slow-scan phone
- D. Frequency modulated phone

T8A07
What is the name of the voice emission most used on amateur HF bands?
- A. Single-sideband phone
- B. Pulse-modulated phone
- C. Slow-scan phone
- D. Frequency modulated phone

T8A08
What is meant by the upper-sideband (USB)?
- A. The part of a single-sideband signal which is above the carrier frequency
- B. The part of a single-sideband signal which is below the carrier frequency
- C. Any frequency above 10 MHz
- D. The carrier frequency of a single-sideband signal

T8A09
What emissions are produced by a transmitter using a reactance modulator?
- A. CW
- B. Test
- C. Single-sideband, suppressed-carrier phone
- D. Phase-modulated phone

T8A10
What other emission does phase modulation most resemble?
- A. Amplitude modulation
- B. Pulse modulation
- C. Frequency modulation
- D. Single-sideband modulation

T8B11
What causes splatter interference?
- A. Keying a transmitter too fast
- B. Signals from a transmitter's output circuit are being sent back to its input circuit
- C. Overmodulation of a transmitter
- D. The transmitting antenna is the wrong length

T8B03
What kind of emission would your FM transmitter produce if its microphone failed to work?
- A. An unmodulated carrier
- B. A phase-modulated carrier
- C. An amplitude-modulated carrier
- D. A frequency-modulated carrier

T8B05
Why is FM voice best for local VHF/UHF radio communications?
- A. The carrier is not detectable
- B. It is more resistant to distortion caused by reflected signals
- C. It has high-fidelity audio which can be understood even when the signal is somewhat weak
- D. Its RF carrier stays on frequency better than the AM modes

T8B10
What is the result of overdeviation in an FM transmitter?
- A. Increased transmitter power
- B. Out-of-channel emissions
- C. Increased transmitter range
- D. Poor carrier suppression

ANSWERS TO PREVIOUS TEST

T8A01	B
T8A02	C
T8A03	B
T8A04	B
T8A05	D
T8A11	B
T8B01	A
T8B02	A
T8B04	B

YOUR ANSWERS TO THIS TEST

- T8A06 _____
- T8A07 _____
- T8A08 _____
- T8A09 _____
- T8A10 _____
- T8B11 _____
- T8B03 _____
- T8B05 _____
- T8B10 _____

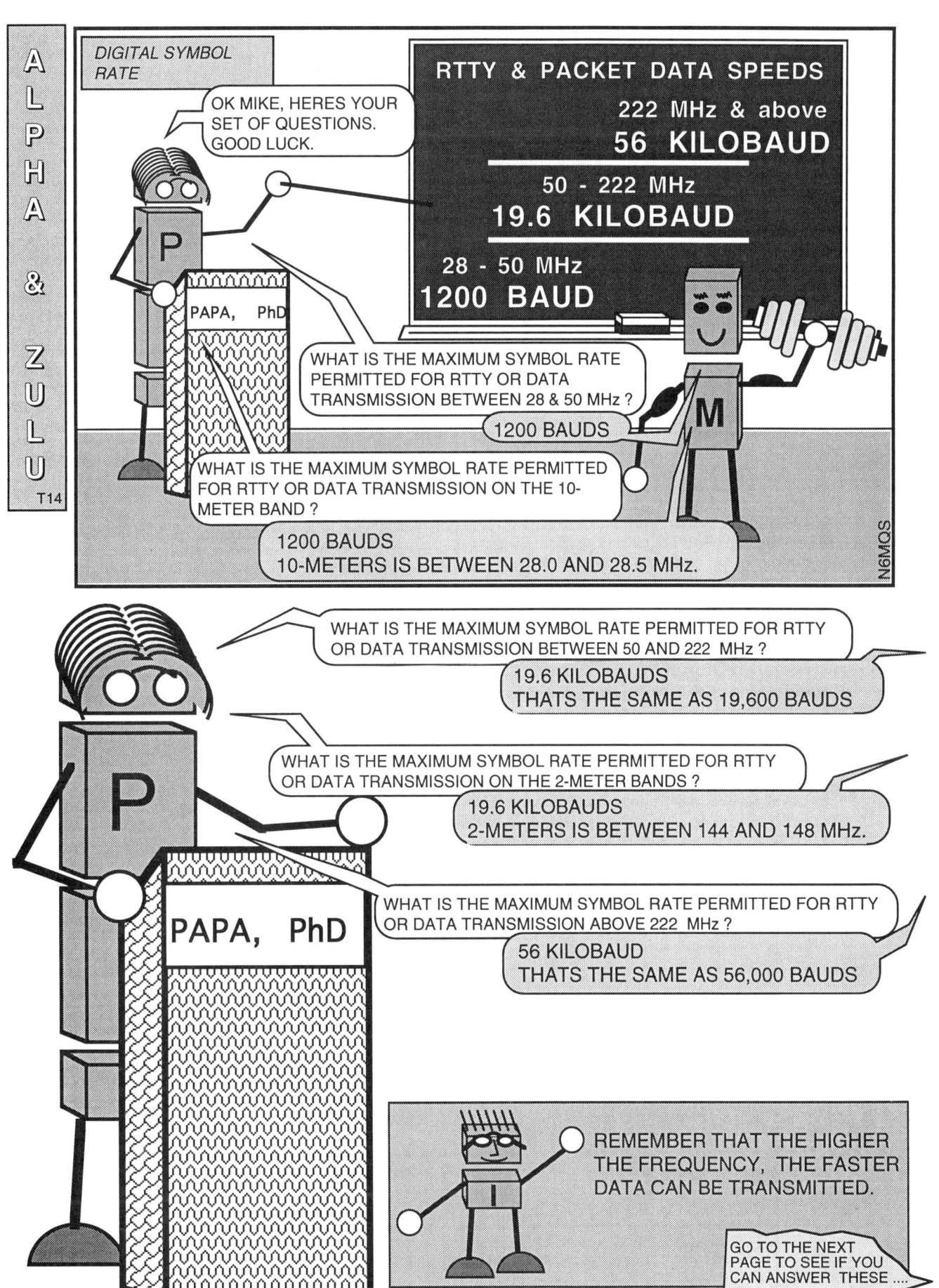

Riding the Airwaves with Alpha & Zulu

SEE IF YOU CAN ANSWER THESE TECHNICIAN QUESTIONS WITHOUT LOOKING BACK TO THE LAST 'TOON'. WRITE YOUR ANSWERS IN THE ANSWER BOX BELOW. GOOD LUCK. DA-DI-DAH

PRACTICE TEST PAGE T14

T1C04 [97.307f4]
What is the maximum symbol rate permitted for packet transmissions on the 10-meter band?
- A. 300 bauds
- B. 1200 bauds
- C. 19.6 kilobauds
- D. 56 kilobauds

T1C06 [97.307f4]
What is the maximum symbol rate permitted for RTTY or data transmissions between 28 and 50 MHz?
- A. 56 kilobauds
- B. 19.6 kilobauds
- C. 1200 bauds
- D. 300 bauds

T1C05 [97.307f5]
What is the maximum symbol rate permitted for packet transmissions on the 2-meter band?
- A. 300 bauds
- B. 1200 bauds
- C. 19.6 kilobauds
- D. 56 kilobauds

T1C07 [97.307f5]
What is the maximum symbol rate permitted for RTTY or data transmissions between 50 and 222 MHz?
- A. 56 kilobauds
- B. 19.6 kilobauds
- C. 1200 bauds
- D. 300 bauds

T1C09 [97.307f6]
What is the maximum symbol rate permitted for RTTY or data transmissions above 222 MHz?
- A. 300 bauds
- B. 1200 bauds
- C. 19.6 kilobauds
- D. 56 kilobauds

ANSWERS TO PREVIOUS TEST
T8A06
D
T8A07
A
T8A08
A
T8A09
D
T8A10
C
T8B11
C
T8B03
A
T8B05
C
T8B10
B

YOUR ANSWERS TO THIS TEST
T1C04
T1C05
T1C06
T1C07
T1C09

artsci inc

Riding the Airwaves with Alpha & Zulu

SEE IF YOU CAN ANSWER THESE TECHNICIAN QUESTIONS WITHOUT LOOKING BACK TO THE LAST 'TOON'. WRITE YOUR ANSWERS IN THE ANSWER BOX BELOW. GOOD LUCK. DA-DI-DAH

PRACTICE TEST PAGE T15

T1C02 [97.307f3/4]
What is the maximum frequency shift permitted for RTTY or data transmissions below 50 MHz?
  A. 0.1 kHz
  B. 0.5 kHz
  C. 1 kHz
  D. 5 kHz

T1C03 [97.307]
What is the maximum frequency shift permitted for RTTY or data transmissions above 50 MHz?
  A. 0.1 kHz or the sending speed, in bauds, whichever is greater
  B. 0.5 kHz or the sending speed, in bauds, whichever is greater
  C. 5 kHz or the sending speed, in bauds, whichever is greater
  D. The FCC rules do not specify a maximum frequency shift above 50 MHz

T1C08 [97.307f5]
What is the maximum authorized bandwidth of RTTY, data or multiplexed emissions using an unspecified digital code within the frequency range of 50 to 222 MHz?
  A. 20 kHz
  B. 50 kHz
  C. The total bandwidth shall not exceed that of a single-sideband phone emission
  D. The total bandwidth shall not exceed 10 times that of a CW emission

T1C10 [97.307f6]
What is the maximum authorized bandwidth of RTTY, data or multiplexed emissions using an unspecified digital code within the frequency range of 222 to 450 MHz?
  A. 50 kHz
  B. 100 kHz
  C. 150 kHz
  D. 200 kHz

T1C11 [97.307f6]
What is the maximum authorized bandwidth of RTTY, data or multiplexed emissions using an unspecified digital code within the 70 cm amateur band?
  A. 300 kHz
  B. 200 kHz
  C. 100 kHz
  D. 50 kHz

ANSWERS TO PREVIOUS TEST
T1C04
B
T1C05
C
T1C06
C
T1C07
B
T1C09
D

YOUR ANSWERS TO THIS TEST
T1C02
T1C03
T1C08
T1C10
T1C11

Riding the Airwaves with Alpha & Zulu

GOOD LUCK.

PRACTICE TEST PAGE T16

T1A05 [97.301/305e]
Which of the following frequencies may a Technician operator who has passed a Morse code test use?
- A. 7.1 - 7.2 MHz
- B. 14.1 - 14.2 MHz
- C. 21.1 - 21.2 MHz
- D. 28.1 - 29.2 MHz

T1A06 [97.301a]
Which operator licenses authorize privileges on 52.525 MHz?
- A. Extra, Advanced only
- B. Extra, Advanced, General only
- C. Extra, Advanced, General, Technician only
- D. Extra, Advanced, General, Technician, Novice

T1A07 [97.301a]
Which operator licenses authorize privileges on 146.52 MHz?
- A. Extra, Advanced, General, Technician, Novice
- B. Extra, Advanced, General, Technician only
- C. Extra, Advanced, General only
- D. Extra, Advanced only

T1A08 [97.301a]
Which operator licenses authorize privileges on 223.50 MHz?
- A. Extra, Advanced, General, Technician, Novice
- B. Extra, Advanced, General, Technician only
- C. Extra, Advanced, General only
- D. Extra, Advanced only

T1A09 [97.301a]
Which operator licenses authorize privileges on 446.0 MHz?
- A. Extra, Advanced, General, Technician, Novice
- B. Extra, Advanced, General, Technician only
- C. Extra, Advanced, General only
- D. Extra, Advanced only

T1A10 [97.301e]
In addition to passing the Technician written examination (Elements 2 and 3A), what must you do before you are allowed to use amateur frequencies below 30 MHz?
- A. Nothing special is needed; all Technicians may use the HF bands at any time
- B. You must notify the FCC that you intend to operate on the HF bands
- C. You must attend a class to learn about HF communications
- D. You must pass a Morse code test (either Element 1A, 1B or 1C)

T1A11 [97.301e]
If you are a Technician licensee, what must you have to prove that you are authorized to use the Novice amateur frequencies below 30 MHz?
- A. A certificate from the FCC showing that you have notified them that you will be using the HF bands
- B. A certificate from an instructor showing that you have attended a class in HF communications
- C. Written proof of having passed a Morse code test
- D. No special proof is required before using the HF bands

ANSWERS TO PREVIOUS TEST

T1C02	C
T1C03	D
T1C08	A
T1C10	B
T1C11	C

YOUR ANSWERS TO THIS TEST

T1A05
T1A06
T1A07
T1A08
T1A09
T1A10
T1A11

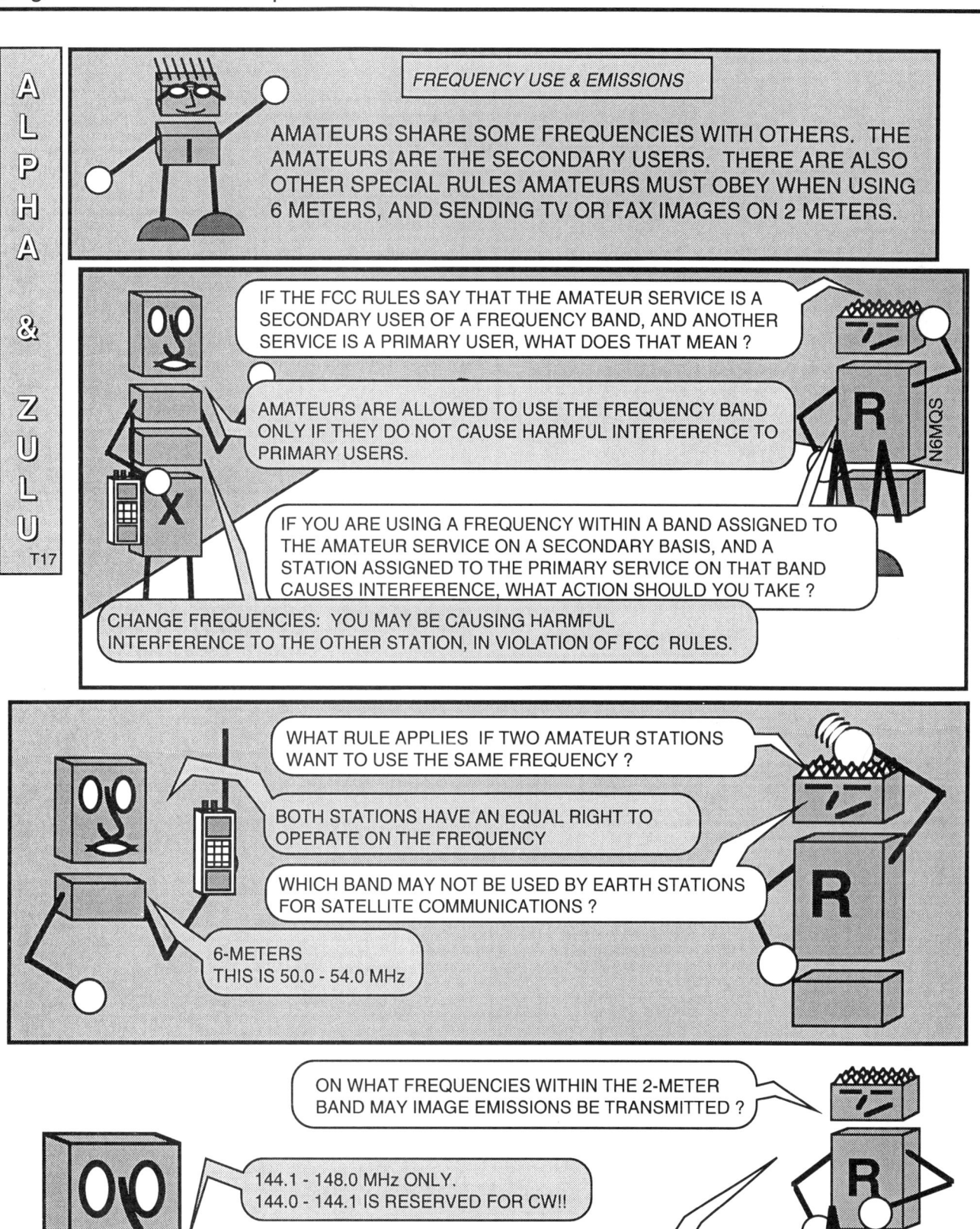

PRACTICE TEST PAGE T17

SEE IF YOU CAN ANSWER THESE TECHNICIAN QUESTIONS WITHOUT LOOKING BACK TO THE LAST 'TOON'. WRITE YOUR ANSWERS IN THE ANSWER BOX BELOW. GOOD LUCK. DA-DI-DAH

T1B04 [97.303]
If the FCC rules say that the amateur service is a secondary user of a frequency band, and another service is a primary user, what does this mean?
A. Nothing special; all users of a frequency band have equal rights to operate
B. Amateurs are only allowed to use the frequency band during emergencies
C. Amateurs are allowed to use the frequency band only if they do not cause harmful interference to primary users
D. Amateurs must increase transmitter power to overcome any interference caused by primary users

T1B05 [97.303]
If you are using a frequency within a band assigned to the amateur service on a secondary basis, and a station assigned to the primary service on that band causes interference, what action should you take?
A. Notify the FCC's regional Engineer in Charge of the interference
B. Increase your transmitter's power to overcome the interference
C. Attempt to contact the station and request that it stop the interference
D. Change frequencies; you may be causing harmful interference to the other station, in violation of FCC rules

T1B08 [97.305c]
On what frequencies within the 6-meter band may phone emissions be transmitted?
A. 50.0 - 54.0 MHz only
B. 50.1 - 54.0 MHz only
C. 51.0 - 54.0 MHz only
D. 52.0 - 54.0 MHz only

T1B09 [97.305c]
On what frequencies within the 2-meter band may image emissions be transmitted?
A. 144.1 - 148.0 MHz only
B. 146.0 - 148.0 MHz only
C. 144.0 - 148.0 MHz only
D. 146.0 - 147.0 MHz only

T1B06 [97.303a]
What rule applies if two amateur stations want to use the same frequency?
A. The station operator with a lesser class of license must yield the frequency to a higher class licensee
B. The station operator with a lower power output must yield the frequency to the station with a higher power output
C. Both station operators have an equal right to operate on the frequency
D. Station operators in ITU Regions 1 and 3 must yield the frequency to stations in ITU Region 2

T1B11 [97.209b2]
Which band may NOT be used by Earth stations for satellite communications?
A. 6 meters
B. 2 meters
C. 70 centimeters
D. 23 centimeters

ANSWERS TO PREVIOUS TEST	
T1A05	C
T1A06	C
T1A07	B
T1A08	A
T1A09	B
T1A10	D
T1A11	C

YOUR ANSWERS TO THIS TEST
T1B04
T1B08
T1B05
T1B09
T1B06
T1B11

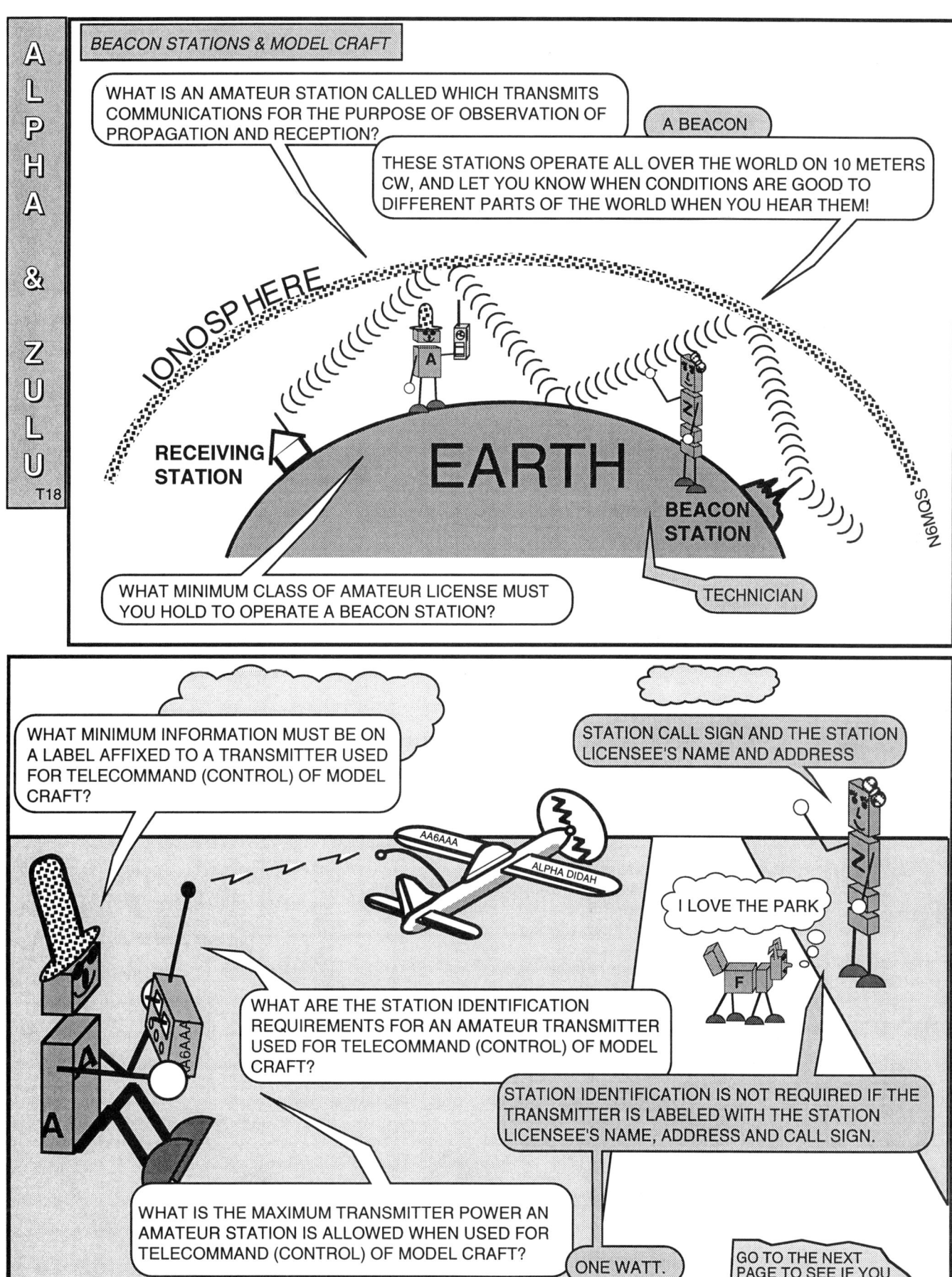

Riding the Airwaves with Alpha & Zulu

SEE IF YOU CAN ANSWER THESE TECHNICIAN QUESTIONS WITHOUT LOOKING BACK TO THE LAST 'TOON'. WRITE YOUR ANSWERS IN THE ANSWER BOX BELOW. GOOD LUCK. DA-DI-DAH

PRACTICE TEST PAGE T18

T1D01 [97.3a9]
What is an amateur station called which transmits communications for the purpose of observation of propagation and reception?
  A. A beacon
  B. A repeater
  C. An auxiliary station
  D. A radio control station

T1D05 [97.203a]
What minimum class of amateur license must you hold to operate a beacon station?
  A. Novice
  B. Technician
  C. General
  D. Amateur Extra

T1D09 [97.215a]
What minimum information must be on a label affixed to a transmitter used for telecommand (control) of model craft?
  A. Station call sign
  B. Station call sign and the station licensee's name
  C. Station call sign and the station licensee's name and address
  D. Station call sign and the station licensee's class of license

T1D10 [97.215a]
What are the station identification requirements for an amateur transmitter used for telecommand (control) of model craft?
  A. Once every ten minutes
  B. Once every ten minutes, and at the beginning and end of each transmission
  C. At the beginning and end of each transmission
  D. Station identification is not required if the transmitter is labeled with the station licensee's name, address and call sign

T1D11 [97.215c]
What is the maximum transmitter power an amateur station is allowed when used for telecommand (control) of model craft?
  A. One milliwatt
  B. One watt
  C. Two watts
  D. Three watts

ANSWERS TO PREVIOUS TEST
T1B04
C
T1B08
B
T1B05
D
T1B09
A
T1B06
C
T1B11
A

YOUR ANSWERS TO THIS TEST
T1D01
T1D05
T1D09
T1D10
T1D11

Riding the Airwaves with Alpha & Zulu

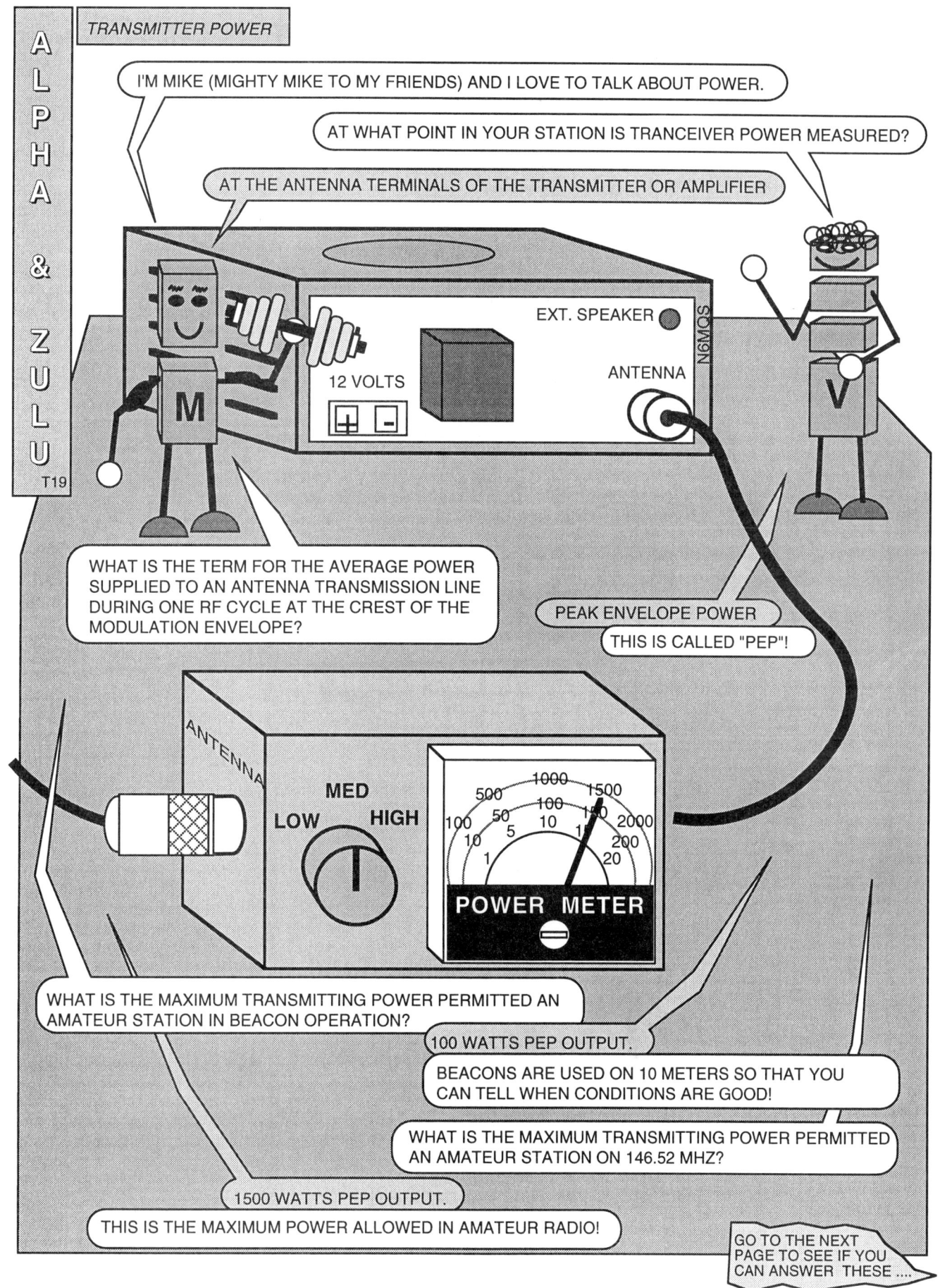

Riding the Airwaves with Alpha & Zulu

SEE IF YOU CAN ANSWER THESE TECHNICIAN QUESTIONS WITHOUT LOOKING BACK TO THE LAST 'TOON'. WRITE YOUR ANSWERS IN THE ANSWER BOX BELOW. GOOD LUCK. DA-DI-DAH

PRACTICE TEST PAGE T19

T1B01 [97.3b6]
At what point in your station is transceiver power measured?
  A. At the power supply terminals inside the transmitter or amplifier
  B. At the final amplifier input terminals inside the transmitter or amplifier
  C. At the antenna terminals of the transmitter or amplifier
  D. On the antenna itself, after the feed line

T1B02 [97.3b6]
What is the term for the average power supplied to an antenna transmission line during one RF cycle at the crest of the modulation envelope?
  A. Peak transmitter power
  B. Peak output power
  C. Average radio-frequency power
  D. Peak envelope power

T1B03 [97.203c]
What is the maximum transmitting power permitted an amateur station in beacon operation?
  A. 10 watts PEP output
  B. 100 watts PEP output
  C. 500 watts PEP output
  D. 1500 watts PEP output

T1B10 [97.313b]
What is the maximum transmitting power permitted an amateur station on 146.52 MHz?
  A. 200 watts PEP output
  B. 500 watts ERP
  C. 1000 watts DC input
  D. 1500 watts PEP output

ANSWERS TO PREVIOUS TEST
T1D01
A
T1D05
B
T1D09
C
T1D10
D
T1D11
B

YOUR ANSWERS TO THIS TEST
T1B01
T1B02
T1B03
T1B10

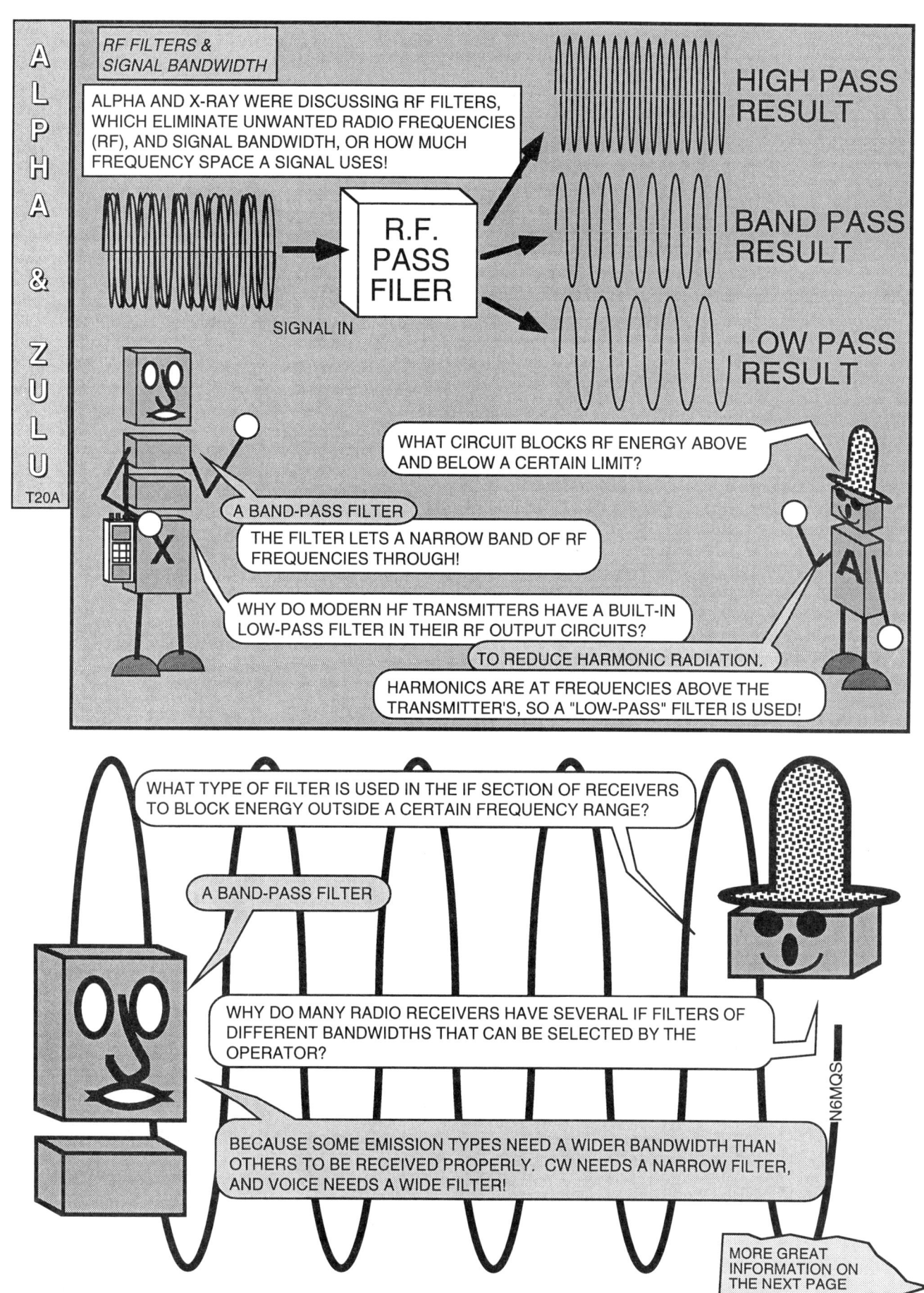

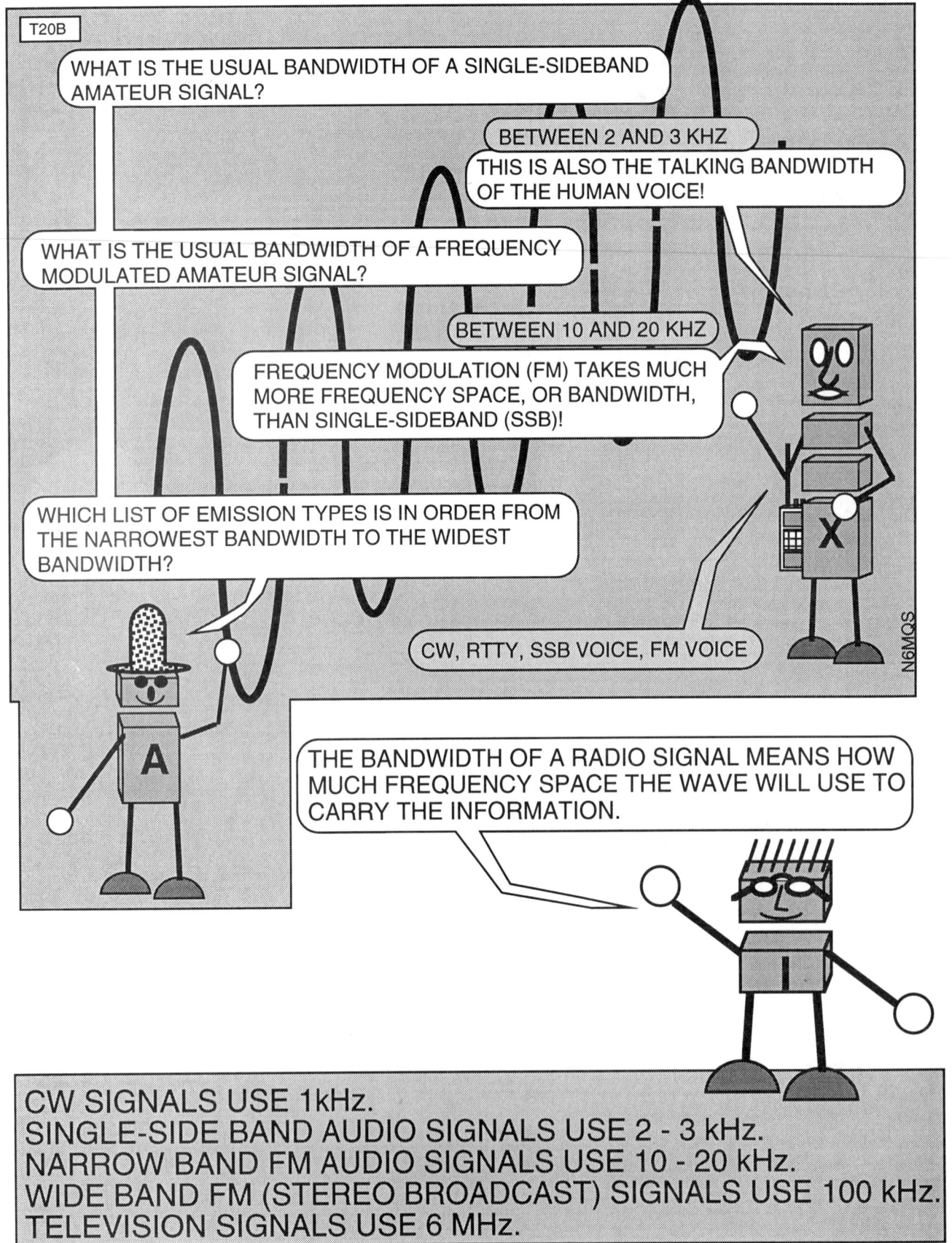

Riding the Airwaves with Alpha & Zulu

SEE IF YOU CAN ANSWER THESE TECHNICIAN QUESTIONS WITHOUT LOOKING BACK TO THE LAST 'TOON'. WRITE YOUR ANSWERS IN THE ANSWER BOX BELOW. GOOD LUCK. DA-DI-DAH

PRACTICE TEST PAGE T20

T7A01
Why do modern HF transmitters have a built-in low-pass filter in their RF output circuits?
- A. To reduce RF energy below a cutoff point
- B. To reduce low-frequency interference to other amateurs
- C. To reduce harmonic radiation
- D. To reduce fundamental radiation

T7A02
What circuit blocks RF energy above and below a certain limit?
- A. A band-pass filter
- B. A high-pass filter
- C. An input filter
- D. A low-pass filter

T7A03
What type of filter is used in the IF section of receivers to block energy outside a certain frequency range?
- A. A band-pass filter
- B. A high-pass filter
- C. An input filter
- D. A low-pass filter

T8B06
Why do many radio receivers have several IF filters of different bandwidths that can be selected by the operator?
- A. Because some frequency bands are wider than others
- B. Because different bandwidths help increase the receiver sensitivity
- C. Because different bandwidths improve S-meter readings
- D. Because some emission types need a wider bandwidth than others to be received properly

T8B07
Which list of emission types is in order from the narrowest bandwidth to the widest bandwidth?
- A. RTTY, CW, SSB voice, FM voice
- B. CW, FM voice, RTTY, SSB voice
- C. CW, RTTY, SSB voice, FM voice
- D. CW, SSB voice, RTTY, FM voice

T8B08
What is the usual bandwidth of a single-sideband amateur signal?
- A. 1 kHz
- B. 2 kHz
- C. Between 3 and 6 kHz
- D. Between 2 and 3 kHz

T8B09
What is the usual bandwidth of a frequency-modulated amateur signal?
- A. Less than 5 kHz
- B. Between 5 and 10 kHz
- C. Between 10 and 20 kHz
- D. Greater than 20 kHz

ANSWERS TO PREVIOUS TEST

T1B01	C
T1B02	D
T1B03	B
T1B10	D

YOUR ANSWERS TO THIS TEST

T7A01	
T7A02	
T7A03	
T8B06	
T8B07	
T8B08	
T8B09	

artsci inc

RADIO MATCHING

THE PURPOSE OF THIS GAME IS TO MATCH THE RADIO TERMS ON THE LEFT SIDE WITH THEIR DESCRIPTION ON THE RIGHT. TO DO THIS, PLACE THE LETTER OF THE DESCRIPTION ON THE LINE IN FRONT OF THE TERM. THERE ARE MORE DESCRIPTIONS THAN TERMS, SO THERE WILL BE ONE DESCRIPTION LEFT OVER. GOOD LUCK.

___ NEUTRAL
___ OHM
___ OMNIDIRECTIONAL
___ PARALLEL CONDUCTOR
___ PEAK ENVELOPE POWER
___ QSO
___ REFLECT
___ REFRACT
___ SKY WAVES
___ STANDING WAVE RATIO
___ TRANSCEIVER
___ WAVELENGTH

A. BOUNCE OFF
B. A RADIO TRANSMITTER AND RECEIVER
C. HAM RADIO CONVERSATION
D. HAVING NO ELECTRICAL CHARGE
E. TWO CONDUCTOR FEED LINE
F. MEASURE OF ELECTRICAL RESISTANCE
G. COMMON ZERO VOLT REFERENCE
H. DISTANCE BETWEEN RF PEAKS
I. MEASURE OF IMPEDANCE MATCHING
J. BENDING OF RF WAVES
K. RADIATES IN ALL DIRECTIONS
L. MEASURE OF TRANSMITTER POWER
M. RADIO WAVES REFLECTED OFF THE IONOSPHERE

Riding the Airwaves with Alpha & Zulu

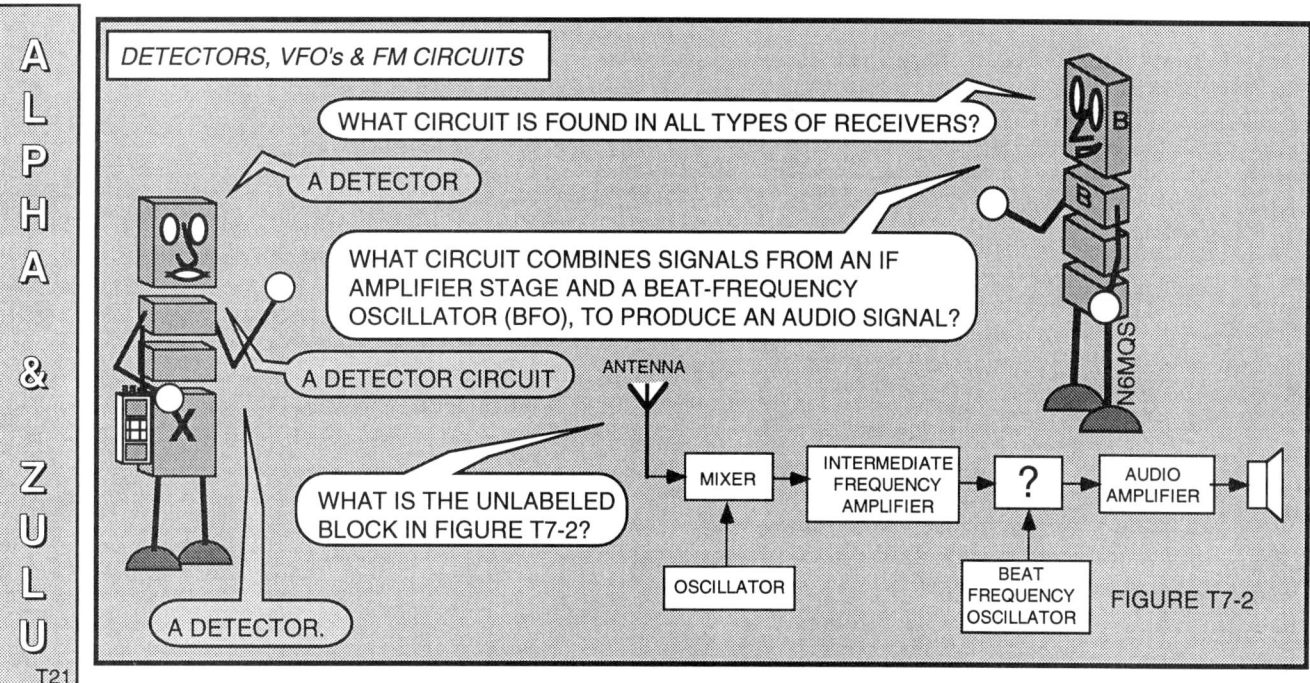
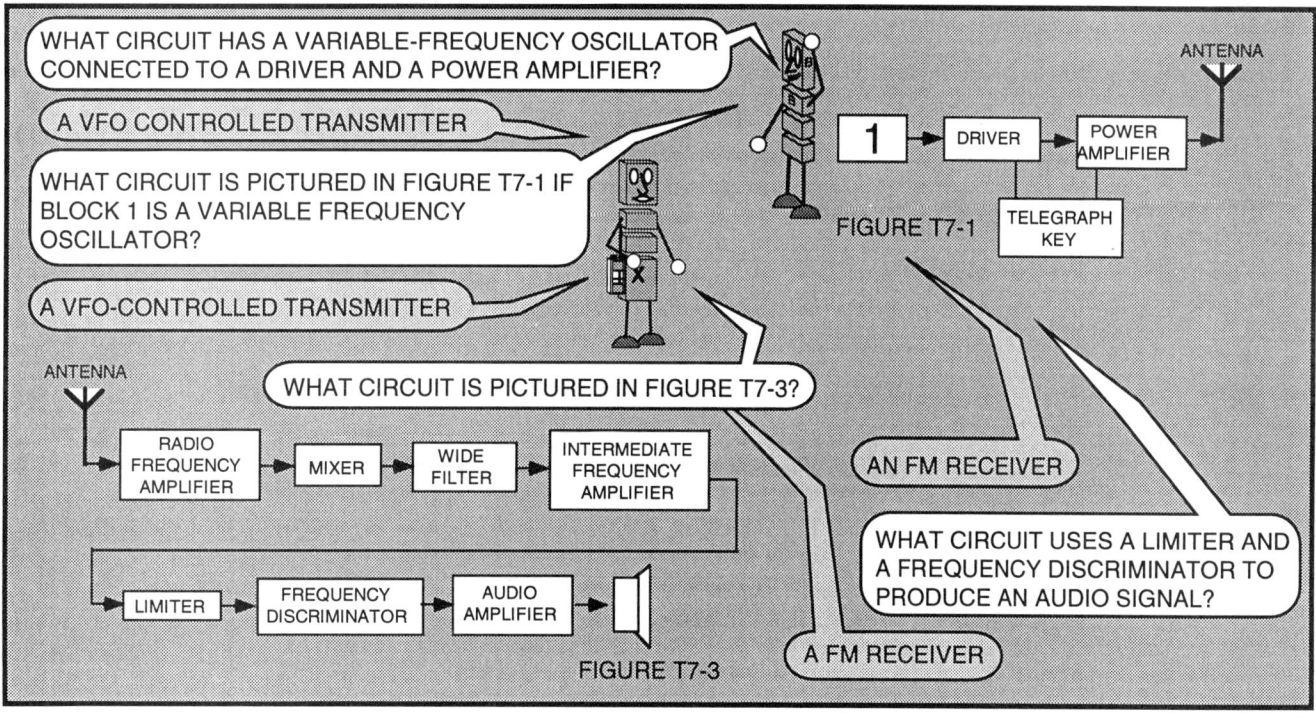
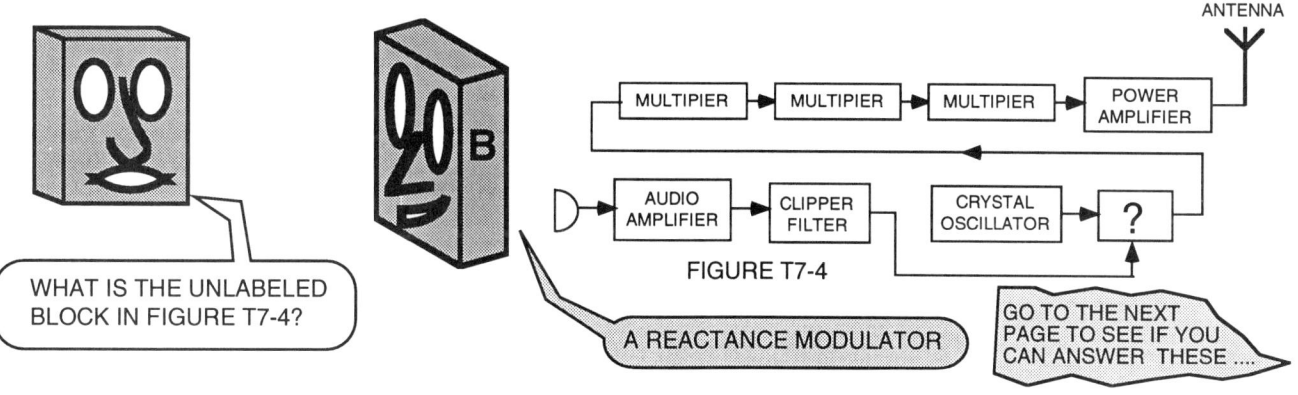

PRACTICE TEST PAGE T21

T7A04
What circuit is found in all types of receivers?
A. An audio filter
B. A beat-frequency oscillator
C. A detector
D. An RF amplifier

T7A05
What circuit has a variable-frequency oscillator connected to a driver and a power amplifier?
A. A packet-radio transmitter
B. A crystal-controlled transmitter
C. A single-sideband transmitter
D. A VFO-controlled transmitter

T7A06
What circuit combines signals from an IF amplifier stage and a beat-frequency oscillator (BFO), to produce an audio signal?
A. An AGC circuit
B. A detector circuit
C. A power supply circuit
D. A VFO circuit

T7A07
What circuit uses a limiter and a frequency discriminator to produce an audio signal?
A. A double-conversion receiver
B. A variable-frequency oscillator
C. A superheterodyne receiver
D. A FM receiver

T7A08
What circuit is pictured in Figure T7-1 if block 1 is a variable-frequency oscillator?
A. A packet-radio transmitter
B. A crystal-controlled transmitter
C. A single-sideband transmitter
D. A VFO-controlled transmitter

T7A09
What is the unlabeled block in Figure T7-2?
A. An AGC circuit
B. A detector
C. A power supply
D. A VFO circuit

T7A10
What circuit is pictured in Figure T7-3?
A. A double-conversion receiver
B. A variable-frequency oscillator
C. A superheterodyne receiver
D. An FM receiver

T7A11
What is the unlabeled block in Figure T7-4?
A. A band-pass filter
B. A crystal oscillator
C. A reactance modulator
D. A rectifier modulator

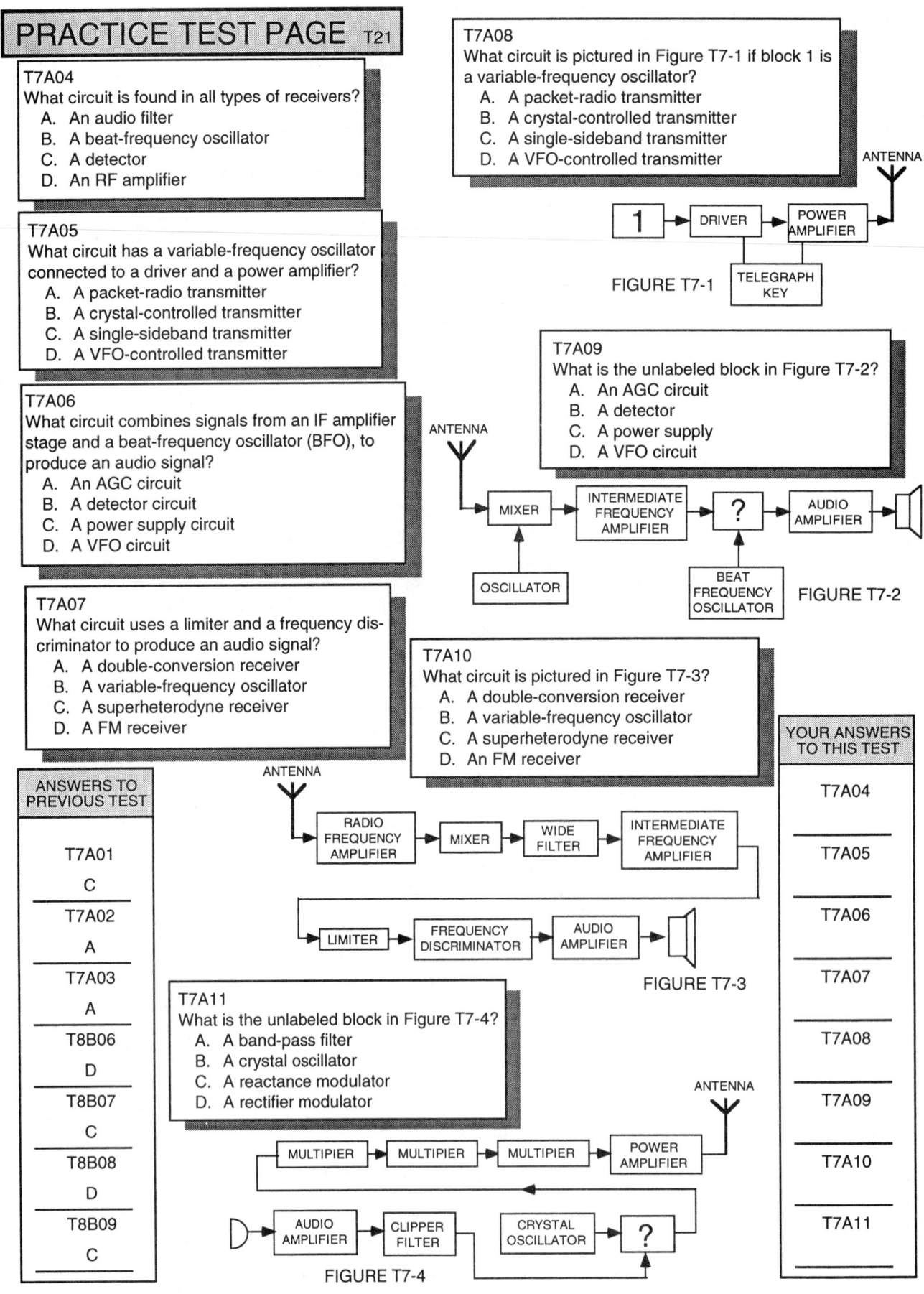

FIGURE T7-1

FIGURE T7-2

FIGURE T7-3

FIGURE T7-4

ANSWERS TO PREVIOUS TEST

T7A01	C
T7A02	A
T7A03	A
T8B06	D
T8B07	C
T8B08	D
T8B09	C

YOUR ANSWERS TO THIS TEST

T7A04
T7A05
T7A06
T7A07
T7A08
T7A09
T7A10
T7A11

SEE IF YOU CAN ANSWER THESE TECHNICIAN QUESTIONS WITHOUT LOOKING BACK TO THE LAST 'TOON'. WRITE YOUR ANSWERS IN THE ANSWER BOX BELOW. GOOD LUCK. DA-DI-DAH

PRACTICE TEST PAGE T22

T4C01
What is a marker generator?
A. A high-stability oscillator that generates reference signals at exact frequency intervals
B. A low-stability oscillator that "sweeps" through a range of frequencies
C. A low-stability oscillator used to inject a signal into a circuit under test
D. A high-stability oscillator which can produce a wide range of frequencies and amplitudes

T4C02
How is a marker generator used?
A. To calibrate the tuning dial on a receiver
B. To calibrate the volume control on a receiver
C. To test the amplitude linearity of a transmitter
D. To test the frequency deviation of a transmitter

T4C03
What device is used to inject a frequency calibration signal into a receiver?
A. A calibrated voltmeter
B. A calibrated oscilloscope
C. A calibrated wavemeter
D. A crystal calibrator

T4C04
What frequency standard may be used to calibrate the tuning dial of a receiver?
A. A calibrated voltmeter
B. Signals from WWV and WWVH
C. A deviation meter
D. A sweep generator

T4C05
How might you check the accuracy of your receiver's tuning dial?
A. Tune to the frequency of a shortwave broadcasting station
B. Tune to a popular amateur net frequency
C. Tune to one of the frequencies of station WWV or WWVH
D. Tune to another amateur station and ask what frequency the operator is using

T4C06
What device produces a stable, low-level signal that can be set to a desired frequency?
A. A wavemeter
B. A reflectometer
C. A signal generator
D. An oscilloscope

T4C07
What is an RF signal generator used for?
A. Measuring RF signal amplitudes
B. Aligning tuned circuits
C. Adjusting transmitter impedance-matching networks
D. Measuring transmission line impedances

ANSWERS TO PREVIOUS TEST

T7A04	C
T7A05	D
T7A06	B
T7A07	D
T7A08	D
T7A09	B
T7A10	D
T7A11	C

YOUR ANSWERS TO THIS TEST

T4C01

T4C02

T4C03

T4C04

T4C05

T4C06

T4C07

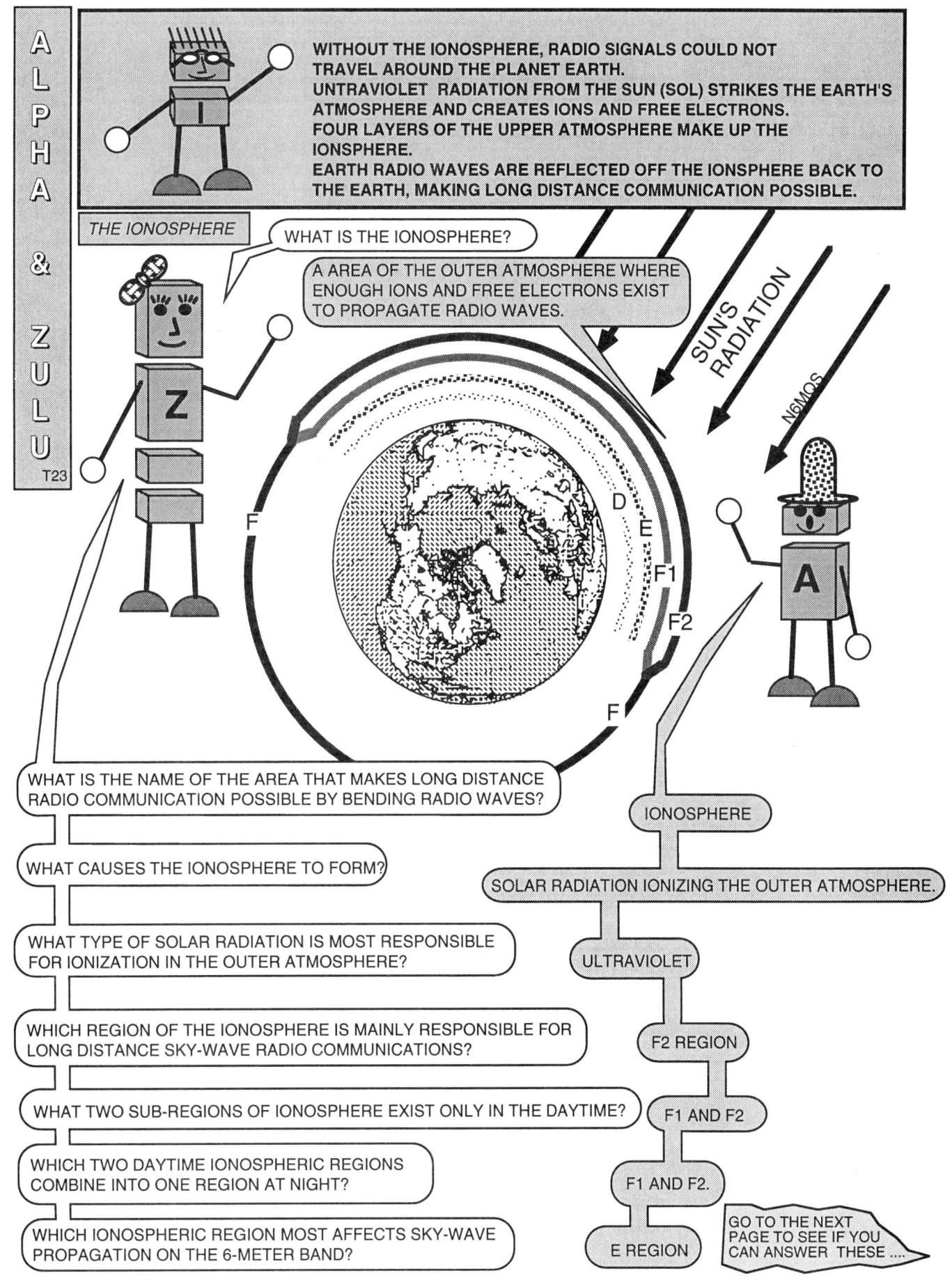

Riding the Airwaves with Alpha & Zulu

SEE IF YOU CAN ANSWER THESE TECHNICIAN QUESTIONS WITHOUT LOOKING BACK TO THE LAST 'TOON'. WRITE YOUR ANSWERS IN THE ANSWER BOX BELOW. GOOD LUCK. DA-DI-DAH

PRACTICE TEST PAGE T23

T3A01
What is the ionosphere?
A. A area of the outer atmosphere where enough ions and free electrons exist to propagate radio waves
B. A area between two air masses of different temperature and humidity, along which radio waves can travel
C. An ionized path in the atmosphere where lightning has struck
D. An area of the atmosphere where weather takes place

T3A02
What is the name of the area that makes long-distance radio communications possible by bending radio waves?
A. Troposphere
B. Stratosphere
C. Magnetosphere
D. Ionosphere

T3A03
What causes the ionosphere to form?
A. Solar radiation ionizing the outer atmosphere
B. Temperature changes ionizing the outer atmosphere
C. Lightning ionizing the outer atmosphere
D. Release of fluorocarbons into the atmosphere

T3A04
What type of solar radiation is most responsible for ionization in the outer atmosphere?
A. Thermal
B. Ionized particle
C. Ultraviolet
D. Microwave

T3A07
Which ionospheric region most affects sky-wave propagation on the 6-meter band?
A. The D region
B. The E region
C. The F1 region
D. The F2 region

T3A09
Which region of the ionosphere is mainly responsible for long-distance sky-wave radio communications?
A. D region
B. E region
C. F1 region
D. F2 region

T3A10
What two sub-regions of ionosphere exist only in the daytime?
A. Troposphere and stratosphere
B. F1 and F2
C. Electrostatic and electromagnetic
D. D and E

T3A11
Which two daytime ionospheric regions combine into one region at night?
A. E and F1
B. D and E
C. F1 and F2
D. E1 and E2

ANSWERS TO PREVIOUS TEST

T4C01	A
T4C02	A
T4C03	D
T4C04	B
T4C05	C
T4C06	C
T4C07	B

YOUR ANSWERS TO THIS TEST

T3A01 _____
T3A02 _____
T3A03 _____
T3A04 _____
T3A07 _____
T3A09 _____
T3A10 _____
T3A11 _____

Riding the Airwaves with Alpha & Zulu

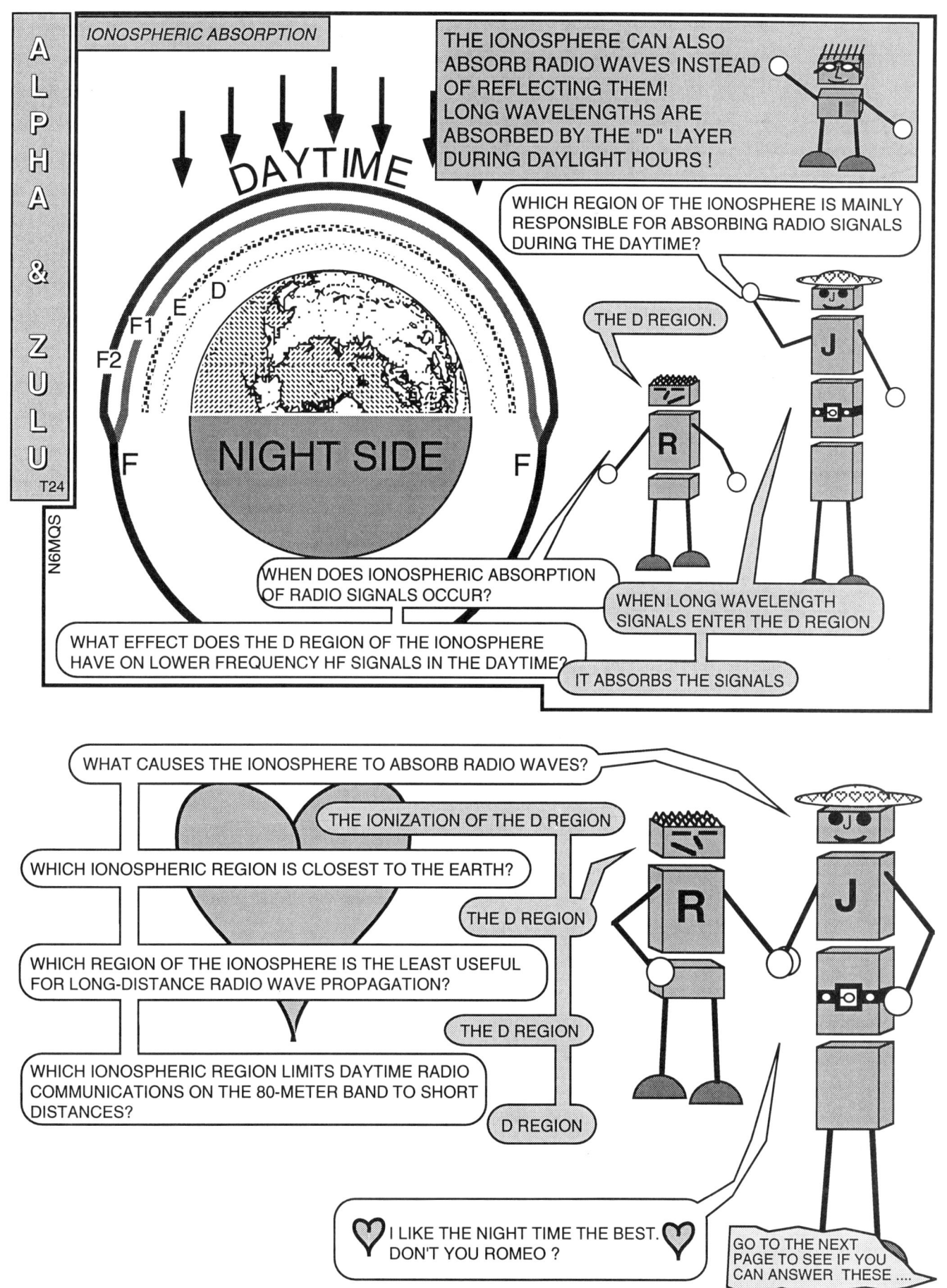

SEE IF YOU CAN ANSWER THESE TECHNICIAN QUESTIONS WITHOUT LOOKING BACK TO THE LAST 'TOON'. WRITE YOUR ANSWERS IN THE ANSWER BOX BELOW. GOOD LUCK. DA-DI-DAH

PRACTICE TEST PAGE T24

T3B01
Which region of the ionosphere is mainly responsible for absorbing radio signals during the daytime?
A. The F2 region
B. The F1 region
C. The E region
D. The D region

T3B02
When does ionospheric absorption of radio signals occur?
A. When tropospheric ducting occurs
B. When long wavelength signals enter the D region
C. When signals travel to the F region
D. When a temperature inversion occurs

T3B03
What effect does the D region of the ionosphere have on lower-frequency HF signals in the daytime?
A. It absorbs the signals
B. It bends the radio waves out into space
C. It refracts the radio waves back to earth
D. It has little or no effect on 80-meter radio waves

T3B04
What causes the ionosphere to absorb radio waves?
A. The weather below the ionosphere
B. The ionization of the D region
C. The presence of ionized clouds in the E region
D. The splitting of the F region

T3A05
Which ionospheric region limits daytime radio communications on the 80-meter band to short distances?
A. D region
B. E region
C. F1 region
D. F2 region

T3A06
Which ionospheric region is closest to the earth?
A. The A region
B. The D region
C. The E region
D. The F region

T3A08
Which region of the ionosphere is the least useful for long-distance radio wave propagation?
A. The D region
B. The E region
C. The F1 region
D. The F2 region

ANSWERS TO PREVIOUS TEST

T3A01	A
T3A02	D
T3A03	A
T3A04	C
T3A07	B
T3A09	D
T3A10	B
T3A11	C

YOUR ANSWERS TO THIS TEST

T3B01

T3B02

T3B03

T3B04

T3A05

T3A06

T3A08

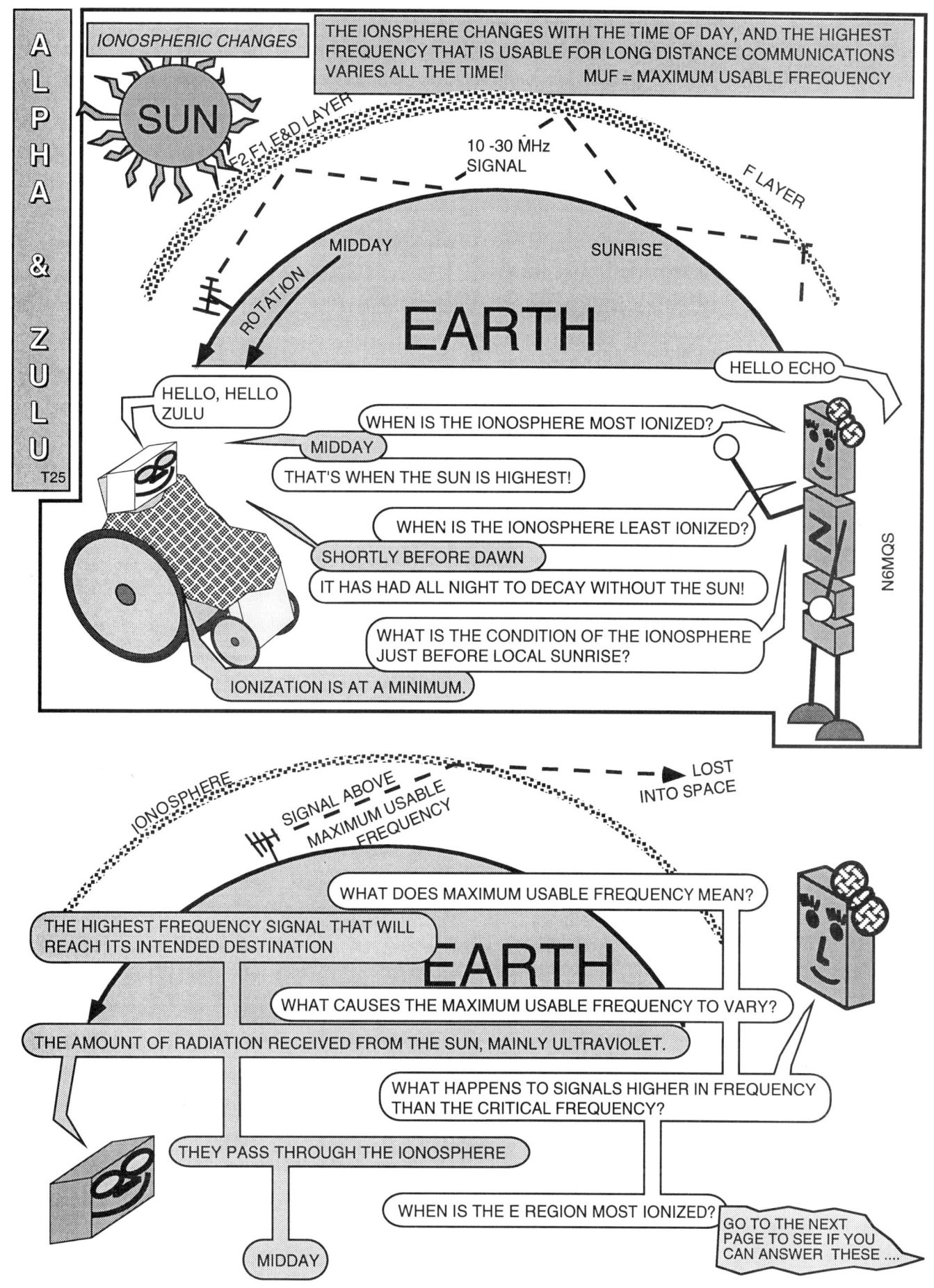

Riding the Airwaves with Alpha & Zulu

SEE IF YOU CAN ANSWER THESE TECHNICIAN QUESTIONS WITHOUT LOOKING BACK TO THE LAST 'TOON'. WRITE YOUR ANSWERS IN THE ANSWER BOX BELOW. GOOD LUCK. DA-DI-DAH

PRACTICE TEST PAGE T25

T3B05
What is the condition of the ionosphere just before local sunrise?
- A. Atmospheric attenuation is at a maximum
- B. The D region is above the E region
- C. The E region is above the F region
- D. Ionization is at a minimum

T3B06
When is the ionosphere most ionized?
- A. Dusk
- B. Midnight
- C. Midday
- D. Dawn

T3B07
When is the ionosphere least ionized?
- A. Shortly before dawn
- B. Just after noon
- C. Just after dusk
- D. Shortly before midnight

T3B08
When is the E region most ionized?
- A. Dawn
- B. Midday
- C. Dusk
- D. Midnight

T3B10
What causes the maximum usable frequency to vary?
- A. The temperature of the ionosphere
- B. The speed of the winds in the upper atmosphere
- C. The amount of radiation received from the sun, mainly ultraviolet
- D. The type of weather just below the ionosphere

T3B11
What does maximum usable frequency mean?
- A. The highest frequency signal that will reach its intended destination
- B. The lowest frequency signal that will reach its intended destination
- C. The highest frequency signal that is most absorbed by the ionosphere
- D. The lowest frequency signal that is most absorbed by the ionosphere

T3B09
What happens to signals higher in frequency than the critical frequency?
- A. They pass through the ionosphere
- B. They are absorbed by the ionosphere
- C. Their frequency is changed by the ionosphere to be below the maximum usable frequency
- D. They are reflected back to their source

ANSWERS TO PREVIOUS TEST	
T3B01	D
T3B02	B
T3B03	A
T3B04	B
T3A05	A
T3A06	B
T3A08	A

YOUR ANSWERS TO THIS TEST	
T3B05	
T3B06	
T3B07	
T3B08	
T3B10	
T3B11	
T3B09	

Riding the Airwaves with Alpha & Zulu

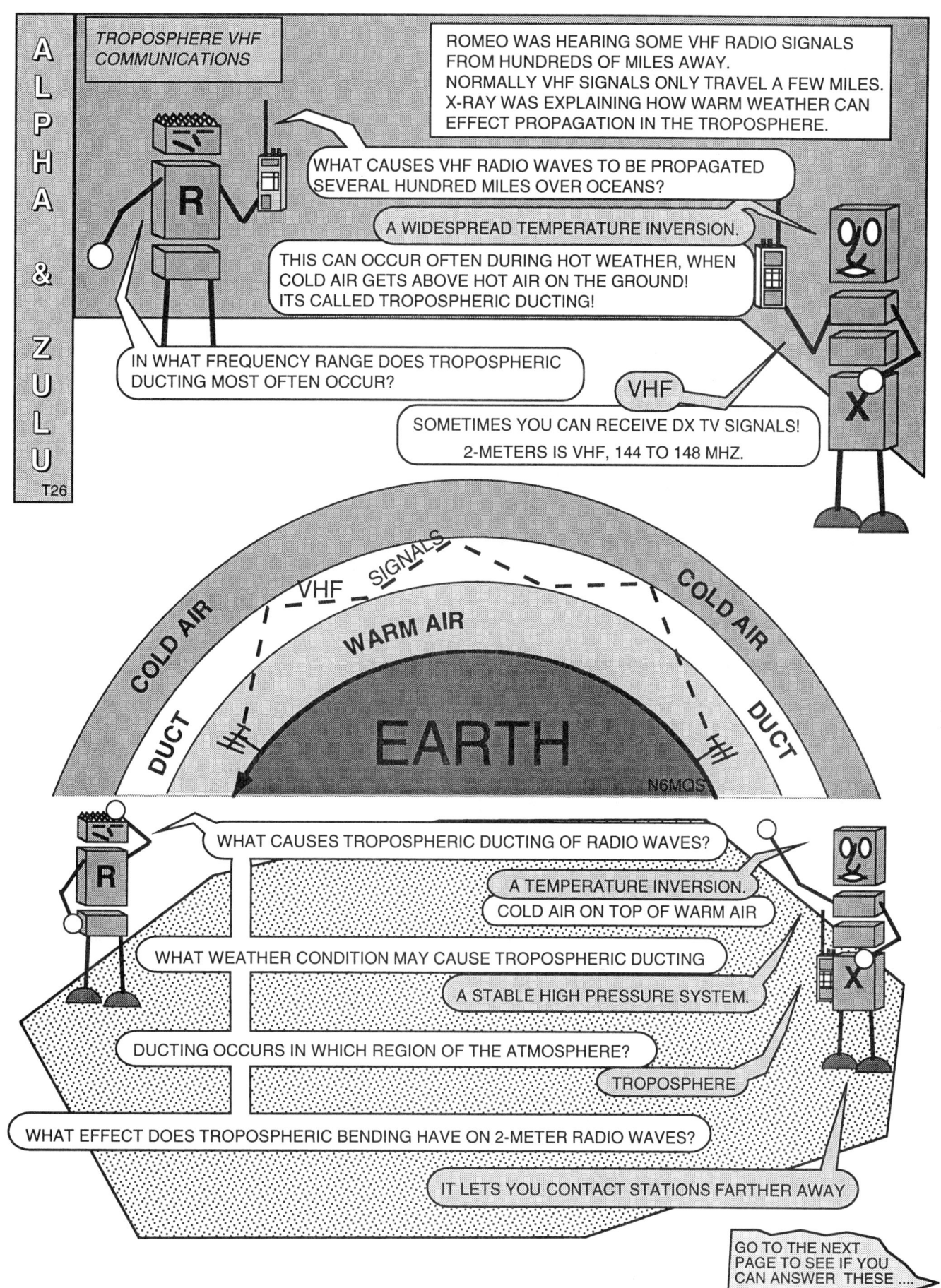

Riding the Airwaves with Alpha & Zulu

SEE IF YOU CAN ANSWER THESE TECHNICIAN QUESTIONS WITHOUT LOOKING BACK TO THE LAST 'TOON'. WRITE YOUR ANSWERS IN THE ANSWER BOX BELOW. GOOD LUCK. DA-DI-DAH

PRACTICE TEST PAGE T26

T3C04
Ducting occurs in which region of the atmosphere?
- A. F2
- B. Ectosphere
- C. Troposphere
- D. Stratosphere

T3C05
What effect does tropospheric bending have on 2-meter radio waves?
- A. It lets you contact stations farther away
- B. It causes them to travel shorter distances
- C. It garbles the signal
- D. It reverses the sideband of the signal

T3C06
What causes tropospheric ducting of radio waves?
- A. A very low pressure area
- B. An aurora to the north
- C. Lightning between the transmitting and receiving stations
- D. A temperature inversion

T3C07
What causes VHF radio waves to be propagated several hundred miles over oceans?
- A. A polar air mass
- B. A widespread temperature inversion
- C. An overcast of cirriform clouds
- D. A high-pressure zone

T3C08
In what frequency range does tropospheric ducting most often occur?
- A. SW
- B. MF
- C. HF
- D. VHF

T3C10
What weather condition may cause tropospheric ducting?
- A. A stable high-pressure system
- B. An unstable low-pressure system
- C. A series of low-pressure waves
- D. Periods of heavy rainfall

ANSWERS TO PREVIOUS TEST	
T3B05	D
T3B06	C
T3B07	A
T3B08	B
T3B10	C
T3B11	A
T3B09	A

YOUR ANSWERS TO THIS TEST
T3C04
T3C05
T3C06
T3C07
T3C08
T3C10

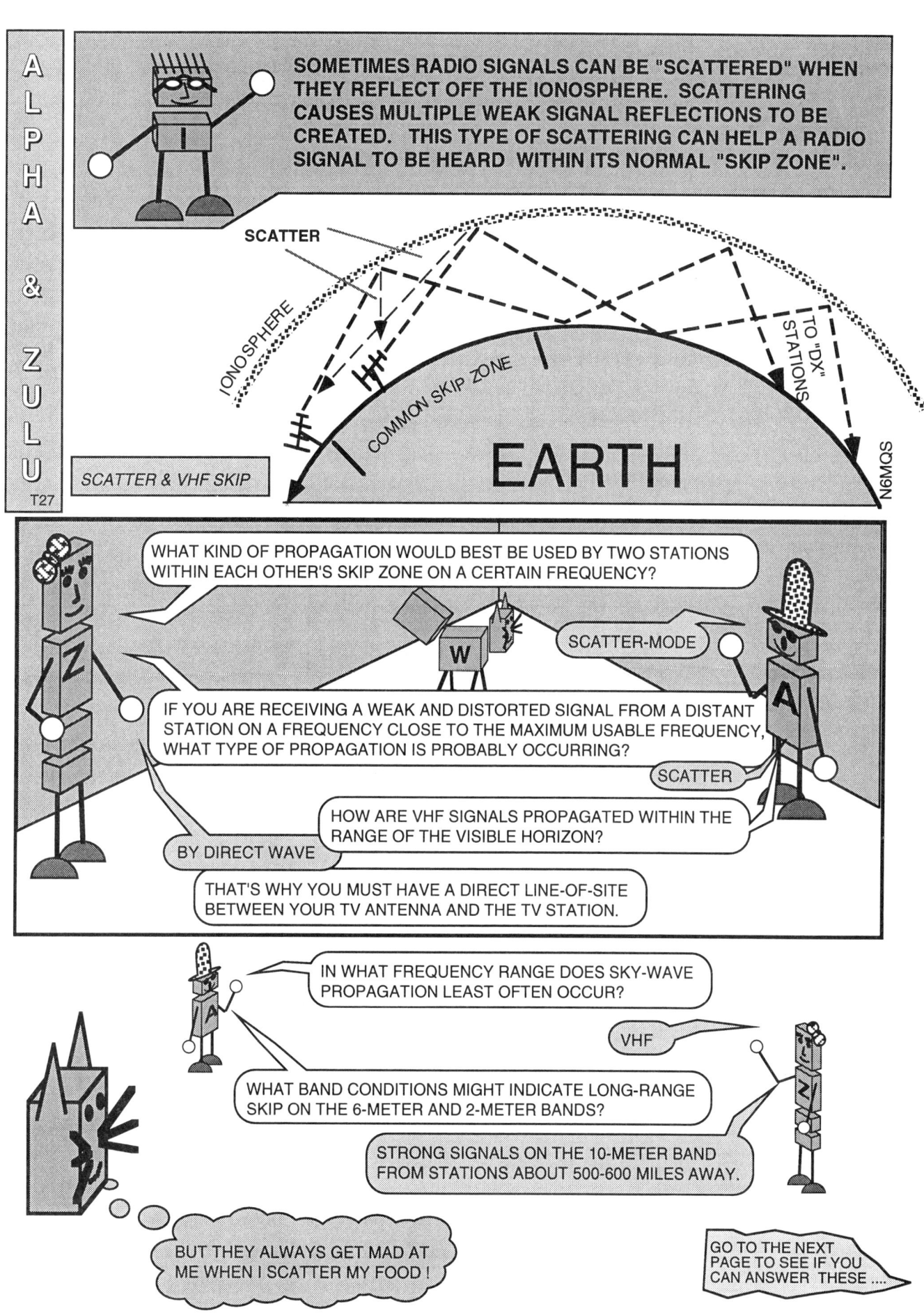

Riding the Airwaves with Alpha & Zulu

SEE IF YOU CAN ANSWER THESE TECHNICIAN QUESTIONS WITHOUT LOOKING BACK TO THE LAST 'TOON'. WRITE YOUR ANSWERS IN THE ANSWER BOX BELOW. GOOD LUCK. DA-DI-DAH

PRACTICE TEST PAGE T27

T3C01
What kind of propagation would best be used by two stations within each other's skip zone on a certain frequency?
 A. Ground-wave
 B. Sky-wave
 C. Scatter-mode
 D. Ducting

T3C02
If you are receiving a weak and distorted signal from a distant station on a frequency close to the maximum usable frequency, what type of propagation is probably occurring?
 A. Ducting
 B. Line-of-sight
 C. Scatter
 D. Ground-wave

T3C03
How are VHF signals propagated within the range of the visible horizon?
 A. By sky wave
 B. By direct wave
 C. By plane wave
 D. By geometric wave

T3C09
In what frequency range does sky-wave propagation least often occur?
 A. LF
 B. MF
 C. HF
 D. VHF

T3C11
What band conditions might indicate long-range skip on the 6-meter and 2-meter bands?
 A. Noise on the 80-meter band
 B. The absence of signals on the 10-meter band
 C. Very long-range skip on the 10-meter band
 D. Strong signals on the 10-meter band from stations about 500-600 miles away

ANSWERS TO PREVIOUS TEST	
T3C04	C
T3C05	A
T3C06	D
T3C07	B
T3C08	D
T3C10	A

YOUR ANSWERS TO THIS TEST
T3C01
T3C02
T3C03
T3C09
T3C11

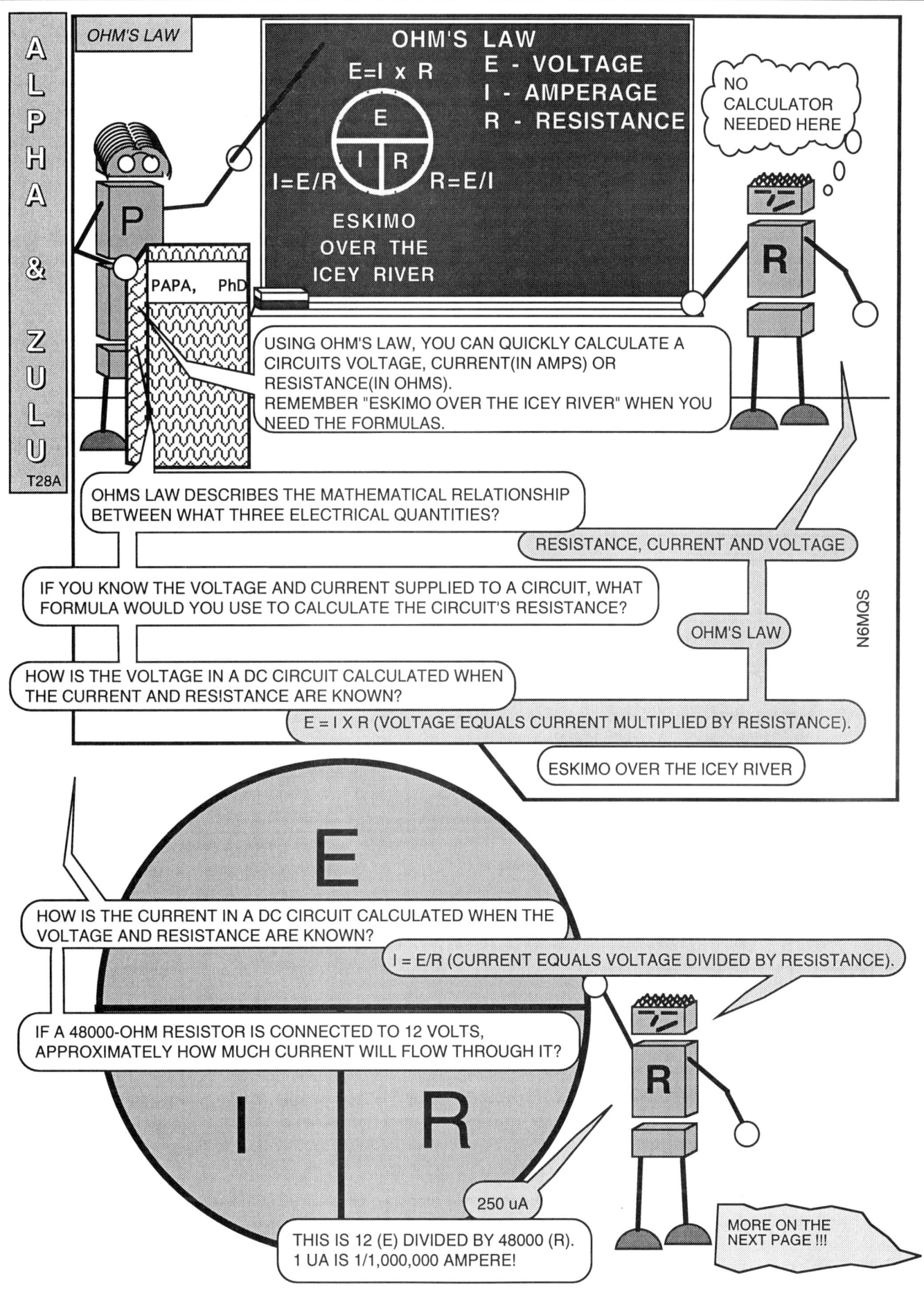

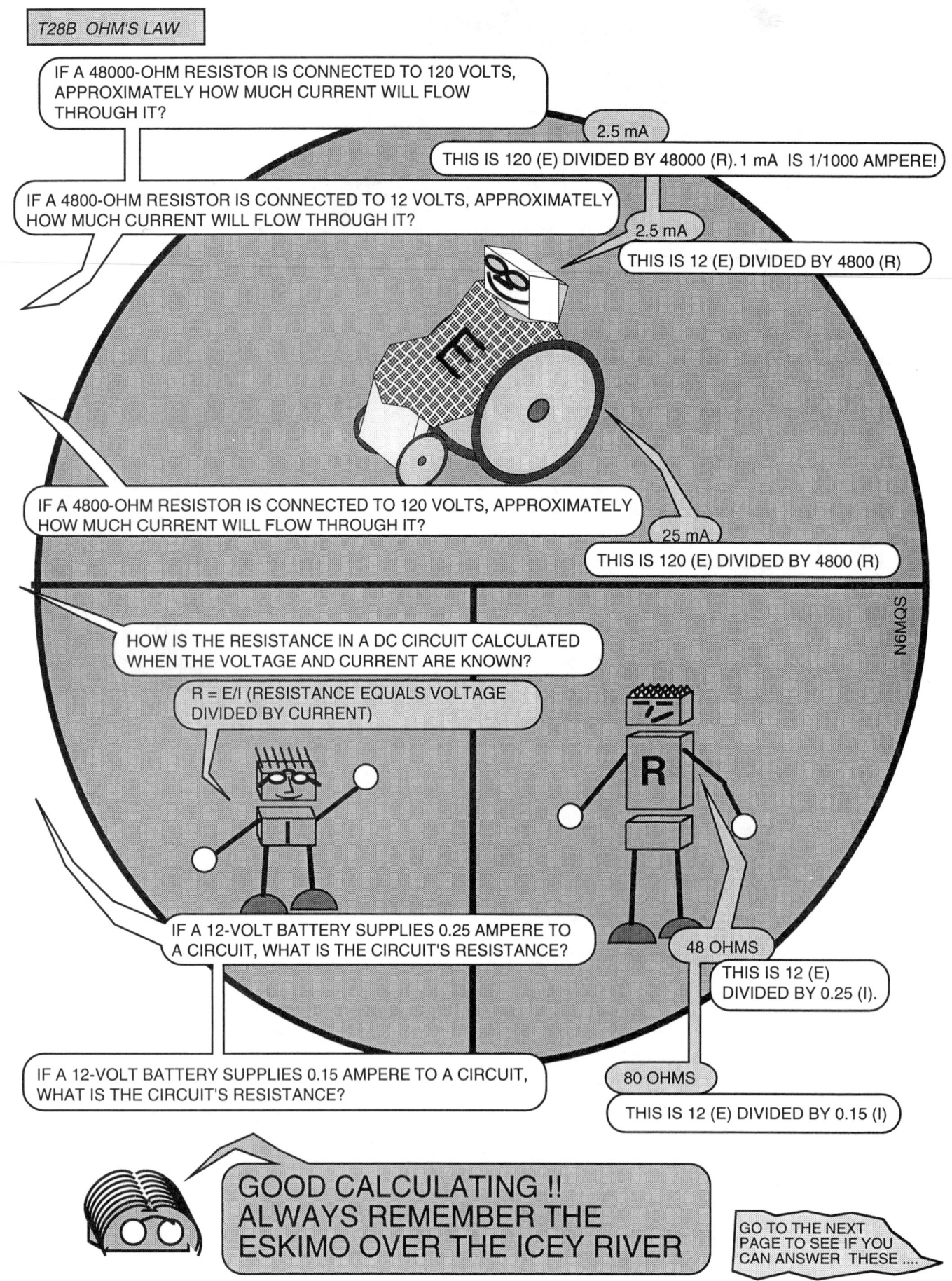

Riding the Airwaves with Alpha & Zulu

"HERE'S ANOTHER TWO PAGE TEST GOOD LUCK."

PRACTICE TEST PAGE T28A

T5B01
Ohm's Law describes the mathematical relationship between what three electrical quantities?
- A. Resistance, voltage and power
- B. Current, resistance and power
- C. Current, voltage and power
- D. Resistance, current and voltage

T5B02
How is the current in a DC circuit calculated when the voltage and resistance are known?
- A. I = R x E [current equals resistance multiplied by voltage]
- B. I = R / E [current equals resistance divided by voltage]
- C. I = E / R [current equals voltage divided by resistance]
- D. I = P / E [current equals power divided by voltage]

T5B03
How is the resistance in a DC circuit calculated when the voltage and current are known?
- A. R = I / E [resistance equals current divided by voltage]
- B. R = E / I [resistance equals voltage divided by current]
- C. R = I x E [resistance equals current multiplied by voltage]
- D. R = P / E [resistance equals power divided by voltage]

T5B04
How is the voltage in a DC circuit calculated when the current and resistance are known?
- A. E = I / R [voltage equals current divided by resistance]
- B. E = R / I [voltage equals resistance divided by current]
- C. E = I x R [voltage equals current multiplied by resistance]
- D. E = P / I [voltage equals power divided by current]

T5B05
If a 12-volt battery supplies 0.25 ampere to a circuit, what is the circuit's resistance?
- A. 0.25 ohm
- B. 3 ohm
- C. 12 ohms
- D. 48 ohms

ANSWERS TO PREVIOUS TEST

T3C01	C
T3C02	C
T3C03	B
T3C09	D
T3C11	D

artsci inc

227

PRACTICE TEST PAGE T28B

T5B06
If a 12-volt battery supplies 0.15 ampere to a circuit, what is the circuit's resistance?
- A. 0.15 ohm
- B. 1.8 ohm
- C. 12 ohms
- D. 80 ohms

T5B07
If a 4800-ohm resistor is connected to 120 volts, approximately how much current will flow through it?
- A. 4 A
- B. 25 mA
- C. 25 A
- D. 40 MA

T5B08
If a 48000-ohm resistor is connected to 120 volts, approximately how much current will flow through it?
- A. 400 A
- B. 40 A
- C. 25 mA
- D. 2.5 mA

T5B09
If a 4800-ohm resistor is connected to 12 volts, approximately how much current will flow through it?
- A. 2.5 mA
- B. 25 mA
- C. 40 A
- D. 400 A

T5B11
If you know the voltage and current supplied to a circuit, what formula would you use to calculate the circuit's resistance?
- A. Ohm's law
- B. Tesla's law
- C. Ampere's law
- D. Kirchhoff's law

T5B10
If a 48000-ohm resistor is connected to 12 volts, approximately how much current will flow through it?
- A. 250 uA
- B. 250 mA
- C. 4000 mA
- D. 4000 A

YOUR ANSWERS TO THIS TEST
T5B01
T5B02
T5B03
T5B04
T5B05
T5B06
T5B07
T5B08
T5B09
T5B11
T5B10

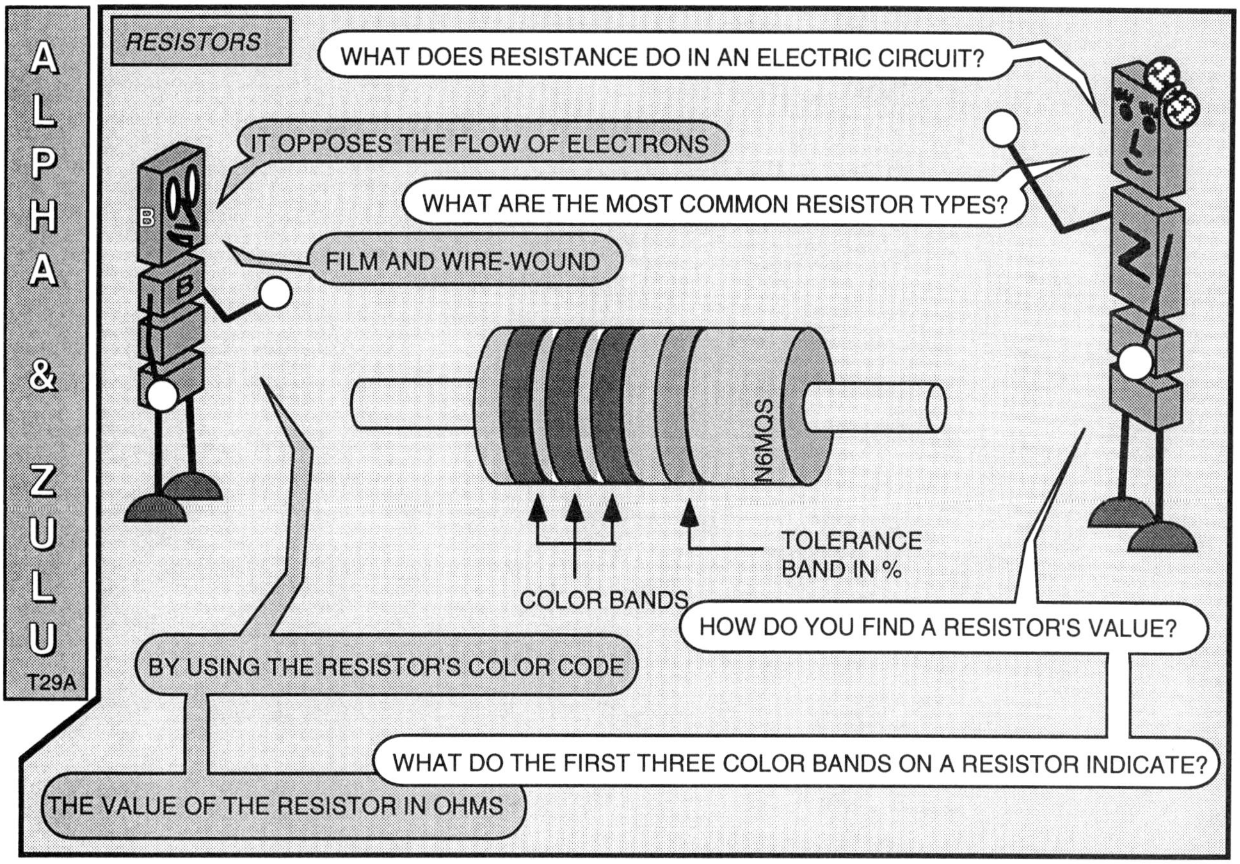

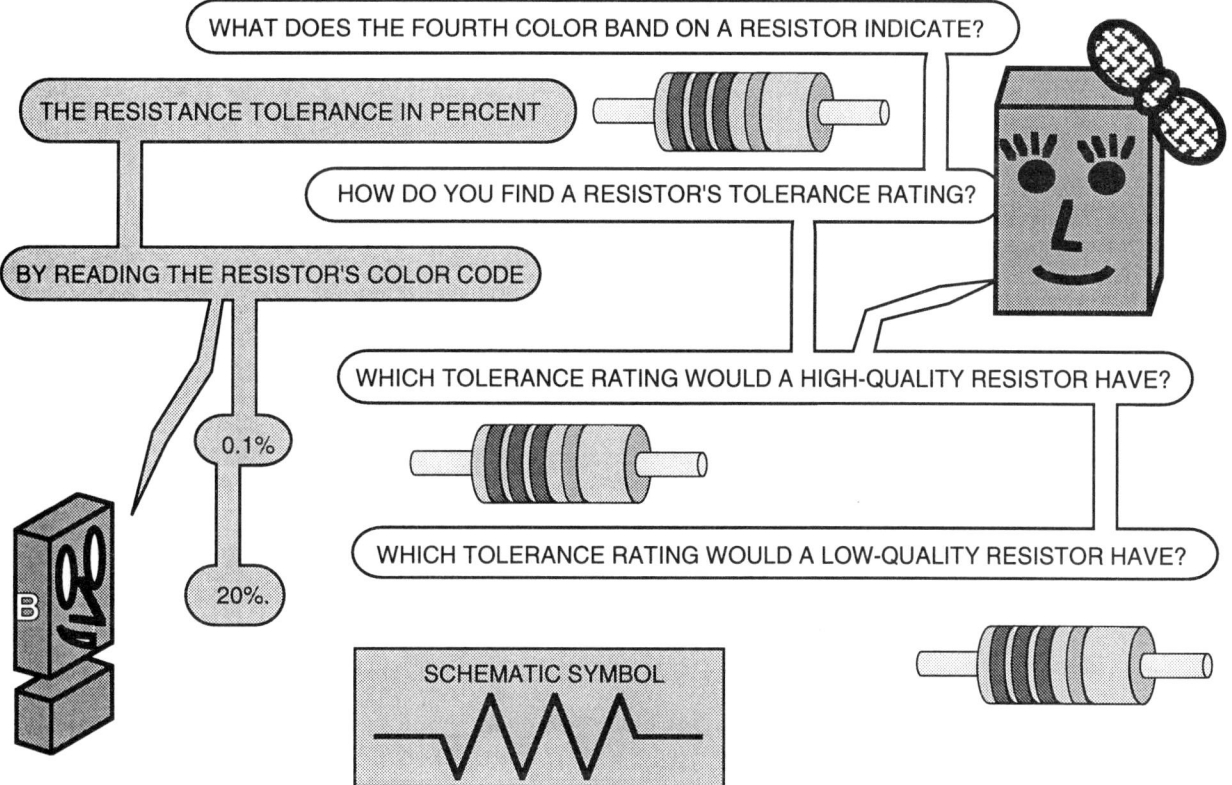

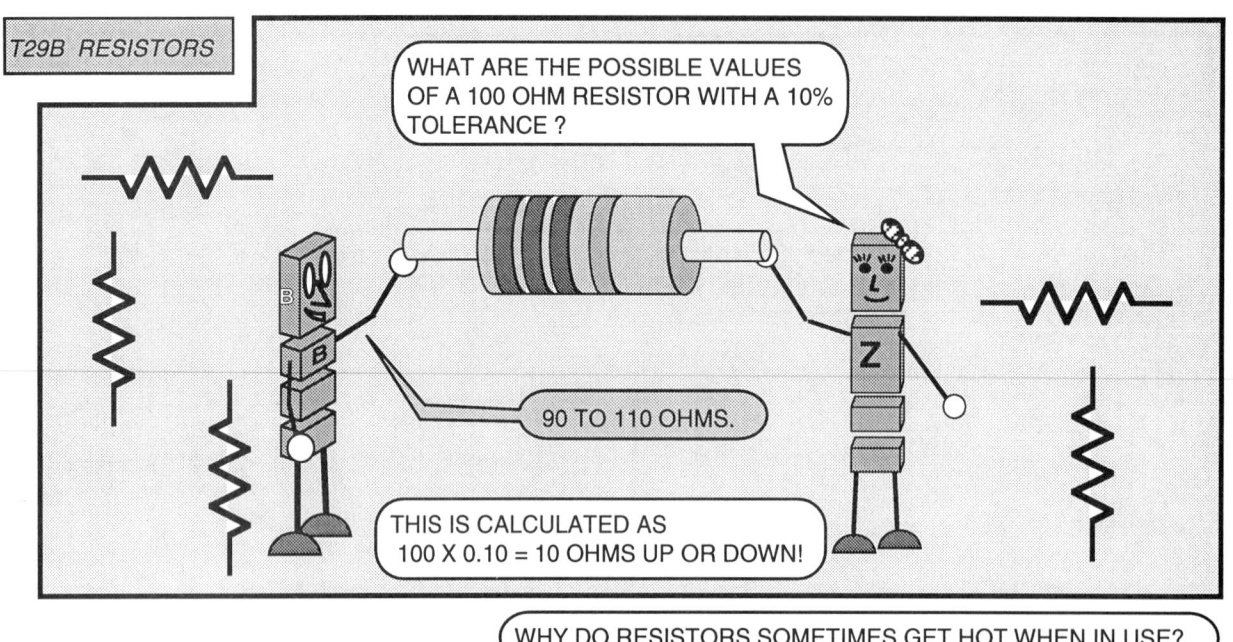
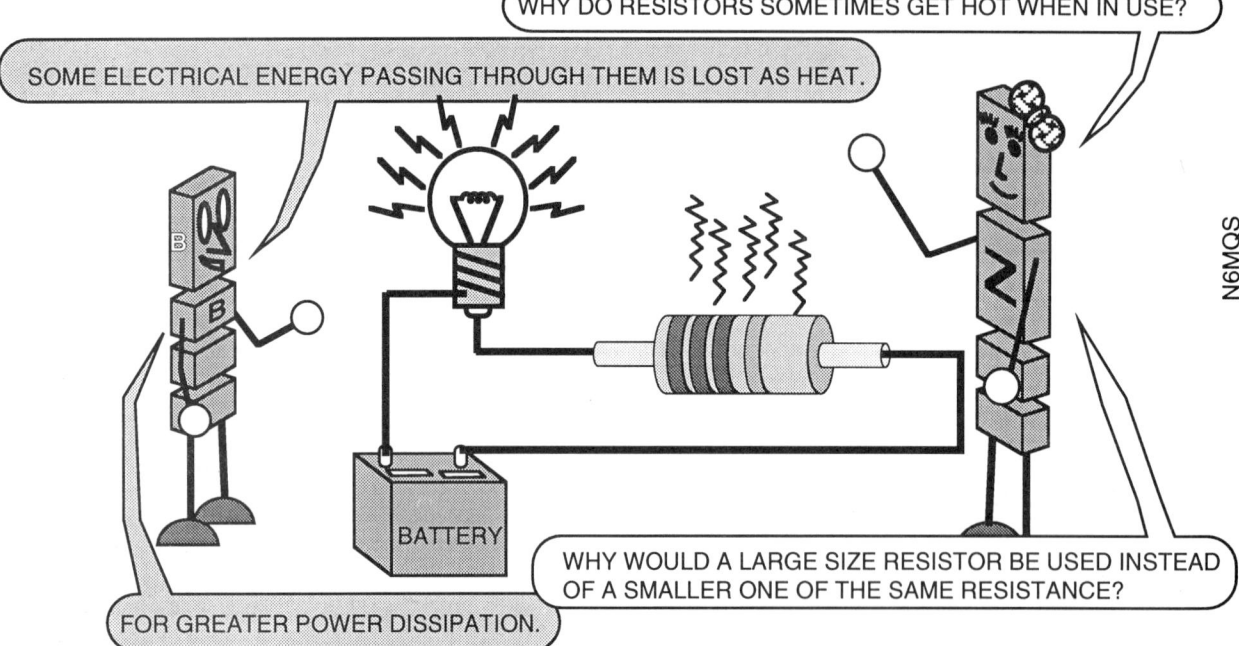
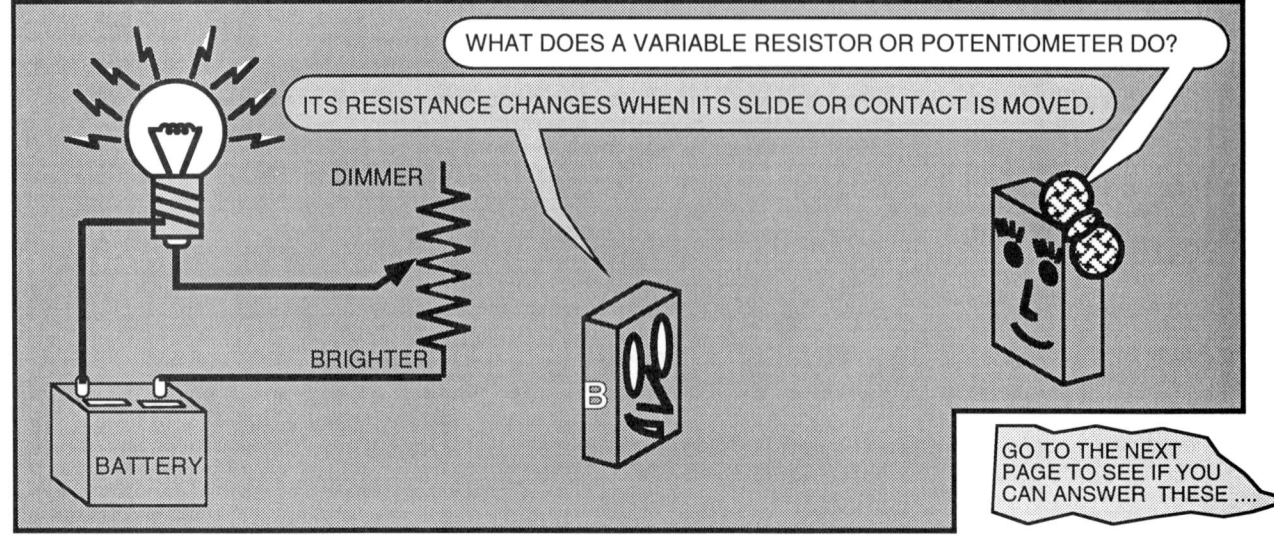

Riding the Airwaves with Alpha & Zulu

WOW, ANOTHER TWO PAGE TEST GOOD LUCK.

PRACTICE TEST PAGE T29A

T5A01
What does resistance do in an electric circuit?
A. It stores energy in a magnetic field
B. It stores energy in an electric field
C. It provides electrons by a chemical reaction
D. It opposes the flow of electrons

T6A01
What are the most common resistor types?
A. Plastic and porcelain
B. Film and wire-wound
C. Electrolytic and metal-film
D. Iron core and brass core

T6A02
What does a variable resistor or potentiometer do?
A. Its resistance changes when AC is applied to it
B. It transforms a variable voltage into a constant voltage
C. Its resistance changes when its slide or contact is moved
D. Its resistance changes when it is heated

T6A03
How do you find a resistor's tolerance rating?
A. By using a voltmeter
B. By reading the resistor's color code
C. By using Thevenin's theorem for resistors
D. By reading its Baudot code

T6A04
What do the first three color bands on a resistor indicate?
A. The value of the resistor in ohms
B. The resistance tolerance in percent
C. The power rating in watts
D. The resistance material

T6A05
What does the fourth color band on a resistor indicate?
A. The value of the resistor in ohms
B. The resistance tolerance in percent
C. The power rating in watts
D. The resistance material

ANSWERS TO PREVIOUS TEST

T5B01	D
T5B02	C
T5B03	B
T5B04	C
T5B05	D
T5B06	D
T5B07	B
T5B08	D
T5B09	A
T5B11	A
T5B10	A

artsci inc

PRACTICE TEST PAGE T29B

T6A06
Why do resistors sometimes get hot when in use?
A. Some electrical energy passing through them is lost as heat
B. Their reactance makes them heat up
C. Hotter circuit components nearby heat them up
D. They absorb magnetic energy which makes them hot

T6A07
Why would a large size resistor be used instead of a smaller one of the same resistance?
A. For better response time
B. For a higher current gain
C. For greater power dissipation
D. For less impedance in the circuit

T6A08
What are the possible values of a 100-ohm resistor with a 10% tolerance?
A. 90 to 100 ohms
B. 10 to 100 ohms
C. 90 to 110 ohms
D. 80 to 120 ohms

T6A09
How do you find a resistor's value?
A. By using a voltmeter
B. By using the resistor's color code
C. By using Thevenin's theorem for resistors
D. By using the Baudot code

T6A10
Which tolerance rating would a high-quality resistor have?
A. 0.1%
B. 5%
C. 10%
D. 20%

T6A11
Which tolerance rating would a low-quality resistor have?
A. 0.1%
B. 5%
C. 10%
D. 20%

YOUR ANSWERS TO THIS TEST

T5A01
T6A01
T6A02
T6A03
T6A04
T6A05
T6A06
T6A07
T6A08
T6A09
T6A10
T6A11

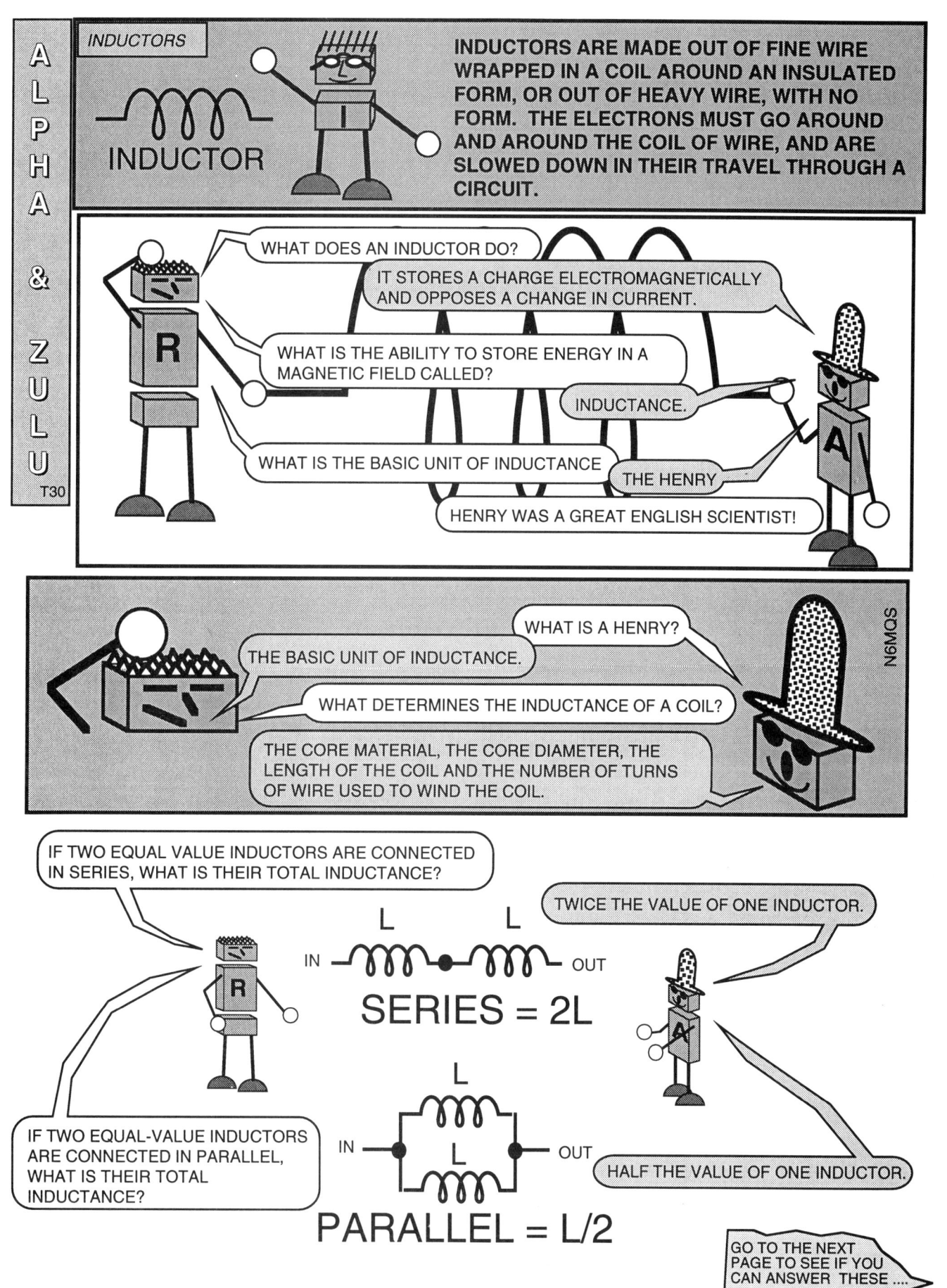

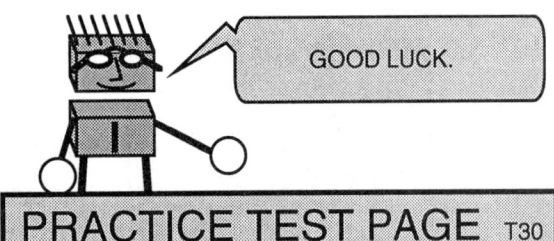

PRACTICE TEST PAGE T30

"GOOD LUCK."

T5A02
What is the ability to store energy in a magnetic field called?
A. Admittance
B. Capacitance
C. Resistance
D. Inductance

T5A03
What is the basic unit of inductance?
A. The coulomb
B. The farad
C. The henry
D. The ohm

T5A04
What is a henry?
A. The basic unit of admittance
B. The basic unit of capacitance
C. The basic unit of inductance
D. The basic unit of resistance

T6B03 (D)
What determines the inductance of a coil?
A. The core material, the core diameter, the length of the coil and whether the coil is mounted horizontally or vertically
B. The core diameter, the number of turns of wire used to wind the coil and the type of metal used for the wire
C. The core material, the number of turns used to wind the core and the frequency of the current through the coil
D. The core material, the core diameter, the length of the coil and the number of turns of wire used to wind the coil

T5A08
If two equal-value inductors are connected in series, what is their total inductance?
A. Half the value of one inductor
B. Twice the value of one inductor
C. The same as the value of either inductor
D. The value of one inductor times the value of the other

T5A09
If two equal-value inductors are connected in parallel, what is their total inductance?
A. Half the value of one inductor
B. Twice the value of one inductor
C. The same as the value of either inductor
D. The value of one inductor times the value of the other

T6B02
What does an inductor do?
A. It stores a charge electrostatically and opposes a change in voltage
B. It stores a charge electrochemically and opposes a change in current
C. It stores a charge electromagnetically and opposes a change in current
D. It stores a charge electromechanically and opposes a change in voltage

ANSWERS TO PREVIOUS TEST

T5A01	D
T6A01	B
T6A02	C
T6A03	B
T6A04	A
T6A05	B
T6A06	A
T6A07	C
T6A08	C
T6A09	B
T6A10	A
T6A11	D

YOUR ANSWERS TO THIS TEST

T5A02 ___
T5A03 ___
T5A04 ___
T6B03 ___
T5A08 ___
T5A09 ___
T6B02 ___

artsci inc

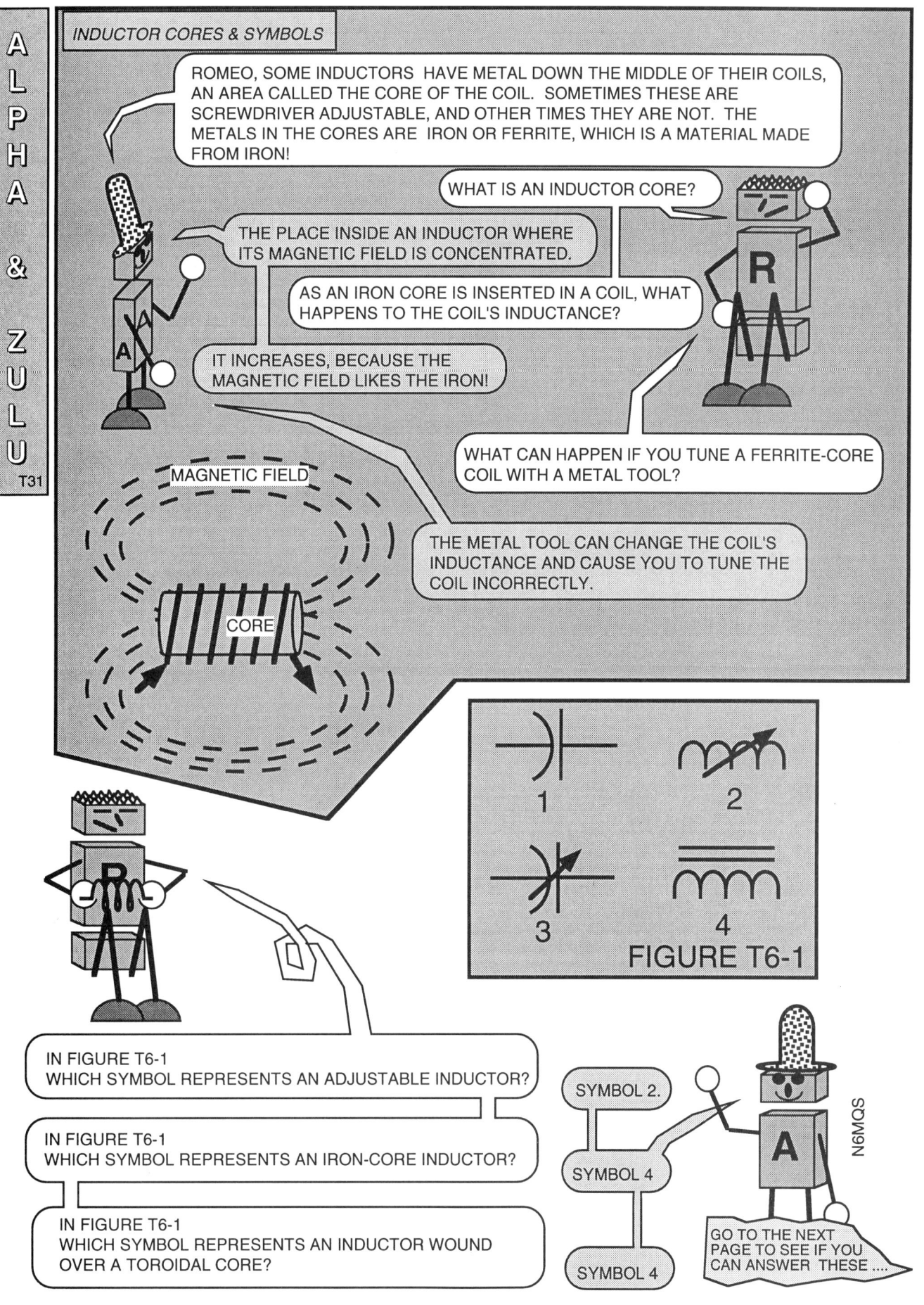

Riding the Airwaves with Alpha & Zulu

SEE IF YOU CAN ANSWER THESE TECHNICIAN QUESTIONS WITHOUT LOOKING BACK TO THE LAST 'TOON'. WRITE YOUR ANSWERS IN THE ANSWER BOX BELOW. GOOD LUCK. DA-DI-DAH

PRACTICE TEST PAGE T31

T6B01
What is an inductor core?
A. The place where a coil is tapped for resonance
B. A tight coil of wire used in a transformer
C. Insulating material placed between the wires of a transformer
D. The place inside an inductor where its magnetic field is concentrated

T6B04
As an iron core is inserted in a coil, what happens to the coil's inductance?
A. It increases
B. It decreases
C. It stays the same
D. It disappears

T6B05
What can happen if you tune a ferrite-core coil with a metal tool?
A. The metal tool can change the coil's inductance and cause you to tune the coil incorrectly
B. The metal tool can become magnetized so much that you might not be able to remove it from the coil
C. The metal tool can pick up enough magnetic energy to become very hot
D. The metal tool can pick up enough magnetic energy to become a shock hazard

T6B06
In Figure T6-1 which symbol represents an adjustable inductor?
A. Symbol 1
B. Symbol 2
C. Symbol 3
D. Symbol 4

T6B07
In Figure T6-1 which symbol represents an iron-core inductor?
A. Symbol 1
B. Symbol 2
C. Symbol 3
D. Symbol 4

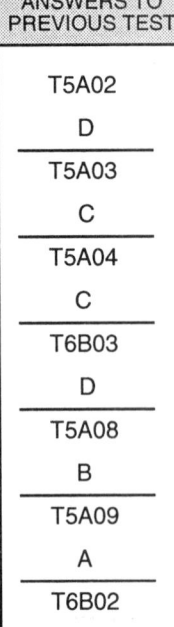

ANSWERS TO PREVIOUS TEST
T5A02
D
T5A03
C
T5A04
C
T6B03
D
T5A08
B
T5A09
A
T6B02
C

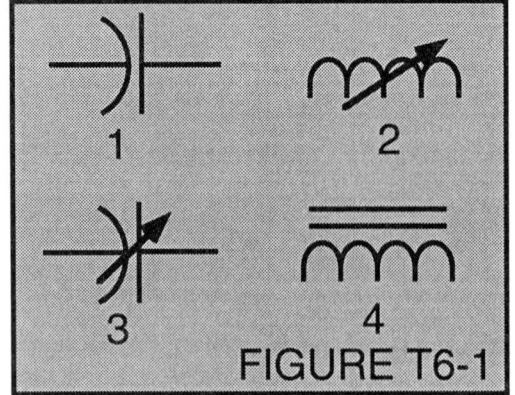

FIGURE T6-1

T6B08
In Figure T6-1 which symbol represents an inductor wound over a toroidal core?
A. Symbol 1
B. Symbol 2
C. Symbol 3
D. Symbol 4

YOUR ANSWERS TO THIS TEST
T6B01
T6B04
T6B05
T6B06
T6B07
T6B08

artsci inc

Riding the Airwaves with Alpha & Zulu

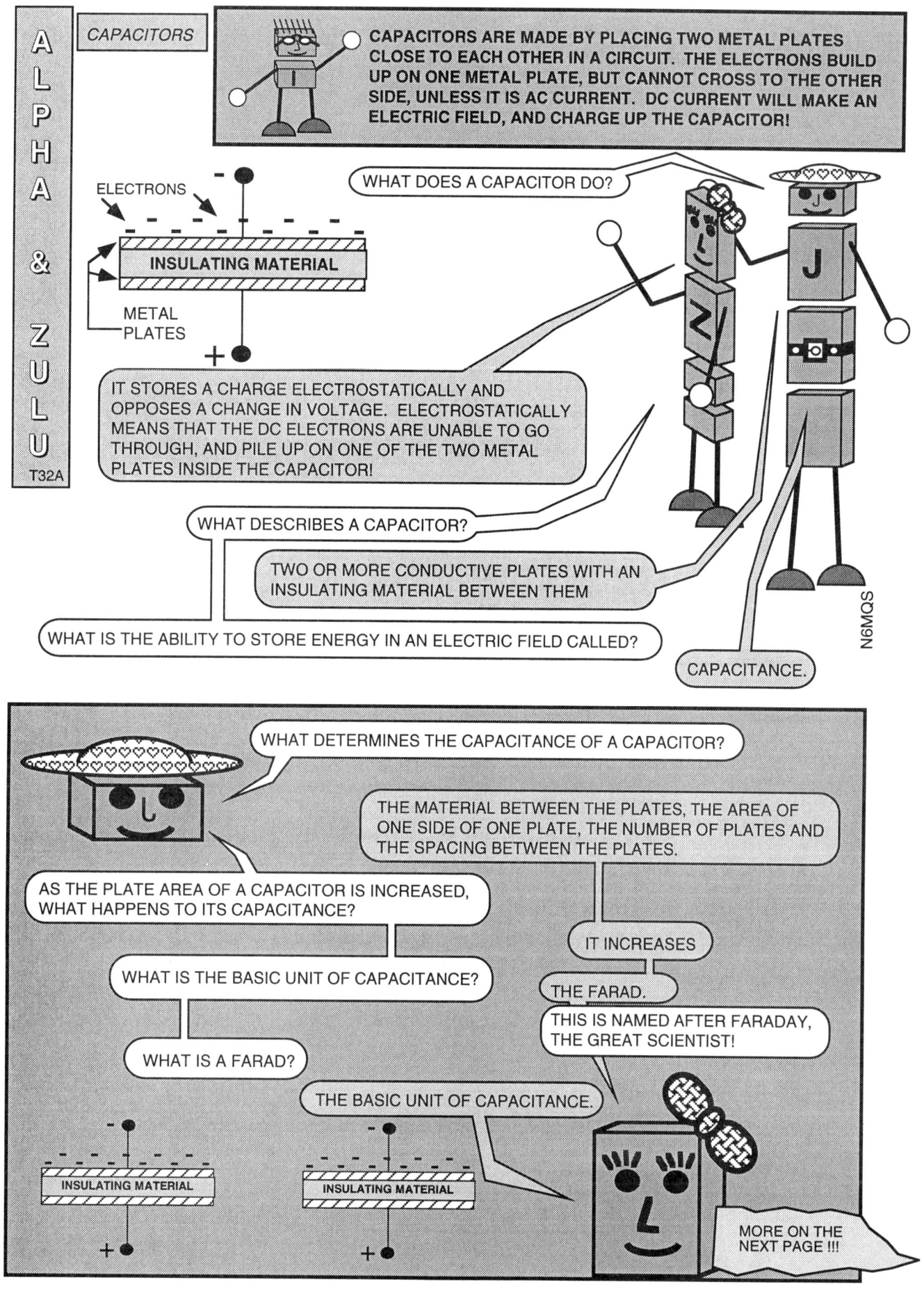

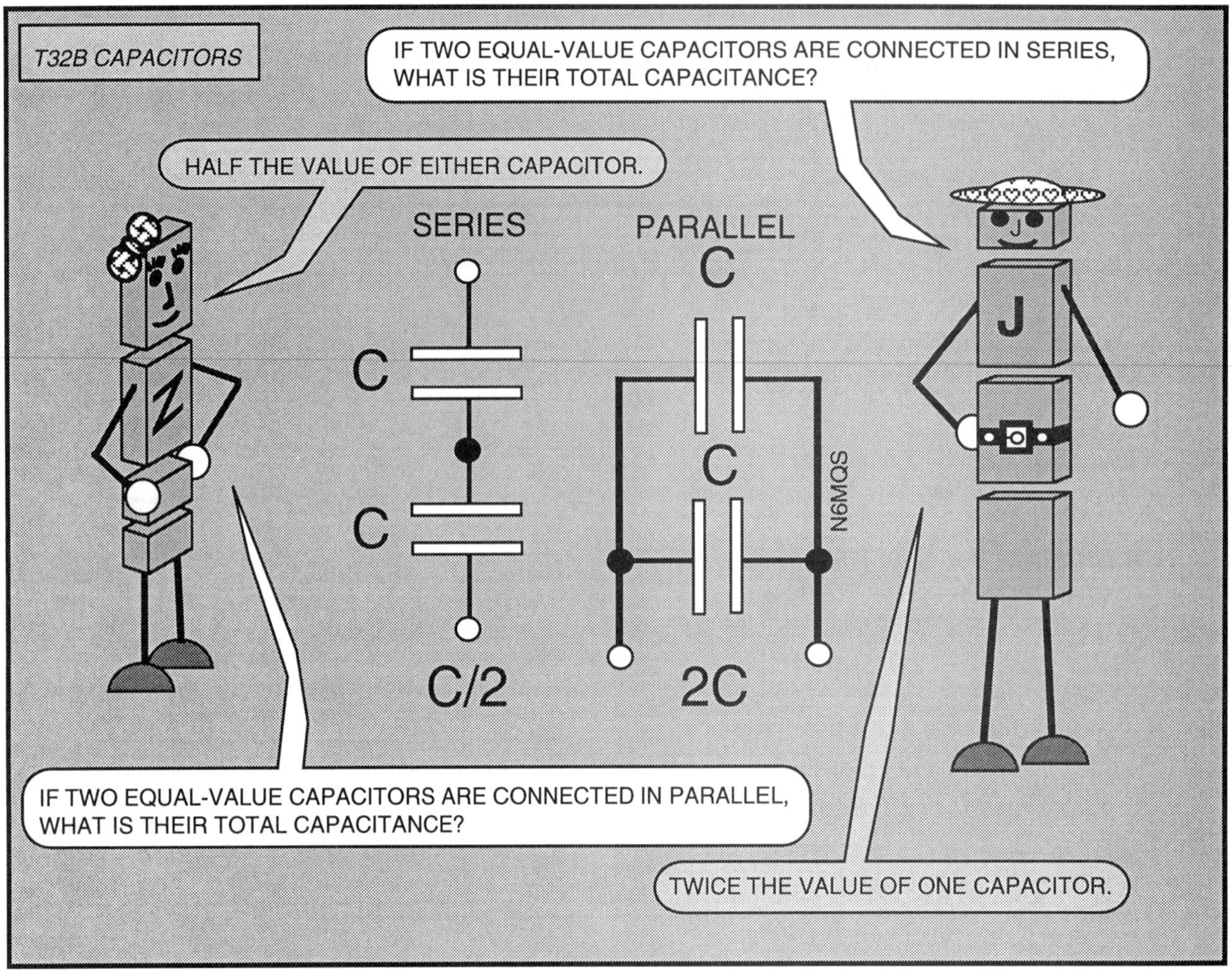

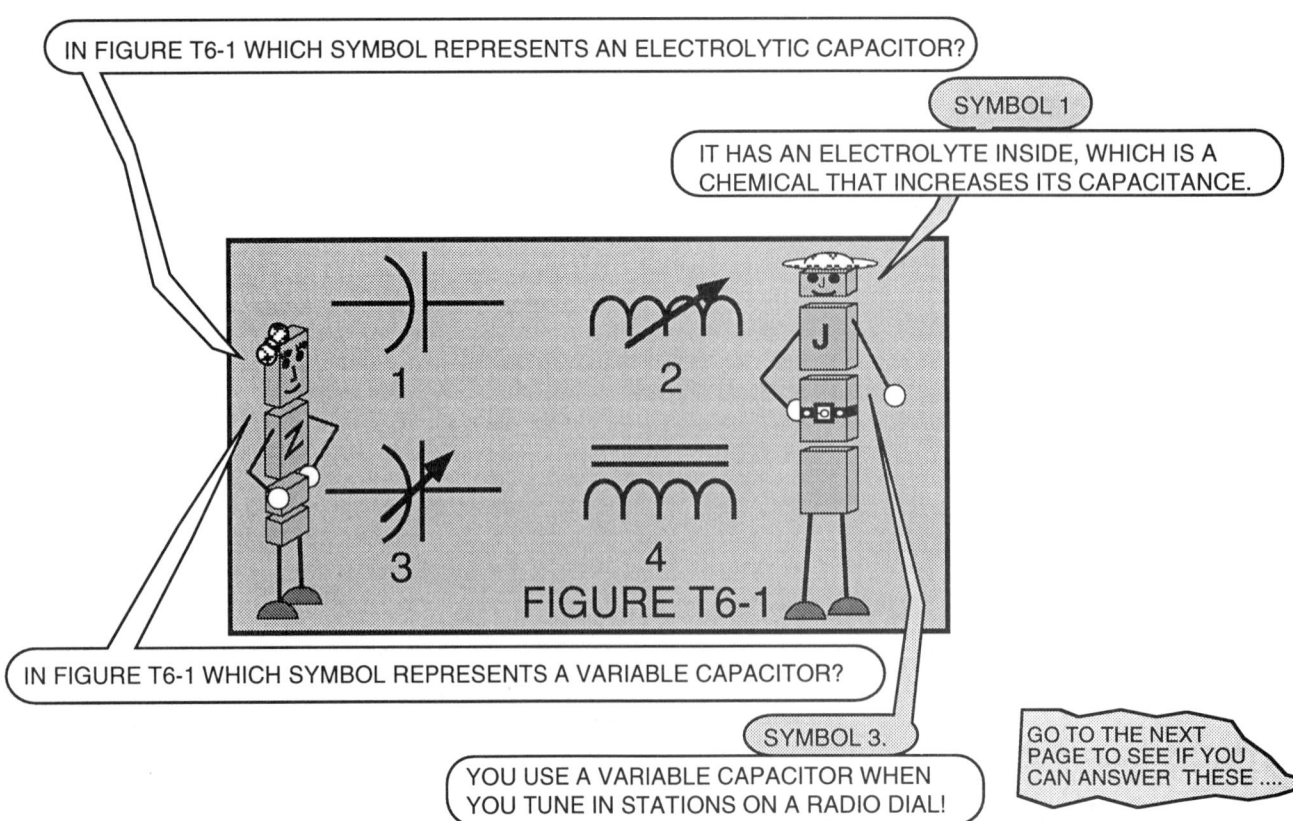

HERE'S A TWO PAGE TEST
GOOD LUCK. DA-DI-DAH

PRACTICE TEST PAGE T32A

T5A05
What is the ability to store energy in an electric field called?
A. Inductance
B. Resistance
C. Tolerance
D. Capacitance

T5A06
What is the basic unit of capacitance?
A. The farad
B. The ohm
C. The volt
D. The henry

T5A07
What is a farad?
A. The basic unit of resistance
B. The basic unit of capacitance
C. The basic unit of inductance
D. The basic unit of admittance

T6B09
In Figure T6-1 which symbol represents an electrolytic capacitor?
A. Symbol 1
B. Symbol 2
C. Symbol 3
D. Symbol 4

T5A10
If two equal-value capacitors are connected in series, what is their total capacitance?
A. Twice the value of one capacitor
B. The same as the value of either capacitor
C. Half the value of either capacitor
D. The value of one capacitor times the value of the other

T5A11
If two equal-value capacitors are connected in parallel, what is their total capacitance?
A. Twice the value of one capacitor
B. Half the value of one capacitor
C. The same as the value of either capacitor
D. The value of one capacitor times the value of the other

T6B10
In Figure T6-1 which symbol represents a variable capacitor?
A. Symbol 1
B. Symbol 2
C. Symbol 3
D. Symbol 4

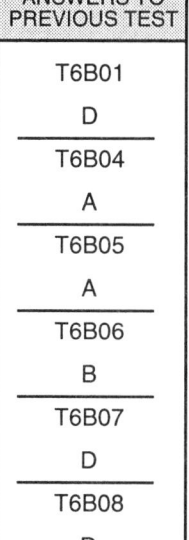

ANSWERS TO PREVIOUS TEST

T6B01	D
T6B04	A
T6B05	A
T6B06	B
T6B07	D
T6B08	D

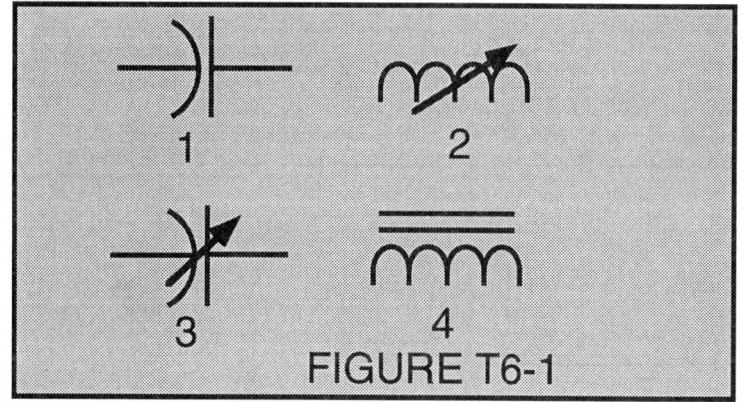

FIGURE T6-1

PRACTICE TEST PAGE T32B

T6B11
What describes a capacitor?
A. Two or more layers of silicon material with an insulating material between them
B. The wire used in the winding and the core material
C. Two or more conductive plates with an insulating material between them
D. Two or more insulating plates with a conductive material between them

T6B12
What does a capacitor do?
A. It stores a charge electrochemically and opposes a change in current
B. It stores a charge electrostatically and opposes a change in voltage
C. It stores a charge electromagnetically and opposes a change in current
D. It stores a charge electromechanically and opposes a change in voltage

T6B13
What determines the capacitance of a capacitor?
A. The material between the plates, the area of one side of one plate, the number of plates and the spacing between the plates
B. The material between the plates, the number of plates and the size of the wires connected to the plates
C. The number of plates, the spacing between the plates and whether the dielectric material is N type or P type
D. The material between the plates, the area of one plate, the number of plates and the material used for the protective coating

T6B14
As the plate area of a capacitor is increased, what happens to its capacitance?
A. It decreases
B. It increases
C. It stays the same
D. It disappears

YOUR ANSWERS TO THIS TEST
T5A05
T5A06
T5A07
T6B09
T5A10
T5A11
T6B10
T6B11
T6B12
T6B13
T6B14

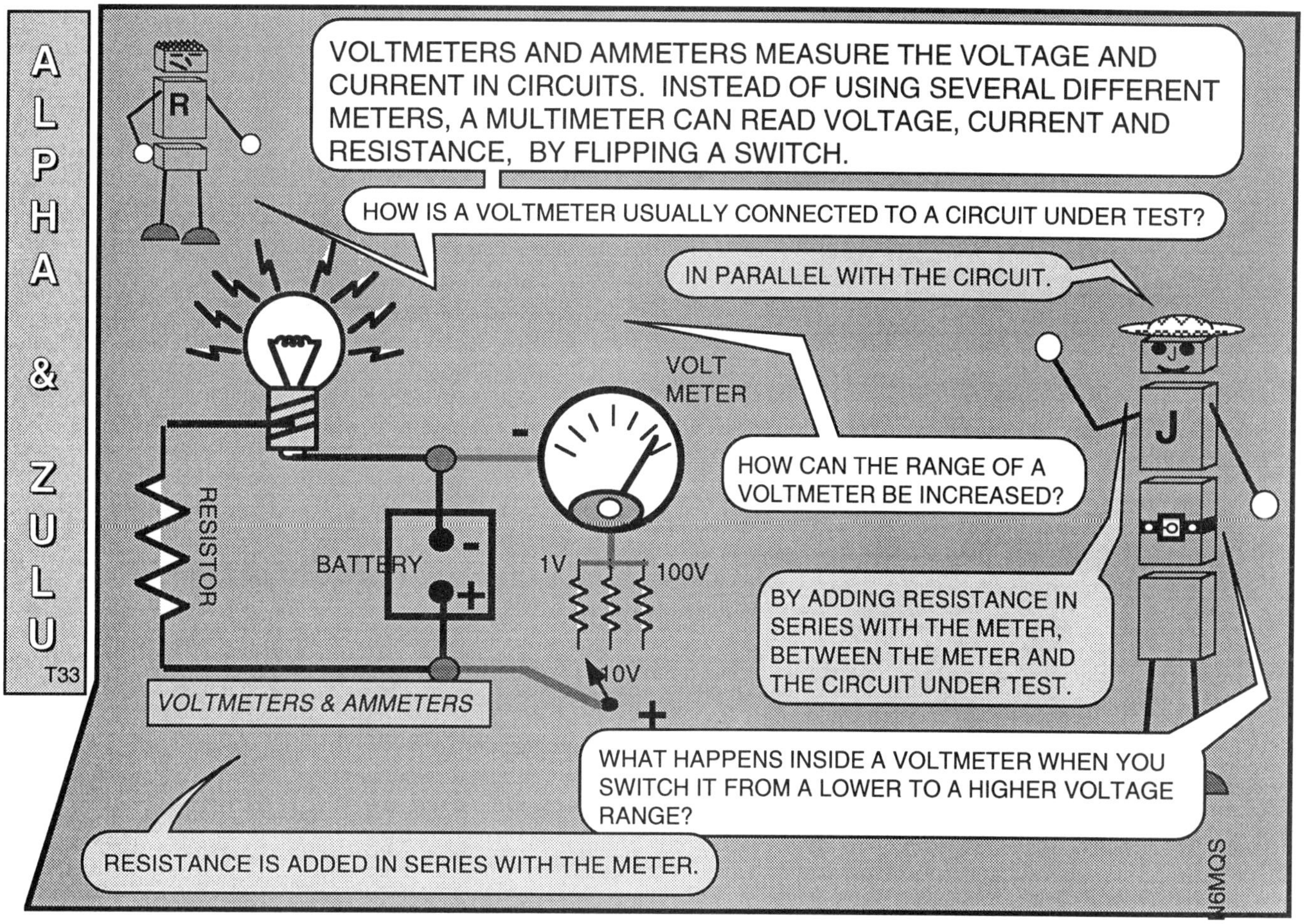

Riding the Airwaves with Alpha & Zulu

ALPHA & ZULU — *VOLTMETERS & AMMETERS* — T33

VOLTMETERS AND AMMETERS MEASURE THE VOLTAGE AND CURRENT IN CIRCUITS. INSTEAD OF USING SEVERAL DIFFERENT METERS, A MULTIMETER CAN READ VOLTAGE, CURRENT AND RESISTANCE, BY FLIPPING A SWITCH.

HOW IS A VOLTMETER USUALLY CONNECTED TO A CIRCUIT UNDER TEST?

IN PARALLEL WITH THE CIRCUIT.

HOW CAN THE RANGE OF A VOLTMETER BE INCREASED?

BY ADDING RESISTANCE IN SERIES WITH THE METER, BETWEEN THE METER AND THE CIRCUIT UNDER TEST.

WHAT HAPPENS INSIDE A VOLTMETER WHEN YOU SWITCH IT FROM A LOWER TO A HIGHER VOLTAGE RANGE?

RESISTANCE IS ADDED IN SERIES WITH THE METER.

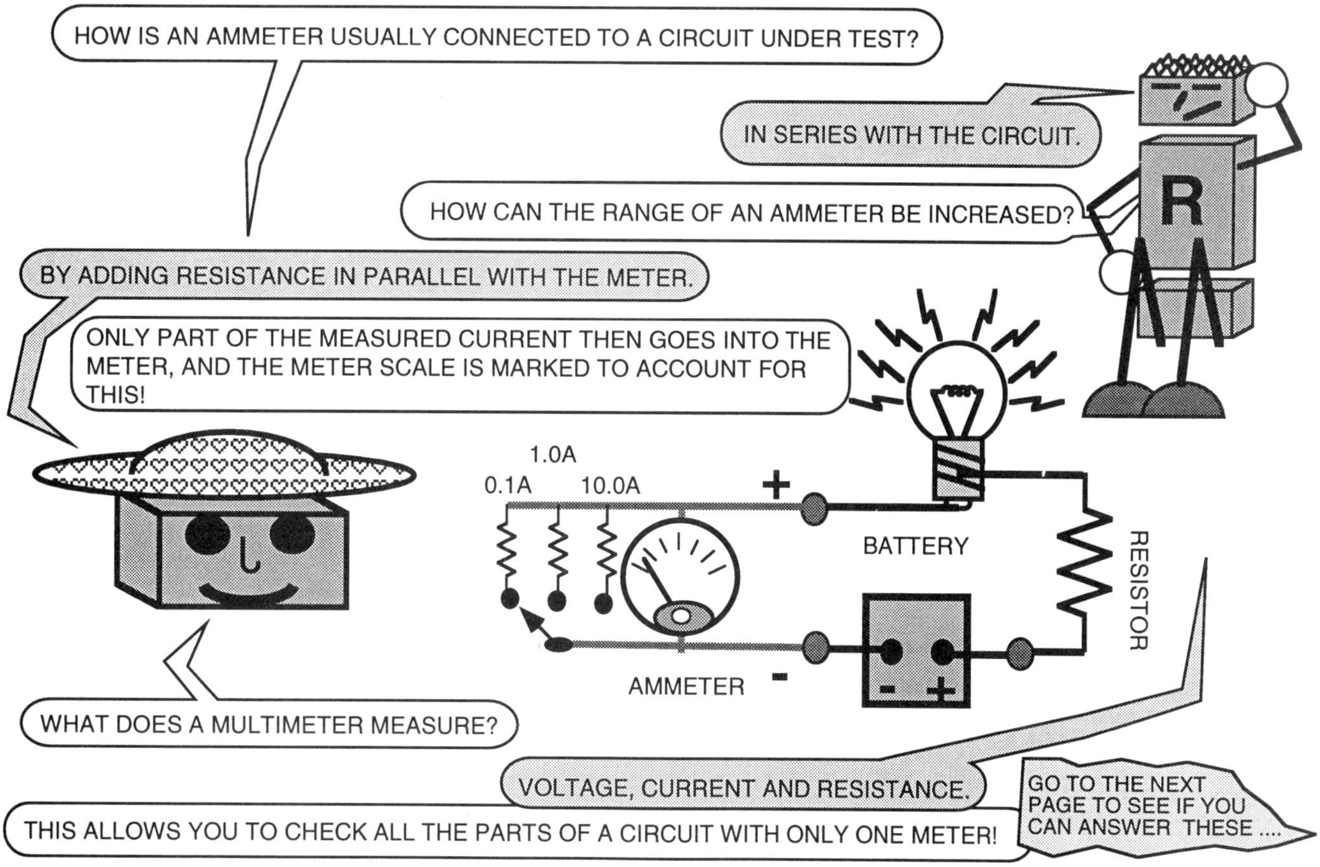

HOW IS AN AMMETER USUALLY CONNECTED TO A CIRCUIT UNDER TEST?

IN SERIES WITH THE CIRCUIT.

HOW CAN THE RANGE OF AN AMMETER BE INCREASED?

BY ADDING RESISTANCE IN PARALLEL WITH THE METER.

ONLY PART OF THE MEASURED CURRENT THEN GOES INTO THE METER, AND THE METER SCALE IS MARKED TO ACCOUNT FOR THIS!

WHAT DOES A MULTIMETER MEASURE?

VOLTAGE, CURRENT AND RESISTANCE.

THIS ALLOWS YOU TO CHECK ALL THE PARTS OF A CIRCUIT WITH ONLY ONE METER!

GO TO THE NEXT PAGE TO SEE IF YOU CAN ANSWER THESE ....

artsci inc

Riding the Airwaves with Alpha & Zulu

SEE IF YOU CAN ANSWER THESE TECHNICIAN QUESTIONS WITHOUT LOOKING BACK TO THE LAST 'TOON'. WRITE YOUR ANSWERS IN THE ANSWER BOX BELOW. GOOD LUCK. DA-DI-DAH

PRACTICE TEST PAGE T33

T4B01
How is a voltmeter usually connected to a circuit under test?
- A. In series with the circuit
- B. In parallel with the circuit
- C. In quadrature with the circuit
- D. In phase with the circuit

T4B02
How can the range of a voltmeter be increased?
- A. By adding resistance in series with the circuit under test
- B. By adding resistance in parallel with the circuit under test
- C. By adding resistance in series with the meter, between the meter and the circuit under test
- D. By adding resistance in parallel with the meter, between the meter and the circuit under test

T4B03
What happens inside a voltmeter when you switch it from a lower to a higher voltage range?
- A. Resistance is added in series with the meter
- B. Resistance is added in parallel with the meter
- C. Resistance is reduced in series with the meter
- D. Resistance is reduced in parallel with the meter

T4B04
How is an ammeter usually connected to a circuit under test?
- A. In series with the circuit
- B. In parallel with the circuit
- C. In quadrature with the circuit
- D. In phase with the circuit

T4B05
How can the range of an ammeter be increased?
- A. By adding resistance in series with the circuit under test
- B. By adding resistance in parallel with the circuit under test
- C. By adding resistance in series with the meter
- D. By adding resistance in parallel with the meter

T4B06
What does a multimeter measure?
- A. SWR and power
- B. Resistance, capacitance and inductance
- C. Resistance and reactance
- D. Voltage, current and resistance

ANSWERS TO PREVIOUS TEST

T5A05	D
T5A06	A
T5A07	B
T6B09	A
T5A10	C
T5A11	A
T6B10	C
T6B11	C
T6B12	B
T6B13	A
T6B14	B

YOUR ANSWERS TO THIS TEST

T4B01	
T4B02	
T4B03	
T4B04	
T4B05	
T4B06	

artsci inc

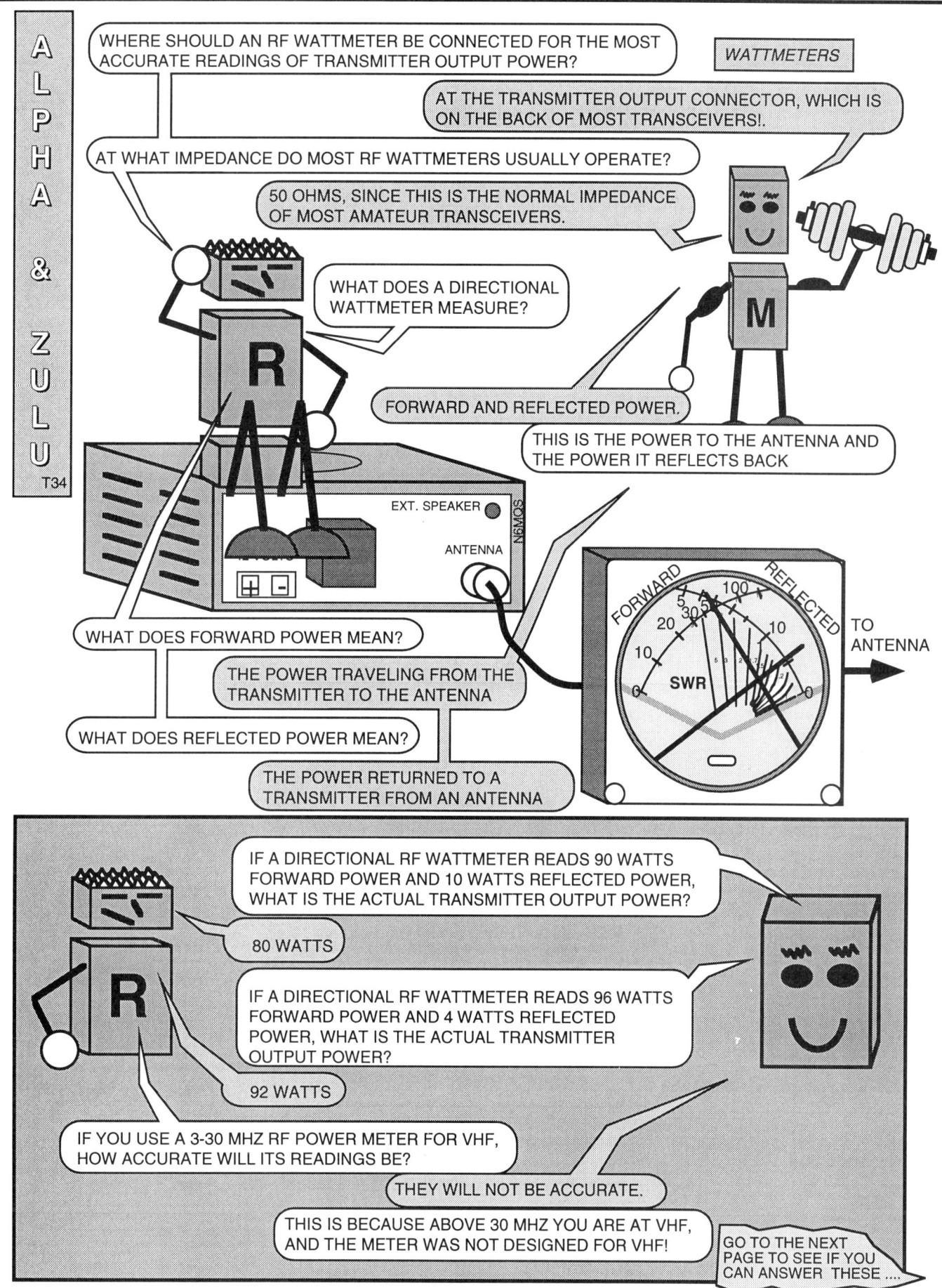

PRACTICE TEST PAGE T34

GOOD LUCK.

T4B07
Where should an RF wattmeter be connected for the most accurate readings of transmitter output power?
A. At the transmitter output connector
B. At the antenna feed point
C. One-half wavelength from the transmitter output
D. One-half wavelength from the antenna feed point

T4B08
At what line impedance do most RF wattmeters usually operate?
A. 25 ohms
B. 50 ohms
C. 100 ohms
D. 300 ohms

T4B09
What does a directional wattmeter measure?
A. Forward and reflected power
B. The directional pattern of an antenna
C. The energy used by a transmitter
D. Thermal heating in a load resistor

T4C10
If you use a 3-30 MHz RF power meter for VHF, how accurate will its readings be?
A. They will not be accurate
B. They will be accurate enough to get by
C. If it properly calibrates to full scale in the set position, they may be accurate
D. They will be accurate providing the readings are multiplied by 4.5

T9B07
What does forward power mean?
A. The power traveling from the transmitter to the antenna
B. The power radiated from the top of an antenna system
C. The power produced during the positive half of an RF cycle
D. The power used to drive a linear amplifier

T9B08
What does reflected power mean?
A. The power radiated down to the ground from an antenna
B. The power returned to a transmitter from an antenna
C. The power produced during the negative half of an RF cycle
D. The power returned to an antenna by buildings and trees

T4B10 (B)
If a directional RF wattmeter reads 90 watts forward power and 10 watts reflected power, what is the actual transmitter output power?
A. 10 watts
B. 80 watts
C. 90 watts
D. 100 watts

T4B11 (C)
If a directional RF wattmeter reads 96 watts forward power and 4 watts reflected power, what is the actual transmitter output power?
A. 80 watts
B. 88 watts
C. 92 watts
D. 100 watts

ANSWERS TO PREVIOUS TEST

T4B01	B
T4B02	C
T4B03	A
T4B04	A
T4B05	D
T4B06	D

YOUR ANSWERS TO THIS TEST

T4B07	
T4B08	
T4B09	
T4C10	
T9B07	
T9B08	
T4B10	
T4B11	

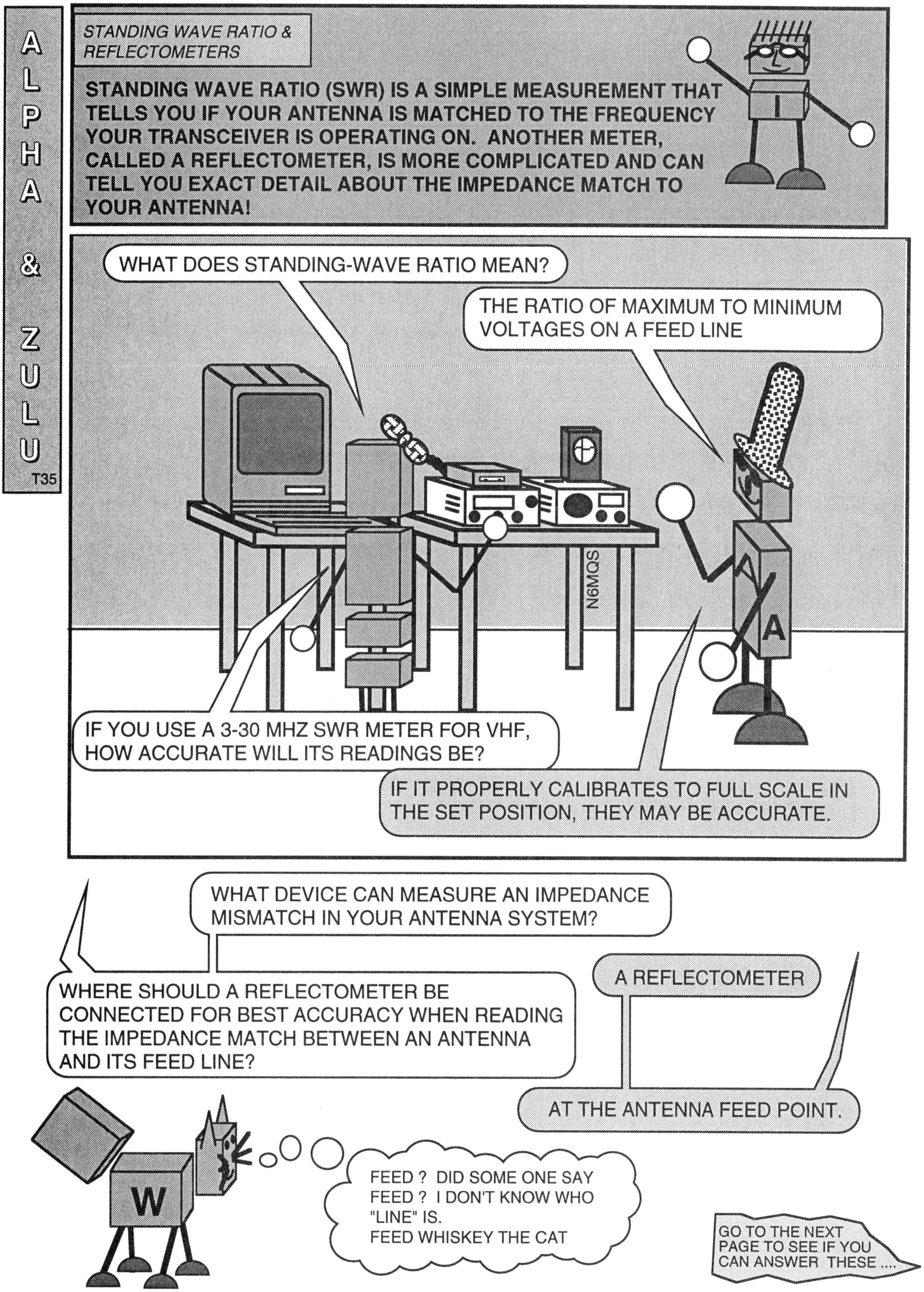

Riding the Airwaves with Alpha & Zulu

ALPHA & ZULU

STANDING WAVE RATIO & REFLECTOMETERS

STANDING WAVE RATIO (SWR) IS A SIMPLE MEASUREMENT THAT TELLS YOU IF YOUR ANTENNA IS MATCHED TO THE FREQUENCY YOUR TRANSCEIVER IS OPERATING ON. ANOTHER METER, CALLED A REFLECTOMETER, IS MORE COMPLICATED AND CAN TELL YOU EXACT DETAIL ABOUT THE IMPEDANCE MATCH TO YOUR ANTENNA!

WHAT DOES STANDING-WAVE RATIO MEAN?

THE RATIO OF MAXIMUM TO MINIMUM VOLTAGES ON A FEED LINE

IF YOU USE A 3-30 MHZ SWR METER FOR VHF, HOW ACCURATE WILL ITS READINGS BE?

IF IT PROPERLY CALIBRATES TO FULL SCALE IN THE SET POSITION, THEY MAY BE ACCURATE.

WHAT DEVICE CAN MEASURE AN IMPEDANCE MISMATCH IN YOUR ANTENNA SYSTEM?

A REFLECTOMETER

WHERE SHOULD A REFLECTOMETER BE CONNECTED FOR BEST ACCURACY WHEN READING THE IMPEDANCE MATCH BETWEEN AN ANTENNA AND ITS FEED LINE?

AT THE ANTENNA FEED POINT.

FEED? DID SOME ONE SAY FEED? I DON'T KNOW WHO "LINE" IS. FEED WHISKEY THE CAT

GO TO THE NEXT PAGE TO SEE IF YOU CAN ANSWER THESE ....

Riding the Airwaves with Alpha & Zulu

SEE IF YOU CAN ANSWER THESE TECHNICIAN QUESTIONS WITHOUT LOOKING BACK TO THE LAST 'TOON'. WRITE YOUR ANSWERS IN THE ANSWER BOX BELOW. GOOD LUCK. DA-DI-DAH

PRACTICE TEST PAGE T35

T9B06
What does standing-wave ratio mean?
- A. The ratio of maximum to minimum inductances on a feed line
- B. The ratio of maximum to minimum resistances on a feed line
- C. The ratio of maximum to minimum impedances on a feed line
- D. The ratio of maximum to minimum voltages on a feed line

T4C11
If you use a 3-30 MHz SWR meter for VHF, how accurate will its readings be?
- A. They will not be accurate
- B. They will be accurate enough to get by
- C. If it properly calibrates to full scale in the set position, they may be accurate
- D. They will be accurate providing the readings are multiplied by 4.5

T4C08
What device can measure an impedance mismatch in your antenna system?
- A. A field-strength meter
- B. An ammeter
- C. A wavemeter
- D. A reflectometer

T4C09
Where should a reflectometer be connected for best accuracy when reading the impedance match between an antenna and its feed line?
- A. At the antenna feed point
- B. At the transmitter output connector
- C. At the midpoint of the feed line
- D. Anywhere along the feed line

ANSWERS TO PREVIOUS TEST

T4B07	A
T4B08	B
T4B09	A
T4C10	A
T9B07	A
T9B08	B
T4B10	B
T4B11	C

YOUR ANSWERS TO THIS TEST

T9B06	
T4C11	
T4C08	
T4C09	

artsci inc

PRACTICE TEST PAGE T36

T9A01
What is a directional antenna?
- A. An antenna which sends and receives radio energy equally well in all directions
- B. An antenna that cannot send and receive radio energy by skywave or skip propagation
- C. An antenna which sends and receives radio energy mainly in one direction
- D. An antenna which sends and receives radio energy equally well in two opposite directions

T9A02
How is a Yagi antenna constructed?
- A. Two or more straight, parallel elements are fixed in line with each other
- B. Two or more square or circular loops are fixed in line with each other
- C. Two or more square or circular loops are stacked inside each other
- D. A straight element is fixed in the center of three or more elements which angle toward the ground

T9A03
What type of beam antenna uses two or more straight elements arranged in line with each other?
- A. A delta loop antenna
- B. A quad antenna
- C. A Yagi antenna
- D. A Zepp antenna

T9A04
How many directly driven elements do most beam antennas have?
- A. None
- B. One
- C. Two
- D. Three

T9A05
What is a parasitic beam antenna?
- A. An antenna where some elements obtain their radio energy by induction or radiation from a driven element
- B. An antenna where wave traps are used to magnetically couple the elements
- C. An antenna where all elements are driven by direct connection to the feed line
- D. An antenna where the driven element obtains its radio energy by induction or radiation from director elements

T9A06
What are the parasitic elements of a Yagi antenna?
- A. The driven element and any reflectors
- B. The director and the driven element
- C. Only the reflectors (if any)
- D. Any directors or any reflectors

T9A07
What is a cubical quad antenna?
- A. Four straight, parallel elements in line with each other, each approximately 1/2-electrical wavelength long
- B. Two or more parallel four-sided wire loops, each approximately one-electrical wavelength long
- C. A vertical conductor 1/4-electrical wavelength high, fed at the bottom
- D. A center-fed wire 1/2-electrical wavelength long

T9A08
What is a delta loop antenna?
- A. A type of cubical quad antenna, except with triangular elements rather than square
- B. A large copper ring or wire loop, used in direction finding
- C. An antenna system made of three vertical antennas, arranged in a triangular shape
- D. An antenna made from several triangular coils of wire on an insulating form

ANSWERS TO PREVIOUS TEST

T9B06	D
T4C11	C
T4C08	D
T4C09	A

YOUR ANSWERS TO THIS TEST

T9A01 _____
T9A02 _____
T9A03 _____
T9A04 _____
T9A05 _____
T9A06 _____
T9A07 _____
T9A08 _____

Riding the Airwaves with Alpha & Zulu

SEE IF YOU CAN ANSWER THESE TECHNICIAN QUESTIONS WITHOUT LOOKING BACK TO THE LAST 'TOON'. WRITE YOUR ANSWERS IN THE ANSWER BOX BELOW. GOOD LUCK. DA-DI-DAH

PRACTICE TEST PAGE T37

T9B01
What does horizontal wave polarization mean?
A. The magnetic lines of force of a radio wave are parallel to the earth's surface
B. The electric lines of force of a radio wave are parallel to the earth's surface
C. The electric lines of force of a radio wave are perpendicular to the earth's surface
D. The electric and magnetic lines of force of a radio wave are perpendicular to the earth's surface

T9B02
What does vertical wave polarization mean?
A. The electric lines of force of a radio wave are parallel to the earth's surface
B. The magnetic lines of force of a radio wave are perpendicular to the earth's surface
C. The electric lines of force of a radio wave are perpendicular to the earth's surface
D. The electric and magnetic lines of force of a radio wave are parallel to the earth's surface

T9B03
What electromagnetic-wave polarization does a Yagi antenna have when its elements are parallel to the earth's surface?
A. Circular
B. Helical
C. Horizontal
D. Vertical

T9B04
What electromagnetic-wave polarization does a half-wavelength antenna have when it is perpendicular to the earth's surface?
A. Circular
B. Horizontal
C. Parabolical
D. Vertical

T9B05
What electromagnetic-wave polarization does most man-made electrical noise have in the HF and VHF spectrum?
A. Horizontal
B. Left-hand circular
C. Right-hand circular
D. Vertical

T9A09
What type of non-directional antenna is easy to make at home and works well outdoors?
A. A Yagi
B. A delta loop
C. A cubical quad
D. A ground plane

T9A10
What type of antenna is made when a magnetic-base whip antenna is placed on the roof of a car?
A. A Yagi
B. A delta loop
C. A cubical quad
D. A ground plane

T9A11
If a magnetic-base whip antenna is placed on the roof of a car, in what direction does it send out radio energy?
A. It goes out equally well in all horizontal directions
B. Most of it goes in one direction
C. Most of it goes equally in two opposite directions
D. Most of it is aimed high into the air

ANSWERS TO PREVIOUS TEST	
T9A01	C
T9A02	A
T9A03	C
T9A04	B
T9A05	A
T9A06	D
T9A07	B
T9A08	A

YOUR ANSWERS TO THIS TEST:
T9B01
T9B02
T9B03
T9B04
T9B05
T9A09
T9A10
T9A11

Riding the Airwaves with Alpha & Zulu

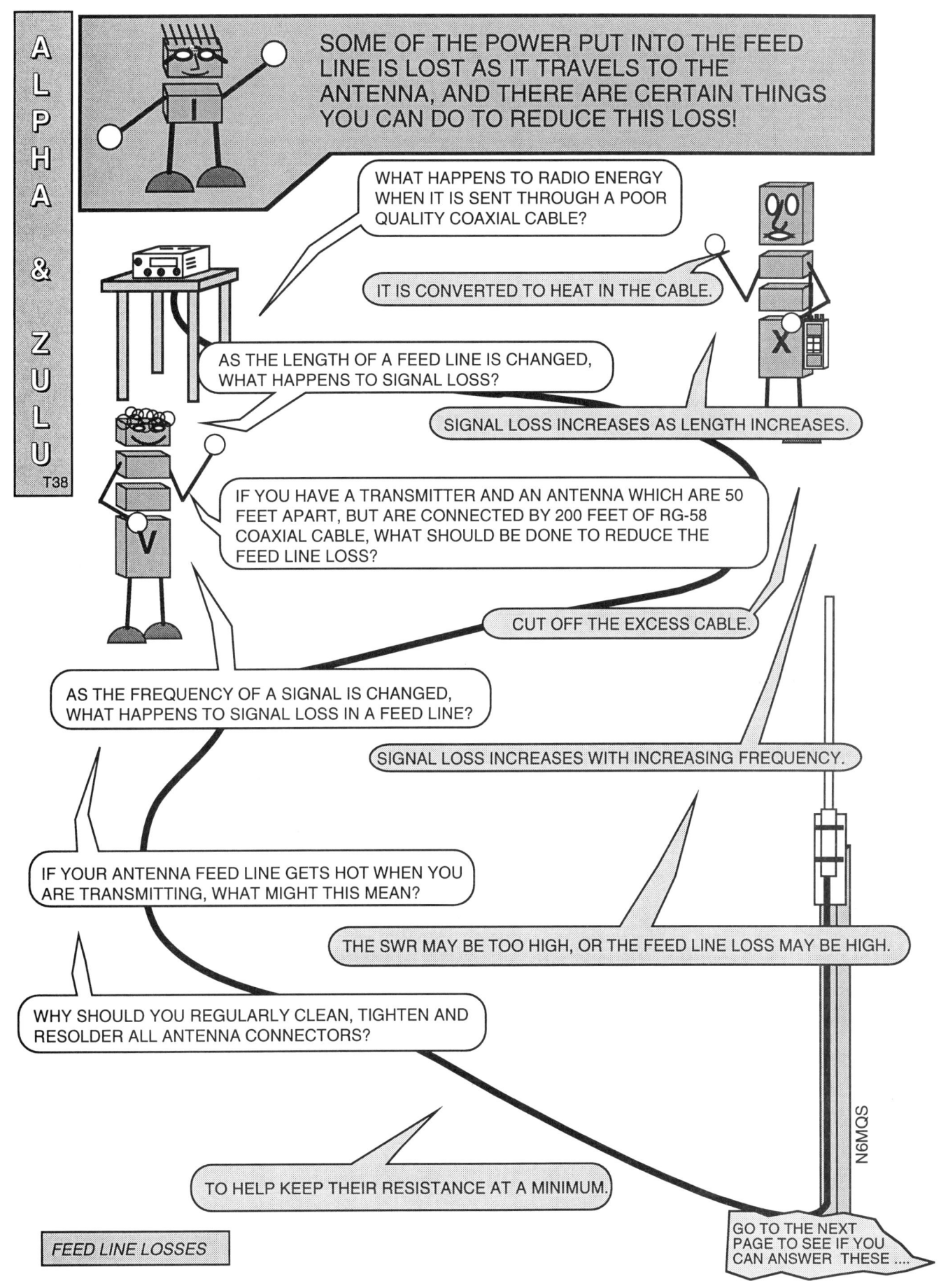

PRACTICE TEST PAGE T38

SEE IF YOU CAN ANSWER THESE TECHNICIAN QUESTIONS WITHOUT LOOKING BACK TO THE LAST 'TOON'. WRITE YOUR ANSWERS IN THE ANSWER BOX BELOW. GOOD LUCK. DA-DI-DAH

T9B09
What happens to radio energy when it is sent through a poor quality coaxial cable?
A. It causes spurious emissions
B. It is returned to the transmitter's chassis ground
C. It is converted to heat in the cable
D. It causes interference to other stations near the transmitting frequency

T9C05
If you have a transmitter and an antenna which are 50 feet apart, but are connected by 200 feet of RG-58 coaxial cable, what should be done to reduce feed line loss?
A. Cut off the excess cable so the feed line is an even number of wavelengths long
B. Cut off the excess cable so the feed line is an odd number of wavelengths long
C. Cut off the excess cable
D. Roll the excess cable into a coil which is as small as possible

T9C06
As the length of a feed line is changed, what happens to signal loss?
A. Signal loss is the same for any length of feed line
B. Signal loss increases as length increases
C. Signal loss decreases as length increases
D. Signal loss is the least when the length is the same as the signal's wavelength

T9C07
As the frequency of a signal is changed, what happens to signal loss in a feed line?
A. Signal loss is the same for any frequency
B. Signal loss increases with increasing frequency
C. Signal loss increases with decreasing frequency
D. Signal loss is the least when the signal's wavelength is the same as the feed line's length

T9C08
If your antenna feed line gets hot when you are transmitting, what might this mean?
A. You should transmit using less power
B. The conductors in the feed line are not insulated very well
C. The feed line is too long
D. The SWR may be too high, or the feed line loss may be high

T9C11
Why should you regularly clean, tighten and re-solder all antenna connectors?
A. To help keep their resistance at a minimum
B. To keep them looking nice
C. To keep them from getting stuck in place
D. To increase their capacitance

ANSWERS TO PREVIOUS TEST	
T9B01	B
T9B02	C
T9B03	C
T9B04	D
T9B05	D
T9A09	D
T9A10	D
T9A11	A

YOUR ANSWERS TO THIS TEST
T9B09
T9C05
T9C06
T9C07
T9C08
T9C11

Riding the Airwaves with Alpha & Zulu

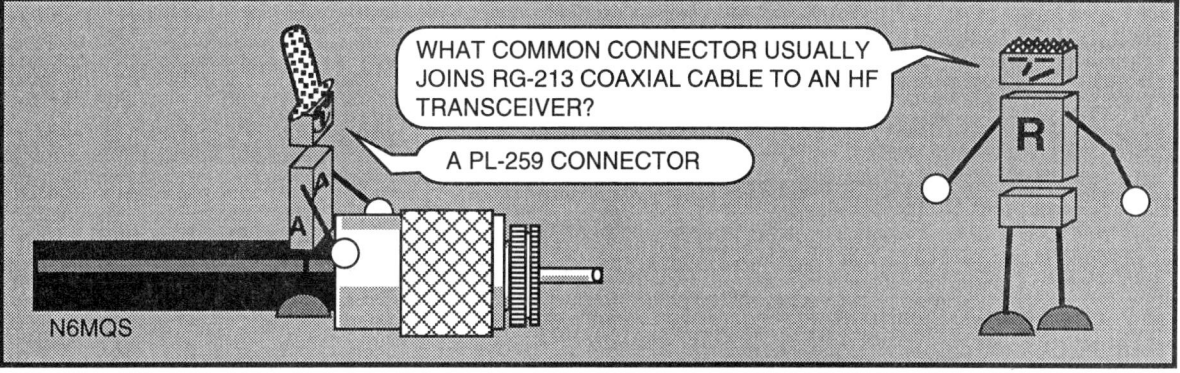
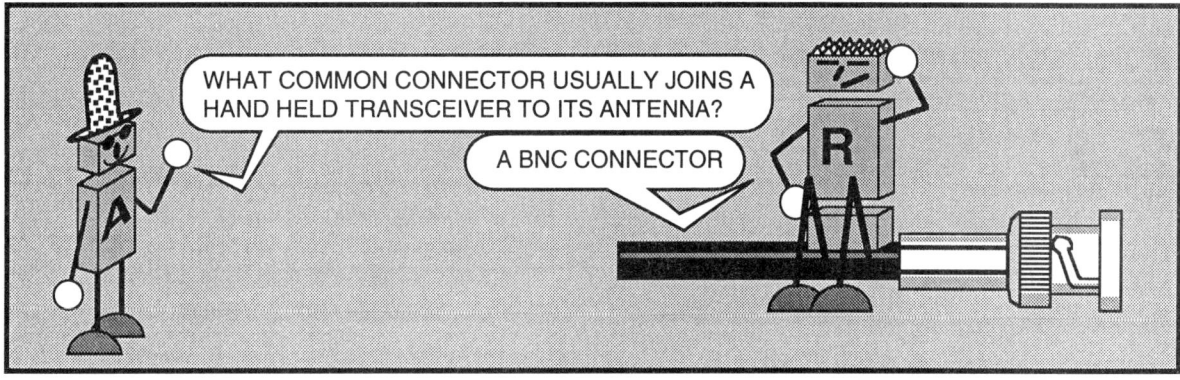

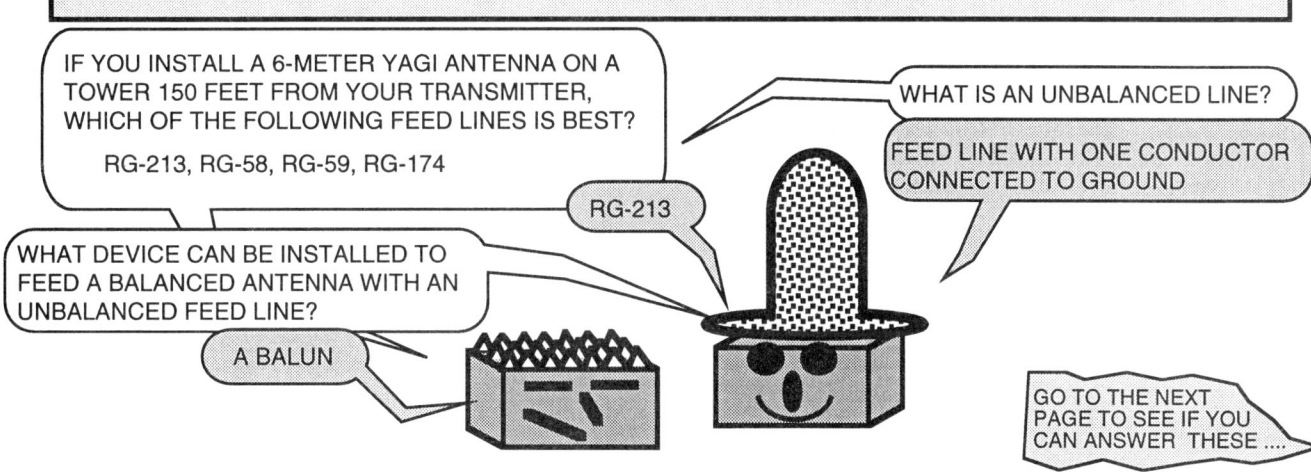

SEE IF YOU CAN ANSWER THESE TECHNICIAN QUESTIONS WITHOUT LOOKING BACK TO THE LAST 'TOON'. WRITE YOUR ANSWERS IN THE ANSWER BOX BELOW. GOOD LUCK. DA-DI-DAH

PRACTICE TEST PAGE T39

T9B10
What is an unbalanced line?
- A. Feed line with neither conductor connected to ground
- B. Feed line with both conductors connected to ground
- C. Feed line with one conductor connected to ground
- D. Feed line with both conductors connected to each other

T9C01
What common connector usually joins RG-213 coaxial cable to an HF transceiver?
- A. An F-type cable connector
- B. A PL-259 connector
- C. A banana plug connector
- D. A binding post connector

T9B11
What device can be installed to feed a balanced antenna with an unbalanced feed line?
- A. A balun
- B. A loading coil
- C. A triaxial transformer
- D. A wavetrap

T9C02
What common connector usually joins a hand held transceiver to its antenna?
- A. A BNC connector
- B. A PL-259 connector
- C. An F-type cable connector
- D. A binding post connector

T9C03
Which of these common connectors has the lowest loss at UHF?
- A. An F-type cable connector
- B. A type-N connector
- C. A BNC connector
- D. A PL-259 connector

T9C04
If you install a 6-meter Yagi antenna on a tower 150 feet from your transmitter, which of the following feed lines is best?
- A. RG-213
- B. RG-58
- C. RG-59
- D. RG-174

ANSWERS TO PREVIOUS TEST
T9B09
C
T9C05
C
T9C06
B
T9C07
B
T9C08
D
T9C11
A

YOUR ANSWERS TO THIS TEST
T9B10
T9B11
T9C01
T9C02
T9C03
T9C04

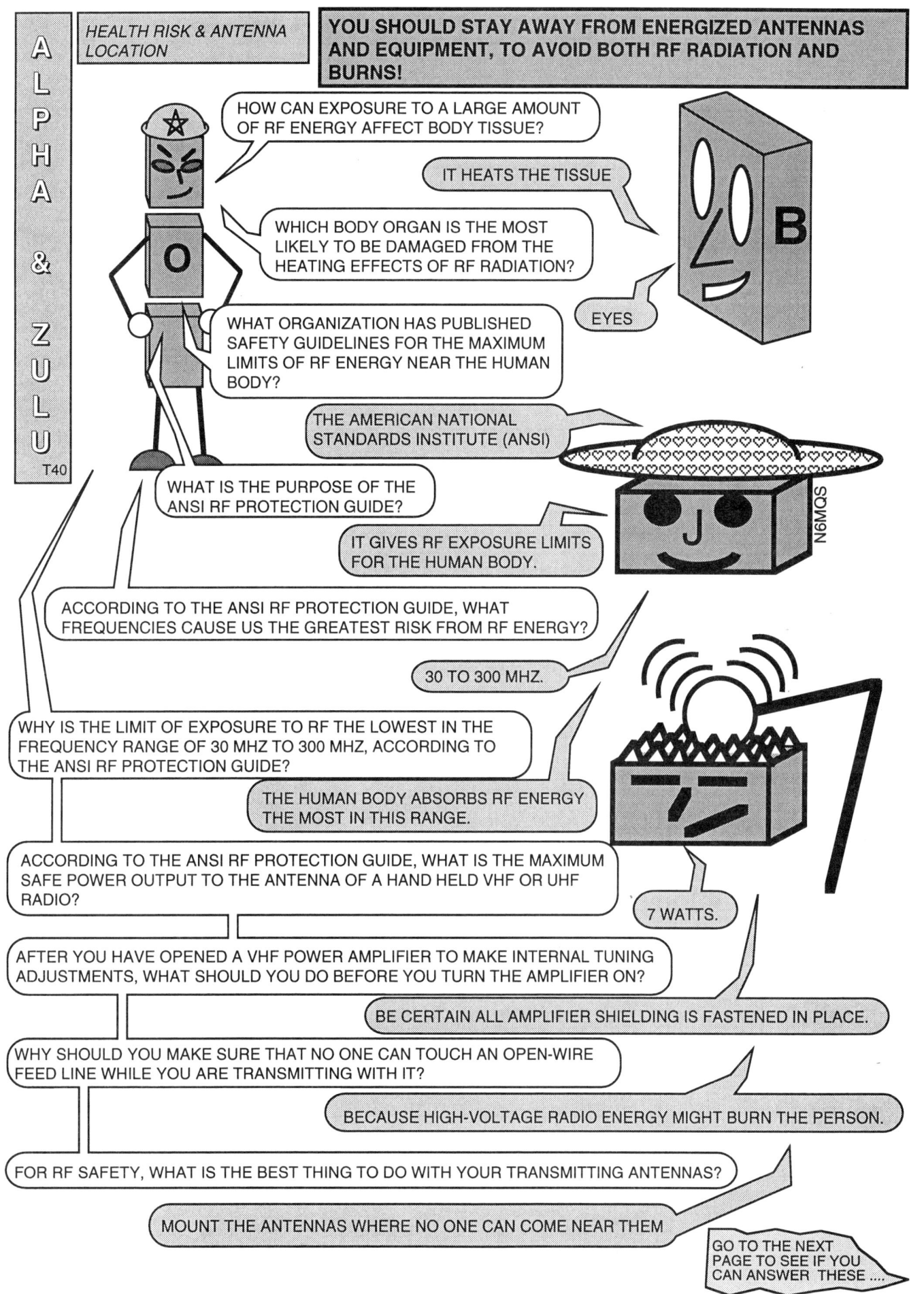

HERE'S ANOTHER TWO PAGE TEST GOOD LUCK.

PRACTICE TEST PAGE T40A

T4D09
How can exposure to a large amount of RF energy affect body tissue?
A. It causes radiation poisoning
B. It heats the tissue
C. It paralyzes the tissue
D. It produces genetic changes in the tissue

T4D10
Which body organ is the most likely to be damaged from the heating effects of RF radiation?
A. Eyes
B. Hands
C. Heart
D. Liver

T4D11
What organization has published safety guidelines for the maximum limits of RF energy near the human body?
A. The Institute of Electrical and Electronics Engineers (IEEE)
B. The Federal Communications Commission (FCC)
C. The Environmental Protection Agency (EPA)
D. The American National Standards Institute (ANSI)

T4D13
According to the ANSI RF protection guide, what frequencies cause us the greatest risk from RF energy?
A. 3 to 30 MHz
B. 300 to 3000 MHz
C. Above 1500 MHz
D. 30 to 300 MHz

T4D12
What is the purpose of the ANSI RF protection guide?
A. It lists all RF frequency allocations for interference protection
B. It gives RF exposure limits for the human body
C. It sets transmitter power limits for interference protection
D. It sets antenna height limits for aircraft protection

ANSWERS TO PREVIOUS TEST

T9B10	C
T9B11	A
T9C01	B
T9C02	A
T9C03	B
T9C04	A

PRACTICE TEST PAGE T40B

T4D14
Why is the limit of exposure to RF the lowest in the frequency range of 30 MHz to 300 MHz, according to the ANSI RF protection guide?
   A. There are more transmitters operating in this range
   B. There are fewer transmitters operating in this range
   C. Most transmissions in this range are for a longer time
   D. The human body absorbs RF energy the most in this range

T4D16
After you have opened a VHF power amplifier to make internal tuning adjustments, what should you do before you turn the amplifier on?
   A. Remove all amplifier shielding to ensure maximum cooling
   B. Make sure that the power interlock switch is bypassed so you can test the amplifier
   C. Be certain all amplifier shielding is fastened in place
   D. Be certain no antenna is attached so that you will not cause any interference

T4D15
According to the ANSI RF protection guide, what is the maximum safe power output to the antenna of a hand held VHF or UHF radio?
   A. 125 milliwatts
   B. 7 watts
   C. 10 watts
   D. 25 watts

T9C09
Why should you make sure that no one can touch an open-wire feed line while you are transmitting with it?
   A. Because contact might cause a short circuit and damage the transmitter
   B. Because contact might break the feed line
   C. Because contact might cause spurious emissions
   D. Because high-voltage radio energy might burn the person

T9C10
For RF safety, what is the best thing to do with your transmitting antennas?
   A. Use vertical polarization
   B. Use horizontal polarization
   C. Mount the antennas where no one can come near them
   D. Mount the antenna close to the ground

YOUR ANSWERS TO THIS TEST
T4D09
T4D10
T4D11
T4D13
T4D12
T4D14
T4D15
T4D16
T9C09
T9C10

artsci inc

RADIO CROSSTALK

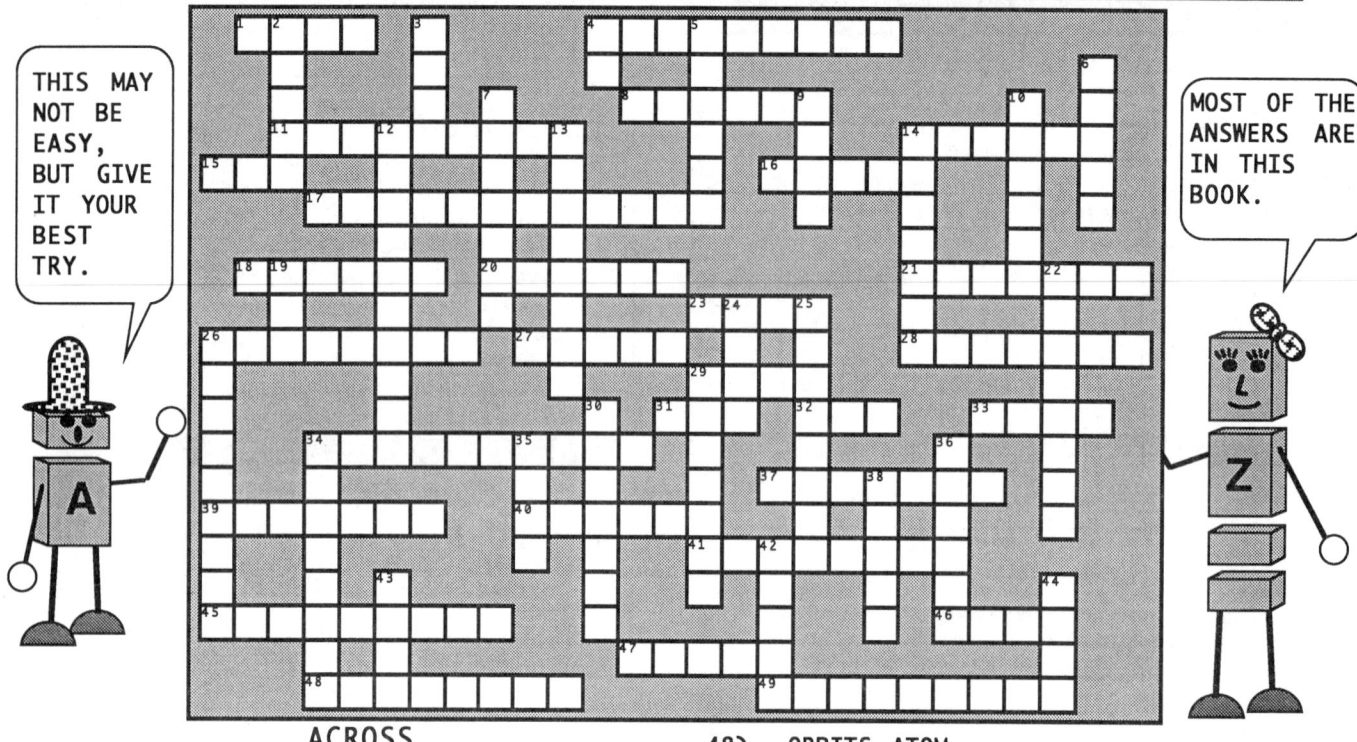

ACROSS

1) MEASURE OF RESISTANCE
4) MATERIAL WITH LOW RESISTANCE
8) TWO WAY COMMUNICATIONS AT THE SAME TIME
11) TRANSMIT TO THE PUBLIC
14) MEASURE OF ELECTRICAL CURRENT
15) TOOL USED FOR MORSE CODE
16) DEVICE TO BALANCE UNBALANCED LOADS
17) UNWANTED SIGNALS
18) CONNECT TO EARTH
20) COAX SHELL
21) ATOMIC CENTER
23) DIRECTIONAL ANTENNA
26) REMOVES LOW FREQUENCIES
27) SHORT WAVE _____
28) INCREASE VOLTAGE, CURRENT
29) RANGE OF FREQUENCIES
31) AMATEUR RADIO OPERATOR
32) 1270 MHZ HAM BAND
33) PROTECTS ELECTRIC CIRCUITS
34) LONGWIRE ORIENTATION
37) INTENTIONAL INTERFERENCE
39) THE SUNS POWER SUPPLY
40) THREE ELEMENT TUBE
41) AUTOMATICALLY RETRANSMITS
45) SMALLER THAN AN ATOM
46) POPULAR BEAM ANTENNA
47) POSITIVE TUBE ELEMENT
48) ORBITS ATOM
49) INDUCTIVE MEASURE

DOWN

2) RADIO FUN
3) FOUR SIDE FULL WAVE ANTENNA
4) BAND THAT NEEDS NO LICENSE
5) COMMON HALFWAVE ANTENNA
6) _____ INTO THE MICROPHONE
7) LARGE GATHERING OF HAMS
9) EXCEPTIONAL PHONETIC LETTER
10) ANTENNA AND MERMAID NAME
12) DEVICE THAT REDUCES VOLUME
13) RTTY KEYBOARD AND MONITOR
14) SOURCE OF RADIO WAVES
19) HAM TERM FOR RADIO EQUIPMENT
22) TRANSMITTED RF
24) MUST BE PASSED TO RECEIVE LICENSE
25) CHANGE AMPLITUDE OR FREQUENCEY OF AN RF SIGNAL
26) SPURIOUS SIGNALS AT A MULTIPLE OF THE TRANSMITTED FREQUENCY
29) PORTABLE POWER PACKAGE
30) CONVERT ELECTRIC ENERGY TO RF WAVES
34) COMMON ANTENNA LENGTH
35) ORGANIZED HAM GROUPS PASSING ALONG SPECIFIC INFORMATION
36) ABILITY TO WORK, POWER
38) LENGTH, ABOUT A YARD
42) RATE OF ENERGY CONSUMPTION
43) DITS AND DAHS
44) AC POWER WAVEFORM

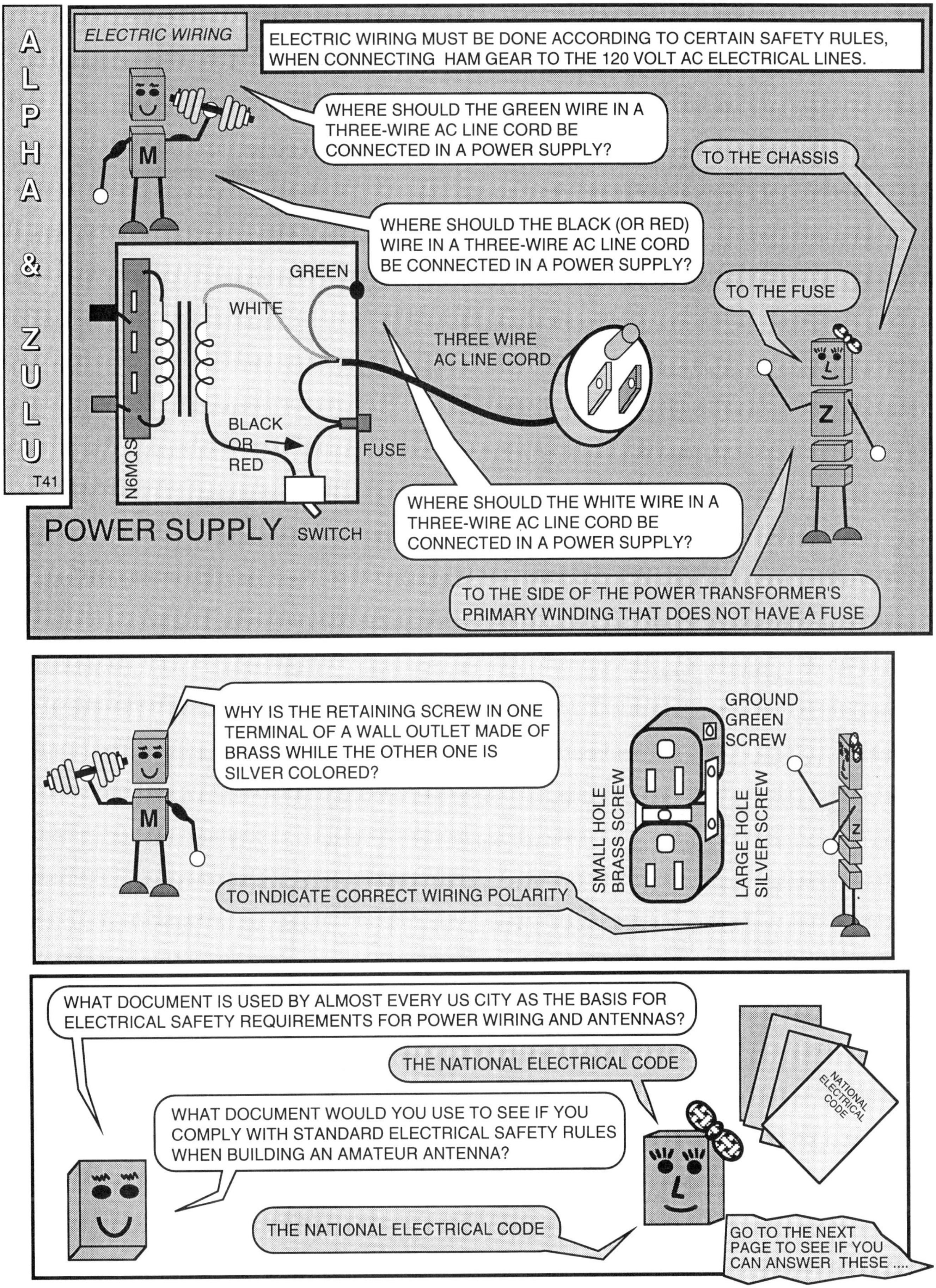

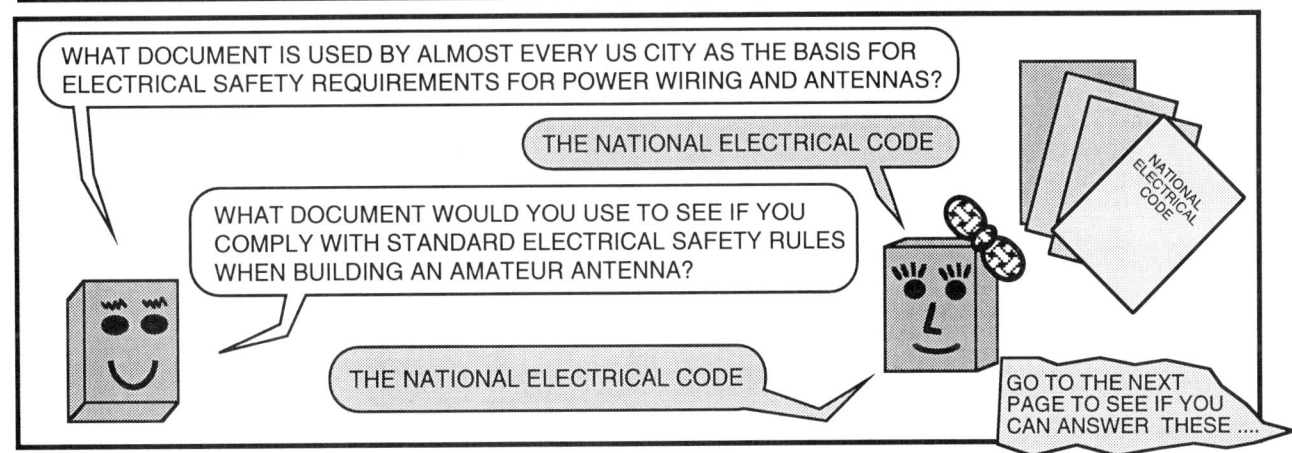

Riding the Airwaves with Alpha & Zulu

SEE IF YOU CAN ANSWER THESE TECHNICIAN QUESTIONS WITHOUT LOOKING BACK TO THE LAST 'TOON'. WRITE YOUR ANSWERS IN THE ANSWER BOX BELOW. GOOD LUCK. DA-DI-DAH

PRACTICE TEST PAGE T41

T4A01
Where should the green wire in a three-wire AC line cord be connected in a power supply?
A. To the fuse
B. To the "hot" side of the power switch
C. To the chassis
D. To the white wire

T4A03
Where should the white wire in a three-wire AC line cord be connected in a power supply?
A. To the side of the power transformer's primary winding that has a fuse
B. To the side of the power transformer's primary winding that does not have a fuse
C. To the chassis
D. To the black wire

T4A02
Where should the black (or red) wire in a three-wire AC line cord be connected in a power supply?
A. To the white wire
B. To the green wire
C. To the chassis
D. To the fuse

T4A04
What document is used by almost every US city as the basis for electrical safety requirements for power wiring and antennas?
A. The Code of Federal Regulations
B. The Proceedings of the IEEE
C. The ITU Radio Regulations
D. The National Electrical Code

T4A05
What document would you use to see if you comply with standard electrical safety rules when building an amateur antenna?
A. The Code of Federal Regulations
B. The Proceedings of the IEEE
C. The National Electrical Code
D. The ITU Radio Regulation

T4A07
Why is the retaining screw in one terminal of a wall outlet made of brass while the other one is silver colored?
A. To prevent corrosion
B. To indicate correct wiring polarity
C. To better conduct current
D. To reduce skin effect

ANSWERS TO PREVIOUS TEST	
T4D09	B
T4D10	A
T4D11	D
T4D13	D
T4D12	B
T4D14	D
T4D15	B
T4D16	C
T9C09	D
T9C10	C

YOUR ANSWERS TO THIS TEST
T4A01
T4A02
T4A03
T4A04
T4A05
T4A07

260 artsci inc

PRACTICE TEST PAGE T42

T4A06
Where should fuses be connected on a mobile transceiver's DC power cable?
A. Between the red and black wires
B. In series with just the black wire
C. In series with just the red wire
D. In series with both the red and black wires

T4A08
How much electrical current flowing through the human body is usually fatal?
A. As little as 1/10 of an ampere
B. Approximately 10 amperes
C. More than 20 amperes
D. Current flow through the human body is never fatal

T4A13
What precaution should you take when leaning over a power amplifier?
A. Take your shoes off
B. Watch out for loose jewelry contacting high voltage
C. Shield your face from the heat produced by the power supply
D. Watch out for sharp edges which may snag your clothing

T4A09
Which body organ can be fatally affected by a very small amount of electrical current?
A. The heart
B. The brain
C. The liver
D. The lungs

T4A12
Where should the main power switch for a high-voltage power supply be located?
A. Inside the cabinet, to kill the power if the cabinet is opened
B. On the back side of the cabinet, out of sight
C. Anywhere that can be seen and reached easily
D. A high voltage power supply should not be switch-operated

T4A10
How much electrical current flowing through the human body is usually painful?
A. As little as 1/500 of an ampere
B. Approximately 10 amperes
C. More than 20 amperes
D. Current flow through the human body is never painful

T4A15
What should you do if you discover someone who is being burned by high voltage?
A. Run from the area so you won't be burned too
B. Turn off the power, call for emergency help and give CPR if needed
C. Immediately drag the person away from the high voltage
D. Wait for a few minutes to see if the person can get away from the high voltage on their own, then try to help

T4A14
What is an important safety rule concerning the main electrical box in your home?
A. Make sure the door cannot be opened easily
B. Make sure something is placed in front of the door so no one will be able to get to it easily
C. Make sure others in your home know where it is and how to shut off the electricity
D. Warn others in your home never to touch the switches, even in an emergency

T4A11
What is the minimum voltage which is usually dangerous to humans?
A. 30 volts
B. 100 volts
C. 1000 volts
D. 2000 volts

ANSWERS TO PREVIOUS TEST

T4A01	C
T4A02	D
T4A03	B
T4A04	D
T4A05	C
T4A07	B

YOUR ANSWERS TO THIS TEST

T4A06	
T4A08	
T4A13	
T4A09	
T4A10	
T4A15	
T4A12	
T4A14	
T4A11	

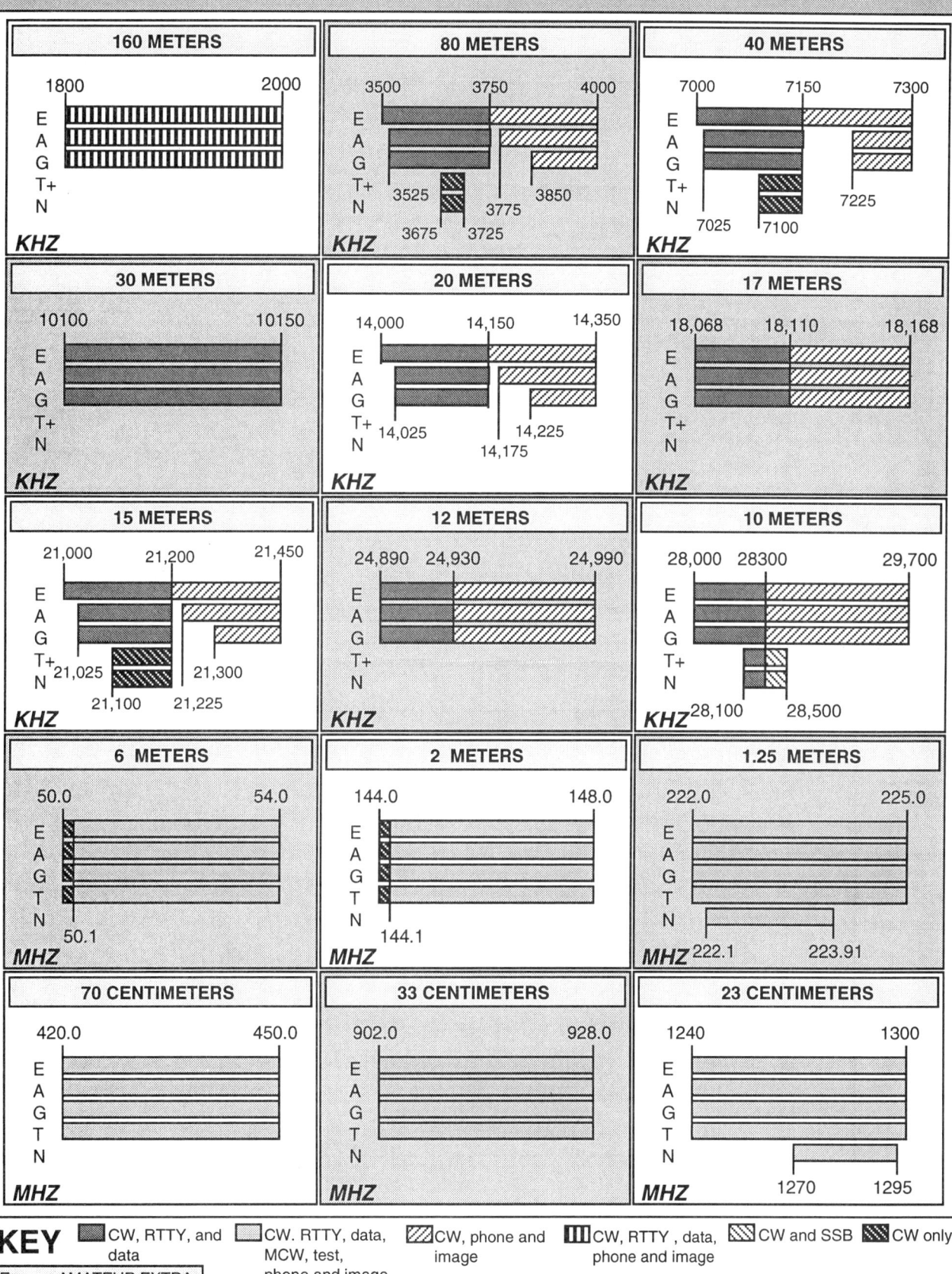

Notes

ANSWERS TO PREVIOUS TEST	
T4A06	D
T4A08	A
T4A13	B
T4A09	A
T4A10	A
T4A15	B
T4A12	C
T4A14	C
T4A11	A

XRAY'S COMMON RADIO EQUATIONS

SEVERAL EQUATIONS ARE DISCUSSED IN THIS BOOK. HERE I HAVE LISTED THEM ALL, ALONG WITH SOME OTHERS THAT YOU WILL COME ACROSS IN THE FUTURE.

OHM'S LAW
OHM'S LAW DEFINES THE RELATIONSHIP BETWEEN, VOTAGE, RESISTANCE, AND CURRENT.

E IS VOLTAGE MEASURED IN VOLTS
I IS CURRENT MEASURED IN AMPERES
R IS RESISTANCE MEASURED IN OHMS

$E = I \times R$
$I = E / R$
$R = E / I$

WATT'S LAW
WATT'S LAW RELATES POWER WITH VOLTAGE AND CURRENT

P IS POWER MEASURED IN WATTS
E IS VOLTAGE MEASURED IN VOLTS
I IS CURRENT MEASURED IN AMPERES

$P = I \times E$
$I = P / E$
$E = P / I$

ELECTRICAL RESISTANCE

RESISTORS IN SERIES

$R_{TOTAL} = R_1 + R_2 + R_3 + \ldots + R_n$

RESISTORS IN PARALLEL

$$R_{TOTAL} = \frac{1}{\frac{1}{R_1} + \frac{1}{R_2} + \frac{1}{R_3} + \ldots + \frac{1}{R_n}}$$

TWO RESISTORS IN PARALLEL

$$R_{TOTAL} = \frac{R_1 \times R_2}{R_1 + R_2}$$

WAVELENGTH, FREQUENCY, AND ANTENNAS

WAVELENGTH (in feet) $= \dfrac{984}{f \text{ (in MHz)}}$

DIPOLE ANTENNAS

ANTENNA LENGTH
LENGTH (in feet) $= \dfrac{468}{f \text{ (in MHz)}}$

QUARTER WAVE VERTICAL ANTENNAS

ANTENNA LENGTH
LENGTH (in feet) $= \dfrac{234}{f \text{ (in MHz)}}$

YAGI ANTENNA ELEMENT LENGTH L (ft)

$L_{driven} = 475 / f \text{ (MHz)}$ $L_{director} = L_{driven} \times 0.95$
$L_{reflector} = L_{driven} \times 1.05$

QUAD ANTENNA CIRCUMFERENCE C (ft)

$C_{driven\ element} = \dfrac{1005}{f \text{ (in MHz)}}$

$C_{director} = \dfrac{975}{f \text{ (in MHz)}}$

$C_{reflector} = \dfrac{1030}{f \text{ (in MHz)}}$

STARTEK
FREQUENCY COUNTERS

NEW MODEL ATH-15

AUTO TRIGGER & HOLD
ULTRA HIGH SPEED
READ & HOLD SIGNAL <.08 SEC
SIGNAL STRENGTH BAR GRAPH

SPECIAL OFFER **$199.** *$235 VALUE*

FEATURES FOR ALL NEW MODEL ATH-15

- *ATH*™ - *AUTO TRIGGER & HOLD*
- AUTOMATIC CLEAN DROPOUT
- *800% FASTER* RESPONSE TIME
- EXTRA BRIGHT LED DIGITS - USABLE IN DAYLIGHT
- 3 to 5 HOUR BATTERY PORTABLE OPERATION
- MAXIMIZED SENSITIVITY - <1mV typ
- NI-CAD BATTERIES & 110VAC ADP/CHARGER INC.
- 1 PPM TCXO STD - 0.2PPM TCXO OPTIONAL
- RANGE 1-1500 MHZ - 6 FAST GATE TIMES
- MANUAL & AUTO HOLD FUNCTION WITH INDICATOR
- 9-12VDC *AUTO-POLARITY* POWER INPUT
- *StarCab*™ QUALITY ALUMINUM CABINET
- FULL YEAR PARTS & LABOR LIMITED WARRANTY

STARTEK Bar Graph counters are *simply the best* for finding frequencies, testing, adjusting, repairing or locating RF devices. Superior sensitivity, longer battery operation, high quality USA construction and sub-compact size are just a few of the reasons to select a **STARTEK** counter.

ALL MODELS MADE IN USA

#*ATH-15*	FREQUENCY COUNTER	
	WITH NI-CADs & 110VAC ADP	$199.
FACTORY INSTALLED OPTIONS:		
#O/S-ATH-15	ONE-SHOT ATH™	40.
#HSTB-15	HIGH STAB TCXO 0.2ppm	100.
ACCESSORIES:		
#CC-90	BLACK VINYL ZIPPER CASE	12.
#TA-90	TELESCOPING BNC ANT	12.
#P-110	PROBE, 200 MHZ, 1X-10X	39.
#M207IC	CABLE FOR MFJ-207/208	10.

SAME DAY SHIPMENT

FACTORY DIRECT ORDER LINES

Orders & Information
305-561-2211
Orders only
800-638-8050
FAX 305-561-9133

TERMS: Shipping-handling charges for Florida add $5 + tax, US & Canada add 5% ($5 min - $10 max), others add 15% of total. COD fee $5, VISA, MC or DISCOVER accepted. Prices & specifications subject to change without notice or obligation.

STARTEK *INTERNATIONAL INC*
398 NE 38th St., Ft. Lauderdale, FL 33334

SELECT YOUR *STARTEK* POCKET COUNTER™ TODAY !

* SPECIAL PRICES

1350	**15-UHS**	**2500**	**3500**	**15-BG**	**35-BG**
1 MHZ - 1300 MHZ	1 MHZ - 1500 MHZ	10 HZ - 2400 MHZ	10 HZ - 3500 MHZ	1.MHZ - 1500 MHZ	1 MHZ - 3200 MHZ
QUALITY & ECONOMY	ULTRA HIGH SENSITIVITY	HI-Z INPUT - LO RANGE	HI-Z INPUT - LO RANGE	ULTRA HIGH SENSITIVITY	ULTRA HIGH SENSITIVITY
(REPLACES #1500A)	(REPLACES #1500HS)	HIGH SENSITIVITY	HIGH SENSITIVITY	2 INCH BAR GRAPH	EXTRA BRIGHT DIGITS
$129	*$159	*$189	$250	*$169	$265

Games solution page

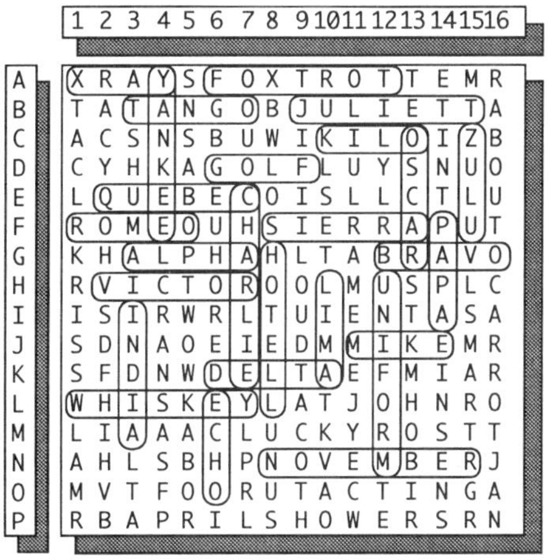

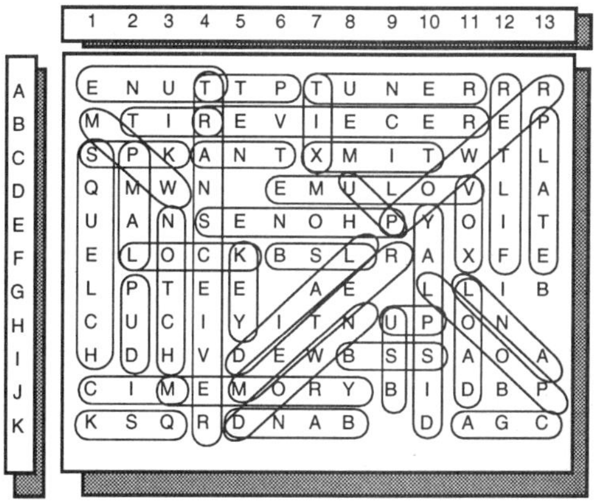

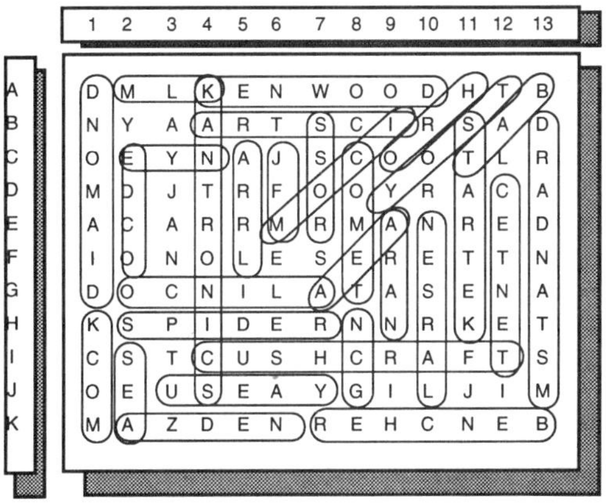

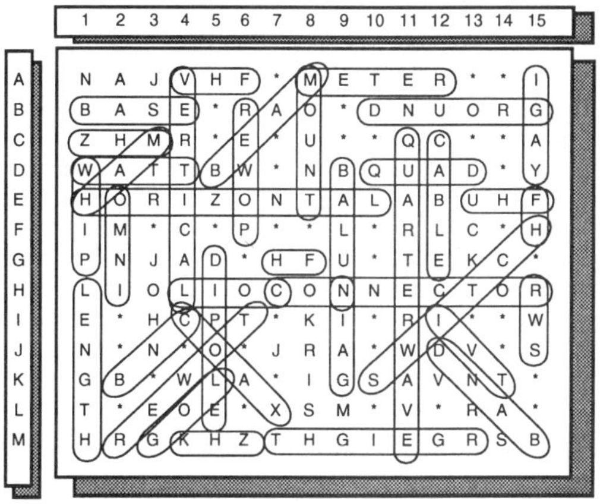

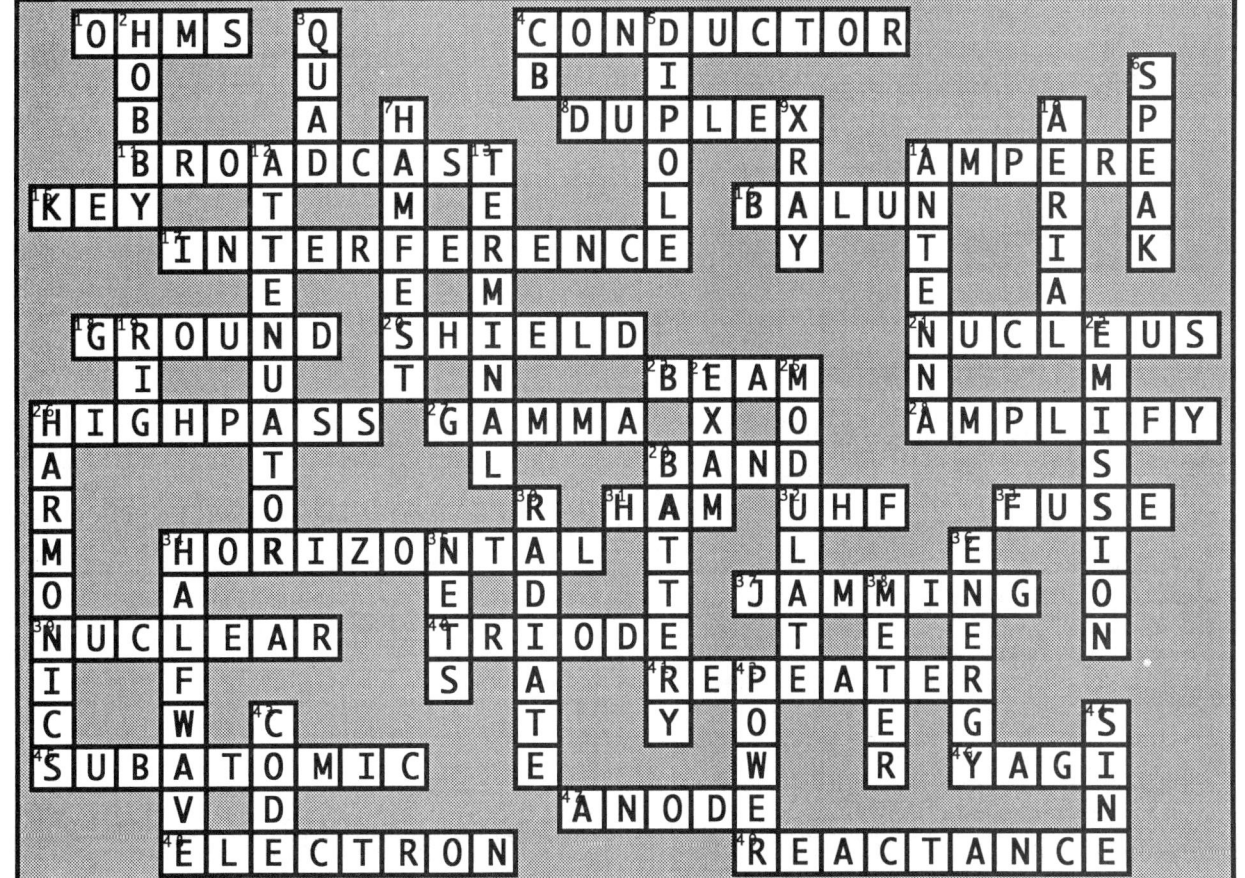

U.S. REPEATER MAPBOOK

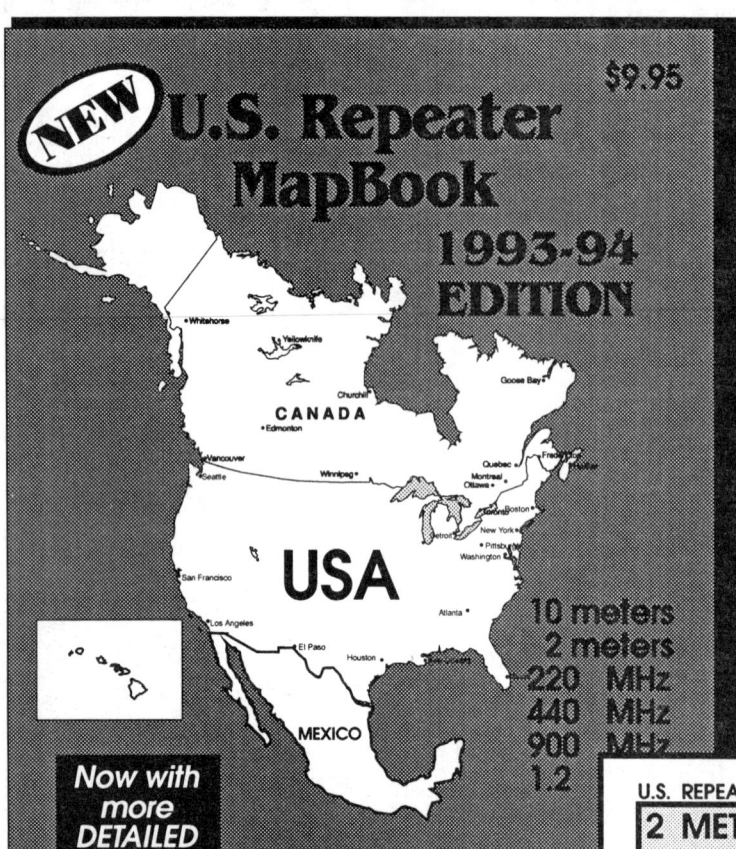

- MORE THAN 160 PAGES!
- MAPS OF ALL 50 US STATES, CANADA, MEXICO, AND THE CARIBBEAN!
- SHOWS POPULAR REPEATERS!
- SHOWS HIGHWAYS, CITIES AND TOURIST INFORMATION NUMBERS.

$9.95

ORDER FROM:
ARTSCI, INC.
P.O. BOX 1848
BURBANK, CA 91507

THE EASY WAY TO FIND REPEATERS!

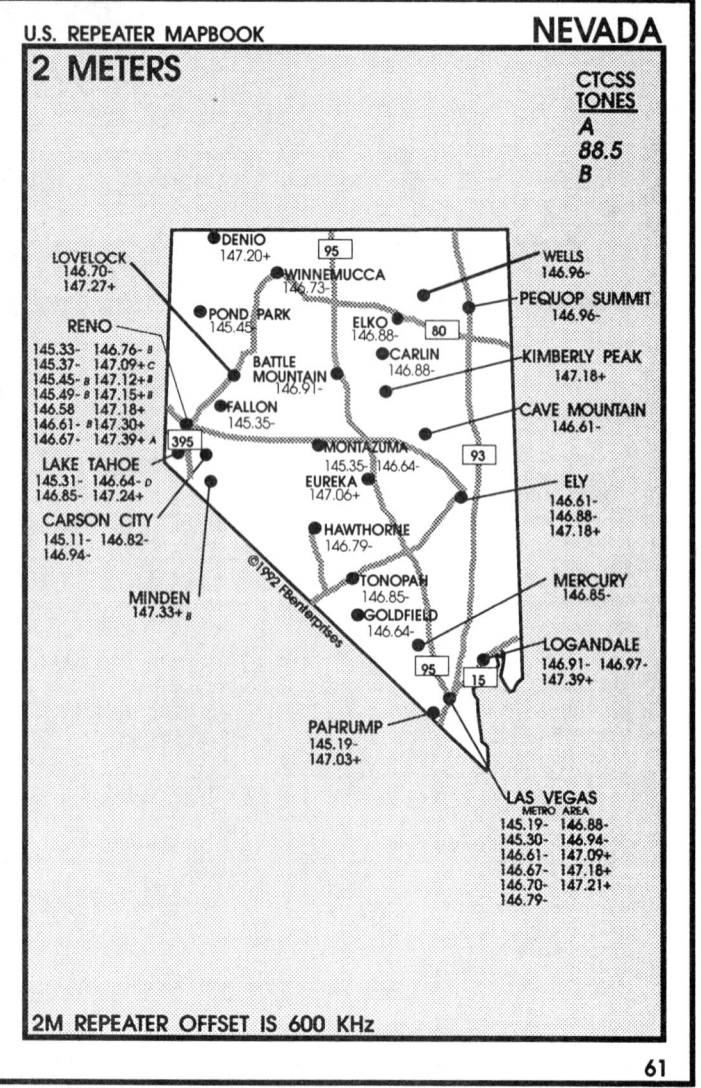

North American Shortwave Frequency Guide

Volume # 1 — $19.95

Frequency	Mode	Service / Times
1.240 MHz	AM	RADIO "THE VOICE OF VIETNAM", (English Service), 1000-1030, 1230-1300, 1330-1400, 2330-2400 UTC, (Spanish Service), 1100-1130 UTC
1.260 MHz	AM	VOICE OF AMERICA (English to Middle East/Europe service) 0100-0330, 1330-1400, 1400-1600, 2100-2200, 2230-2400 UTC (English to VOA Europe) 0300-0330, 0800-1000 UTC
1.296 MHz	AM	BRITISH BROADCASTING CORP., 0300-0330, 0430-0500, 0600-0630, 2200-2330 UTC
1.323 MHz	AM	BRITISH BROADCASTING CORP., 0000-2400 UTC
1.350 MHz	AM	VOICE OF AMERICA, (Service to Middle East) 0000-0500, 1330-2200, 2230-2300 UTC
1.413 MHz	AM	BRITISH BROADCASTING CORP., 0200-0230, 1300-1400, 1730-1830
1.530 MHz	AM	VOICE OF AMERICA (English to American Republics service) 0030-0100 UTC
1.575 MHz	AM	VOICE OF AMERICA (English to Pacific service) 2230-2400, 0030-0100 UTC. (English to VOA Europe) 1530-1600 UTC
	AM	ARMED FORCES RADIO, JAPAN, U.S. AIR FORCE, 0005-2205 UTC
1.580 MHz	AM	VOICE OF AMERICA (English to Caribbean) 0000-0200, 1000-1200 UTC (English to American Republics) 0030-0200 UTC
1.610 MHz		TRAVELERS INFORMATION SERVICE ACROSS U.S.
1.800 MHz	CW	**START OF AMATEUR RADIO 160 METER BAND (Ends 2.000 MHz)**
1.890 MHz	LSB	W1AW ARRL VOICE BULLETINS
2.500 MHz	VOICE	WWV INTERNATIONAL STANDARDS TIME FREQUENCY

©1992 artsci inc.   Shortwave listening guide

- Frequency listings of transmissions on the shortwave band
- English and Spanish Broadcasts

$19.95

TO ORDER: Use the order form in the back of this book

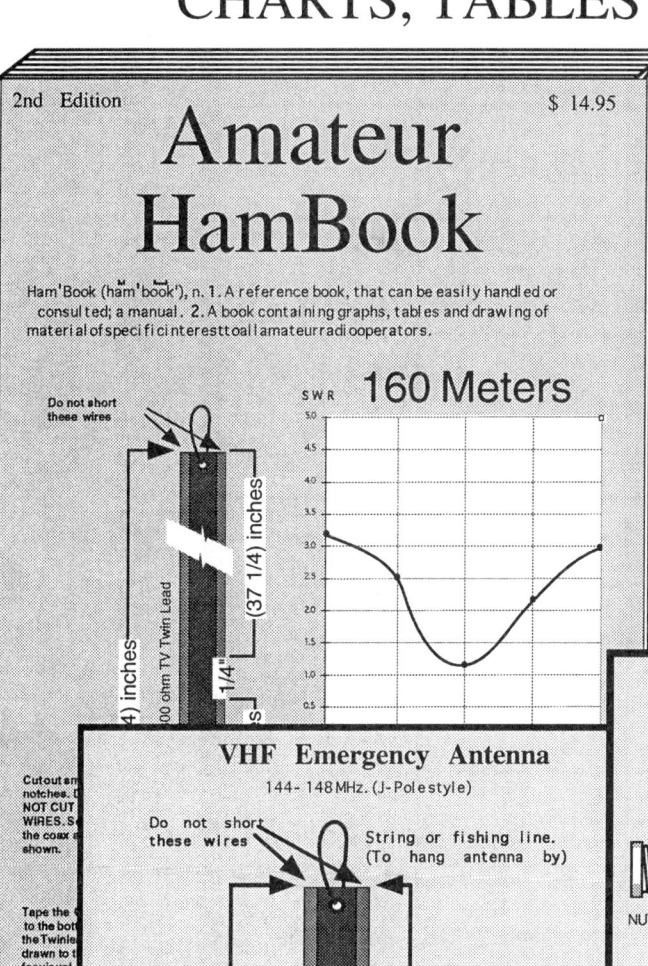

ALL IN ONE!
CHARTS, TABLES & DRAWINGS

- Antenna Construction.
- Latitude & Longitude of U.S. Cities
- Shortwave, Rtty, Satellite Freqs.
- PL hookup plans
- W.A.S. award worksheets
- SWR tables and charts
- Much more.. All in one book.

"THIS IS A GREAT BOOK"

"I COULD NOT HAVE DONE BETTER MYSELF"

"TO ORDER - USE THE FORM IN THE BACK OF THIS BOOK"

$14.95
plus $4.00 shipping

ORDER FORM

TITLE	DESCRIPTION	PRICE	QTY	EXTENSION
Radio/Tech Modifications VOL 5A	Over 200 pages of mods for ICOM, KENWOOD Radios & all models of Scanners	19.95		
Radio/Tech Modifications VOL 5B	Over 200 pages of mods for ALINCO, YAESU, STANDARD and all models of CB equipment.	19.95		
Federal Assignments Volume #3	Scanner Frequency guide for all Federal Government Agencies. Over 300 pages	24.95		
U.S. Repeater Guide #3 1993-1994	VHF & UHF Repeater guide for the USA with State Maps showing popular repeaters.	9.95		
Amateur HamBook #2	Construction plans, coax, antenna, connector, SWR charts. A must have.	14.95		
Lost Users Manuals	Operating Instructions for all popular amateur Mobiles & Ht's.	19.95		
North American Shortwave Directory	Complete Listing of all activity on the HF band 0-30MHz.	19.95		
Ham Radio Resource Guide	For Southern California only. Testing, Club, Repeater, maps & more	6.00		
Quick-N-Easy Shortwave Listening	Beginning guide to getting started with shortwave listening. Antennas, Accessories, Receivers & more.	9.95		
Riding the airwaves with Alpha & Zulu	Novice & No-code license test book using cartoon strips to teach. for ages 8 - 80 !!!	14.95		

MAIL ORDER FORM TO:

ARTSCI INC.
P.O. BOX 1428
BURBANK, CA 91507
(818) 843-4080
FAX: (818) 846-2298

Shipping charge outside the U.S. is $10.00 or more

SUBTOTAL	
SALES TAX 8.25% CA	
SHIPPING	$ 4.00
ORDER TOTAL	

SHIP TO:

NAME
ADDRESS
CITY   ST   ZIP
PHONE (   )

BILLING INFORMATION Give us your phone number in case we have a problem processing your order.

☐ CHECK ENCLOSED

☐ VISA / MasterCharge / DISCOVER / AMERICAN EXPRESS

CARD #

EXP DATE

artsci